AF412907

ADVANCES IN SPACE BIOLOGY AND MEDICINE

Volume 8

Series Editor

Augusto Cogoli

Space Biology Group
ETH Technopark
Zurich, Switzerland

Cell Biology and Biotechnology in Space

Editor

Augusto Cogoli

Space Biology Group
ETH Technopark
Zurich, Switzerland

2002

ELSEVIER

Amsterdam – Boston – London – New York – Oxford – Paris – San Diego –
San Francisco – Singapore – Sydney – Tokyo

ELSEVIER SCIENCE B.V.
Sara Burgerhartstraat 25
P.O. Box 211, 1000 AE Amsterdam, The Netherlands

First edition 2002

Library of Congress Cataloging in Publication Data
A catalog record from the Library of Congress has been applied for.

British Library Cataloguing in Publication Data
A catalogue record from the British Library has been applied for.

ISBN: 0-444-50735-3
ISSN: 1569-2574

Contents

Foreword

The *Advances in Space Biology and Medicine* (ASBM) series was initiated in 1991 by Sjoerd Bonting, then at the SETI Institute at NASA-Ames Research Center in Moffett Field, California, with the publishing house JAI Press. Volume 7 was the last to be edited by Sjoerd. He did a magnificent job and ASBM contributed substantially to the dissemination of the concepts and results of space biology and medicine in the scientific community. ASBM is still the only systematic series of books dedicated to these disciplines. Therefore, I felt greatly honored when, on the occasion of the joint COSPAR/IAF meeting in Washington in the International Space Year 1992, Sjoerd asked me, first, to start as co-editor and, later, to be the series editor of ASBM. I accepted with enthusiasm and collaborated with him to the preparation Vol. 7 which was published by JAI Press in 1999. In the same year JAI Press was taken over by the prestigious publishing house Elsevier. However, it took some time before Elsevier decided to continue the publication of ASBM. I was happily surprised when in November 2000 Hendrik Van Leusen of Elsevier asked me to continue the series. The progress of the construction of the International Space Station (ISS) and the momentum gained by its utilization as a laboratory for basic science and technology were important elements in the decision to continue ASBM. Meanwhile, Sjoerd is enjoying his retirement in Goor, a small town in The Netherlands.

Each of the seven volumes in the series was either a collection of articles from different branches of space biology and medicine, such as cell biology, immunology, radiation biology, plant biology, bone physiology, cardiovascular physiology, vestibular physiology, etc., or it was dedicated to a special event, such as Volumes 3 and 5 which contained a collection of papers dedicated to the first and second European isolation and confinement studies, respectively.

Hendrik suggested restarting ASBM with volumes that were each dedicated to a particular field of life science in space. I agreed as there is an increasing amount of data gathered from space activities as well as from ground-based simulations. It can be expected that a larger number of scientists worldwide will be involved in the experimentation on ISS, biosatellites, sounding rockets, fall towers and several kinds of clinostats or rotating devices that are "randomizing" the gravity vector. In addition, gravity-unloading models like bed-rest studies on human subjects as well as suspended animals are widely used. In future, therefore, ASBM will be coordinated by myself as series editor who will select an editor for each volume. The target will be one volume per year. Angelina Jokovic, assistant to Hendrik at Elsevier, is helping in this challenging job.

The first volume published by Elsevier is this Volume 8. With the new start I choose to be also the volume editor. The volume is dedicated to cell biology, my own field of expertise. Volume 9 will be dedicated to developmental biology and will be edited by Jörg Marthy of the Laboratoire Arago, the renowned oceanographic institute in Banyuls sur Mer, France. Other volumes will follow in human physiology, plant biology, radiation biology, etc.

For this volume I have invited ten experts in their respective field. All of them, with the exception of Roger Binot who is coordinating the microgravity application activities at ESA, are directly involved in experimentation in low gravity and also have a high reputation for their scientific activity not related to gravitational and space biology. The articles address basic science such as signal transduction, cell differentiation, structure and function of the cytoskeleton, as well as applied science such as tissue engineering and bioreactor technology. In the case of three cell systems discussed in this book, namely lymphocytes, osteoclasts and chondrocytes, studies at the cellular level in zero gravity may contribute to a better understanding of severe physiological dysfunctions such as immune depression, osteoporosis and cartilage injuries.

While the other articles deal in general with single cells as such, Mauro Maccarrone and colleagues and James Tabony and co-authors focus on a single component of the cell. Lipoxygenase is a key enzyme which dioxygenates unsaturated fatty acids, thus initiating lipoperoxidation of membranes or the synthesis of signaling molecules, or inducing structural and metabolic changes in the cell. Mauro and collaborators have and will investigate the behaviour of the enzyme in microgravity. Some of the most exciting experiments in space biology have been conducted by James and collaborators on ground simulations and on sounding rockets. Thereby, they have observed that the formation of microtubuli from tubulin *in vitro* is altered in microgravity. This has an important impact of the formation and function of the cytoskeleton in the cell. James approaches—in plain and understandable language (to biologists)—the complicated theoretical background of the theory of bifurcations in non-equilibrium systems.

The article by Ruth Hemmersbach and Richard Bräucker deals with gravi-perception in protozoa. Here, an exciting finding is discussed: the identification of a gravity-sensing organelle, Müller's body, in the protozoan Loxodes.

Lymphocytes, T cells in particular, are a good but complicated model of signal transduction and cell differentiation. Immune dysfunction, mainly attributed to T cells, had been observed in several space crew members since the early times of human space flight. The discovery of the strong inhibition of mitogenic activation *in vitro* in 0 g followed in 1983 in Spacelab 1. All this has attracted the interest of several investigators. One of them is Marian Lewis who carried out several ground-based as well as space studies focusing on the cytoskeleton and apopotosis.

Bone demineralization is another consequence of long-duration space flight. One of the prominent investigators of osteoblasts function in 0 g is Millie Hughes-Fulford. Back in her laboratory after serving as payload specialist in Spacelab SLS-1 in 1991, she continued her studies on osteoblast function under altered gravitational conditions. Bone demineralization in space is a good model for the study of osteoporosis, one of the plagues of elderly people. Ranieri Cancedda is a world leader in the study of osteoporosis and participates in the microgravity application program of ESA (MAP) by carrying out investigations in space.

Lisa Freed and Gordana Vunjac-Novakovic conducted investigations on the growth of artificial cartilages either in a ground-based model like the rotating vessel or in real low gravity aboard MIR station. Thereby bioreactors developed by NASA are described. Isabelle Walther is involved in investigations in space with *Saccharomyces cerevisiae* in a sophisticated bioreactor. The aim is to build up technological and scientific know-how for instrumentation for tissue engineering aboard ISS.

Ludmila Buravkova and her co-authors give a broad overview of the basic cell biology and biotechnology project carried out in Russia. This is particularly important as a great many such activities are unknown to the western scientific community.

There is strong pressure on ESA to use ISS not only as a platform for basic science but also to develop a program of commercialization activities. One such activity is biotechnology, tissue engineering in particular. Roger Binot is the co-ordinator at ESA of the Microgravity Application Program (MAP), which started in 2000 with the participation of several European universities and non-aerospace companies. In fact, one of the major goals of MAP is to attract companies who are interested in the manufacture in space of commercially profitable products. In his function as MAP coordinator, Roger is not a bureaucrat but rather a valuable scientific advisor to several MAP teams.

According with the publisher the ten articles in this book underwent minimum editing in order to respect the language and the opinion of the authors.

In conclusion, I am confident that this volume will provide first hand and new information to those interested as well as to those involved in space biology and may help to introduce newcomers to this young and exciting discipline.

Augusto Cogoli
Zurich, 2002

Cell Biology and Biotechnology in Space
A. Cogoli (editor)
© 2002 Elsevier Science B.V. All rights reserved

Lipoxygenase Activity in Altered Gravity

Mauro Maccarrone, Natalia Battista, Monica Bari and Alessandro Finazzi-Agrò

Department of Experimental Medicine and Biochemical Sciences and Biomedical Space Center, University of Rome Tor Vergata, Rome, Italy

Abstract

Lipoxygenases are a family of enzymes which dioxygenate unsaturated fatty acids, thus initiating lipoperoxidation of membranes or the synthesis of signalling molecules, or inducing structural and metabolic changes in the cell. This activity is the basis for the critical role of lipoxygenases in a number of pathophysiological conditions, both in animals and plants. We review the effects of microgravity on the catalytic efficiency of purified soybean (*Glycine max*) lipoxygenase-1, as well as the modulation of the activity and expression of 5-lipoxygenase in human erythroleukemia K562 cells subjected to altered gravity. We also outline the molecular properties of the lipoxygenase family and discuss its possible involvement in space-related processes, such as apoptosis (programmed cell death) and immunodepression. Finally, we discuss the modulation of cyclooxygenase activity and expression in K562 cells exposed to altered gravity, because cyclooxygenase catalyzes the oxidation of arachidonate through a pathway different from that catalyzed by lipoxygenase activity.

The lipoxygenase family

Lipoxygenases (linoleate:oxygen oxidoreductase, EC 1.13.11.12; LOXs) are a family of monomeric non-heme, non-sulphur iron-containing dioxygenases, which catalyze the conversion of polyunsaturated fatty acids into conjugated hydroperoxides. These substrates are essential fatty acids in humans, but are not found in most bacteria and LOXs are also consistently absent from typical prokaryotes. In animal and plant cells LOXs are widely expressed, sometimes at high levels, where they may initiate the synthesis of signalling molecules involved in structural or metabolic changes in cells. Mammalian lipoxygenases have been implicated in several inflammatory conditions such as arthritis, psoriasis and bronchial asthma [1]. They have also been implicated in atherosclerosis [2], brain ageing [3], HIV infection [4], kidney disease [5,6] and terminal differentiation of keratinocytes [7]. In plants, lipoxygenases favour germination, participate in the synthesis of traumatin and jasmonic acid and in the response to abiotic stress [8]. LOXs from animal and plant tissues have been purified, sequenced and characterized, and have been shown to form a closely related family with no similarities to other known proteins. In the phylogenetic tree the plant and animal enzymes form two distinct branches with several subgroups within each kingdom [9]. With arachidonic (eicosatetraenoic, C20:4) acid as substrate, different LOX isozymes can add a hydroperoxy group at carbons 5, 12 or 15, and therefore they are called 5-, 12- or 15-lipoxygenases. Also linoleic (octadecadienoic, C18:2) acid and linolenic (octadecatrienoic, C18:3) acid are substrates of LOXs. Soybean (*Glycine max*) lipoxygenase-1 (LOX-1) is a 15-lipoxygenase widely used as a prototype for studying the homologous family of lipoxygenases from tissues of different species, both in structural [10–12] and kinetic [13–15] investigations. The primary sequence [16] and three-dimensional structure [10,11] of LOX-1 have been determined, showing that the enzyme is a prolate ellipsoid with a dimension of 90 × 65 × 60 Å, with 839 amino acid residues and a molecular mass of 93840 Da. LOX-1 is made of two domains: a 146-residue β-barrel at the N-terminus (domain I) and a 693-residue helical bundle at the C-terminus (domain II). The iron-containing active site is in the center of domain II, liganded to four conserved histidines and to the carboxyl group of the conserved C-terminal isoleucine. The active site can be reached through two cavities (I and II), shown in Fig. 1. Cavity I presents an ideal path for the entrance of molecular oxygen from the outside to the iron, whereas cavity II can

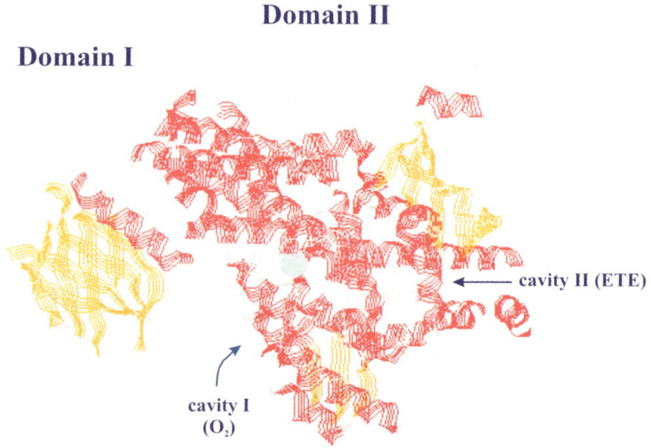

Fig. 1. Schematic diagram of the three-dimensional structure of soybean (*Glycine max*) lipoxygenase-1, showing the small N-terminal domain I and the large C-terminal domain II. The iron-containing active site is located in domain II, and can be reached by molecular oxygen (O_2) through cavity I and by arachidonic acid (eicosatetraenoic acid, ETE) through cavity II. The β-sandwiches are represented in yellow, the α-helices in red, the random coils in grey and the iron in light blue. The three-dimensional structure was modelled through the RASMOL program, using the lipoxygenase-1 sequence (PDB accession number: 2SBL).

accommodate arachidonic acid or even a slightly larger fatty acid [10]. LOX-1 is usually in the inactive ferrous (Fe^{2+}) form and oxidation to the active ferric (Fe^{3+}) enzyme is required for catalysis [17]. Mammalian lipoxygenases do not have the N-terminal domain present in LOX-1 and related plant lipoxygenases, thus showing smaller molecular masses (75–80 kDa compared to 94–104 kDa in plants).

Mammalian 5-lipoxygenase (arachidonate:oxygen 5-oxidoreductase, EC 1.13.11.34; 5-LOX), which is found primarily in polymorphonuclear leukocytes, macrophages and mast cells [18], where it plays a central role in cellular leukotriene synthesis (Fig. 2), has been most thoroughly studied. The interaction of human 5-LOX with cellular proteins has been recently investigated through the two-hybrid approach [19], showing that this enzyme binds to proteins as different as a coactosin-like protein (an element of the cytoskeleton), a transforming growth factor (TGF) type β receptor-I-associated protein 1 (involved in TGF signalling), and a hypothetical helicase (a DNA metabolizing enzyme). Complex protein–protein interactions have been demonstrated in nuclear membrane translocation, activation and acquisition of substrate by 5-LOX in intact cells [9]. In fact, cell activation by different stimuli results in translocation of 5-LOX to the nuclear membrane, where it associates with a "5-LOX activating protein" (FLAP), an 18 kDa integral membrane protein which acts as an arachidonic acid transfer protein and is essential for full leukotriene biosynthesis. FLAP has homology to leukotriene C_4 synthase [20] and other microsomal glutathione transferases [21], but no enzymatic activity itself [9]. Initially, the localization of FLAP was believed to be at the outer cell membrane, but later it became clear that it is associated with the nuclear envelope. Thus, arachidonic acid released from the nuclear membrane of leukocytes, and presented to 5-LOX by FLAP, may be the primary substrate for leukotriene synthesis [3]. This nuclear localization of 5-LOX activity has particular relevance for the role of this enzyme in apoptosis (programmed cell death, PCD), which will be discussed later. Also regulatory, but nonenzymatic, activities of 5-LOX have been reported, made possible by an Src homology 3 (SH3) binding motif, which enables the interaction with growth factor receptor-bound protein 2 (Grb2) and cytoskeletal proteins [19]. In this line, a possible interaction of 5-LOX with the nuclear factor-κB (NF-κB) complex seems to indicate that 5-LOX protein might influence also gene transcription [22]. The capability of 5-LOX to enter into the nucleus [23] supports its regulatory role on the transcription process, though the functional implications of the nonenzymatic functions of 5-LOX protein remain to be elucidated.

Effect of microgravity on the activity of pure soybean LOX-1

Experiments in space clearly show that several cellular processes, such as growth rates, signalling pathways and gene expression, are modified when cells are placed under conditions of weightlessness [24,25]. As yet, there is no coherent explanation for these observations, nor it is known which molecules might act as gravity sensors [24,25]. Recently, microtubule self-organization has been shown to be gravity-dependent [26], suggesting that investigations at the molecular level might fill the gap

4

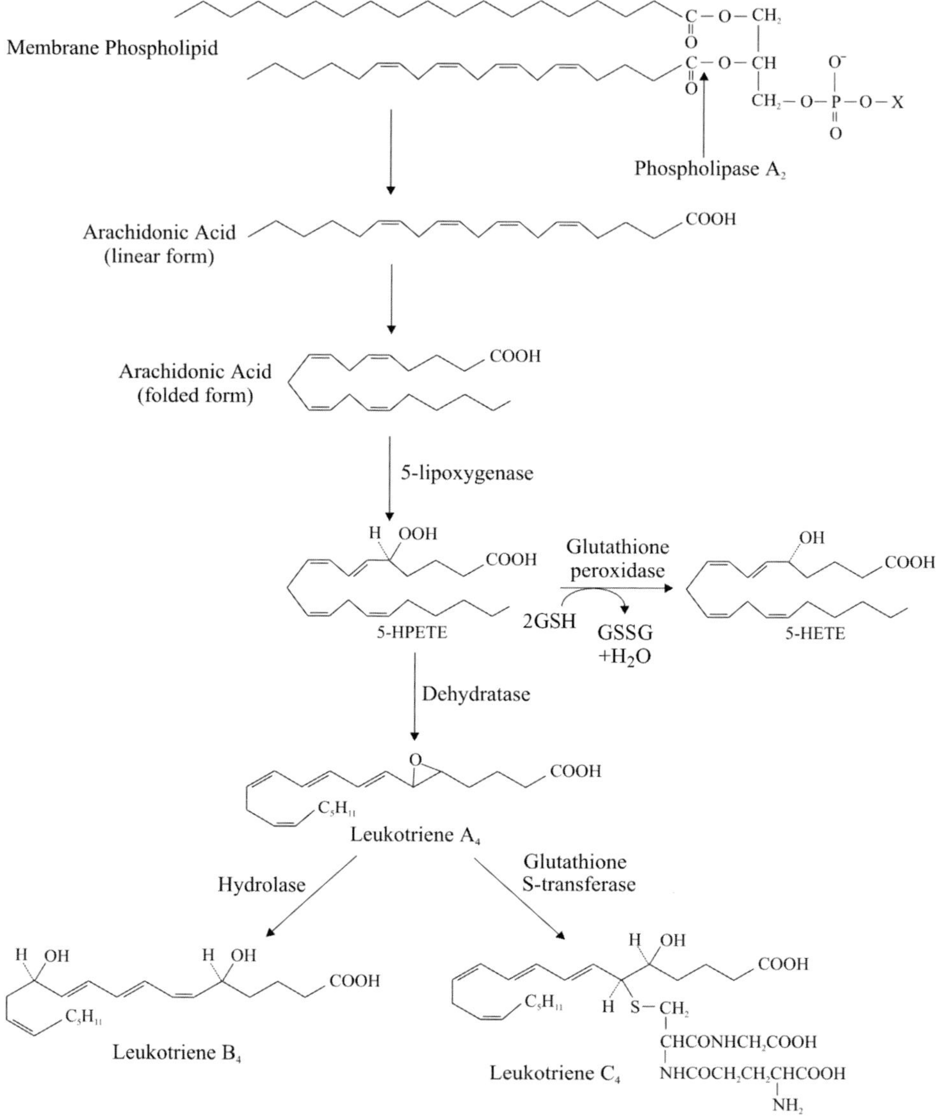

Fig. 2. The arachidonate metabolism catalyzed by 5-lipoxygenase, leading to the synthesis of bioactive molecules, such as 5-hydro(pero)xyeicosatetraenoic acid (5-H(P)ETE) and leukotrienes. The enzymatic activities of glutathione peroxidase, dehydratase, hydrolase and glutathione S-transferase take part in the leukotriene metabolism in humans. GSH, reduced glutathione; GSSG, oxidized glutathione.

between observation and understanding of space effects. Cellular activities are mostly controlled by enzymes, and pathological conditions can arise from alteration of just one of them [27]. Yet, biochemical investigations in microgravity are still very scant, because of the flight costs and the need for special instrumentation to be developed. We took advantage of a fibre optics spectrometer—the EMEC (Effect of Microgravity on Enzymatic Catalysis) module [28] developed for ESA by Officine Galileo (Alenia Difesa, Florence, Italy)—to measure the dioxygenation reaction by pure LOX-1 during the 28th parabolic flight campaign of the European Space Agency [29,30]. The aim was to check whether microgravity might affect enzyme catalysis. A reduced gravity environment was obtained by flying a specially modified Airbus A300 Zero-G through a series of parabolic manoeuvres, which result in approximately 25–30 s at a gravity level = $10^{-2}g$ (the so-called "microgravity phase") [31].

We found that the EMEC module was suitable for measuring the activity of LOX-1, because the Xenon arc lamp was stable over a long enough period to obtain a linear dioxygenation of the substrate by the enzyme (Fig. 3A). The progress curves of the LOX-1-catalyzed reaction measured in the EMEC module on the ground and in flight are shown in Figs. 3B and 3C, respectively. Interestingly, the absorbance at time 0 was almost identical for the respective linoleic acid concentrations in the ground and flight samples, suggesting that flight conditions (e.g., vibrations) did not affect the kinetics. Reaction rates calculated within this time frame showed that LOX-1 activity depended on substrate concentration according to a typical Michaelis–Menten kinetics (Fig. 3D), yielding an apparent Michaelis–Menten constant (K_m) of 10.5±0.5 and 2.6±0.1 μM, on the ground and in flight, respectively (Table 1). The apparent maximum velocity (V_{max}) was 22±1 and 23±1 μM·min^{-1}, on the ground and in flight; therefore, we showed that microgravity reduces the K_m of lipoxygenase activity on linoleate to one-fourth of the 1 g control, without affecting the V_{max} (Table 1). Consequently, the catalytic efficiency of LOX-1 (k_{cat}/K_m) was approximately four-fold higher in flight than on the ground (Table 1). The observation that K_m but not V_{max} was affected suggests that microgravity facilitates the encounter of the enzyme with the substrate to form the complex (ES). The gravity appears to affect the diffusion process which leads to enzyme catalysis. Indeed, in diffusion-controlled reactions or reaction steps, macroscopic concentration patterns can be formed from an initially homogeneous solution by way of nonlinear dynamics processes [32]. Such processes lead to concentration (density) fluctuations, which are subject to a buoyancy force under gravity. This small, directional, gravity-driven molecular transport can affect molecule–molecule interaction, as shown in microtubule self-organization [26,32]. The different dimensions of the molecules involved

Table 1

Kinetic parameters for the dioxygenation of linoleic acid by lipoxygenase-1 (LOX-1)

LOX-1	K_m (μM)	V_{max} (μM min^{-1})	k_{cat} (s^{-1})	k_{cat}/K_m (M^{-1} s^{-1})
On the ground (1 g) in Lambda 18	12.5±0.6	25±1	52	4.2×10^6
On the ground (1 g) in EMEC	10.5±0.5	22±1	46	4.4×10^6
In flight (=10^{-2}g) in EMEC	2.6±0.1*	23±1	48	18.5×10^6*

*Denotes $p < 0.01$ compared to on-ground controls in EMEC ($n = 6$).

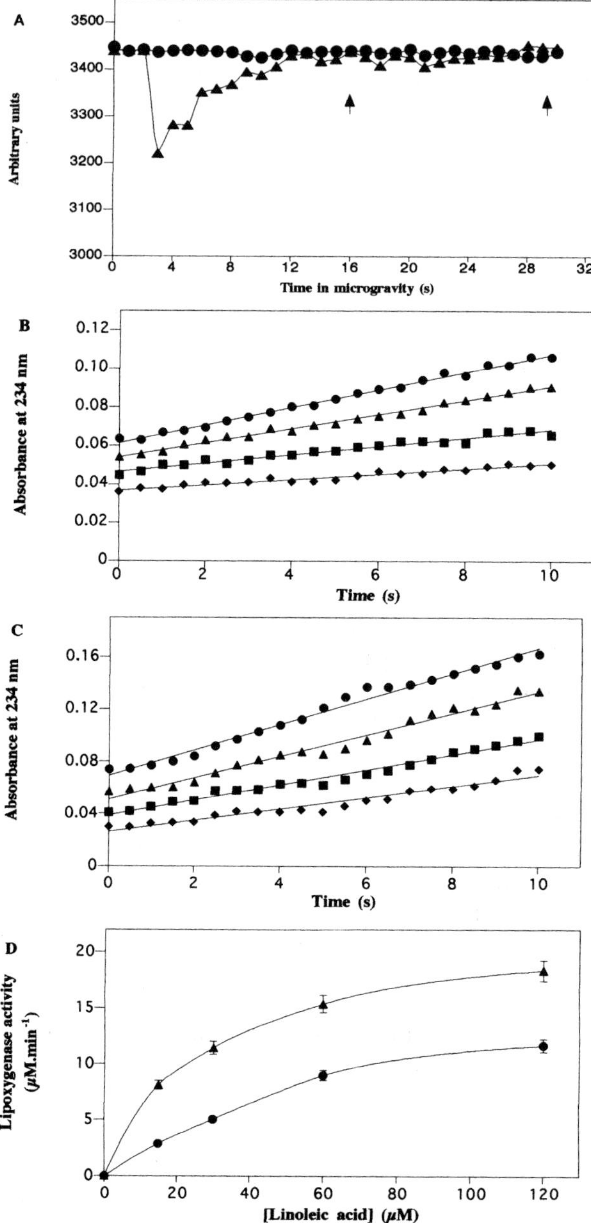

Fig. 3. Activity of soybean (*Glycine max*) lipoxygenase-1 (LOX-1) on the ground and in flight. (A) Response of the Xenon arc lamp of the EMEC module, on the ground (circles) or in flight (triangles). Arrows indicate the time window used to assay LOX-1 activity. (B) Time-course of the dioxygenation of linoleic acid by LOX-1 in the EMEC module on the ground. (C) Time-course of the dioxygenation of linoleic acid by LOX-1 in the EMEC module in flight. In panels B–C, linoleic acid was used at concentrations of 15 μM (diamonds), 30 μM (squares), 60 μM (triangles) or 120 μM (circles). (D) Dependence of LOX-1 activity on linoleic acid concentration, on the ground (circles) or in flight (triangles).

in microtubule self-organization and in enzyme–substrate interactions leave open the possibility that other gravity-dependent factors might control LOX-1 catalysis in low gravity conditions, like the partitioning of linoleic acid at the water/oil interface [33]. This would explain the observation that other enzymes, which work on water soluble, low molecular weight substrates, did not show any effect of gravity on their activity [34]. Since type-1 LOX is the main lipoxygenase in plants and shares with mammalian lipoxygenases several structural and mechanistic properties [9,13], these results could have a broader meaning. They might also form a biochemical background for the immunodepression [24,35] and the bone mass reduction [36] observed in humans during space missions, as discussed later in this review.

Effect of altered gravity on cellular 5-lipoxygenase and cyclooxygenase

Arachidonate metabolites generated by the activity of 5-LOX (the so-called "lipoxygenase pathway") are responsible for lymphocyte maturation [37] and programmed death (apoptosis) of neuronal cells [38]. Therefore, 5-LOX might be relevant for activities in space, because among the most striking effects of space environment are indeed those on T lymphocyte activation [39–41], neuronal cell growth and suspectedly apoptosis [35,36,42].

In the past few years, we studied the possible effects of gravity on the activity and expression of 5-LOX, by subjecting human erythroleukemia K562 cells to simulated hypogravity or hypergravity [43]. We found that exposure of K562 cells to simulated microgravity (by clinorotation) or hypergravity (by centrifugation) for 12 hours did not affect cell viability (not shown), while it significantly affected the activity of 5-lipoxygenase (Table 2). 5-LOX activity reached 35% of the $1 \times g$ control when K562 cells were subjected to $0.00049 \times g$ for 12 hours, whereas hypergravity stimulus at $22 \times g$ for the same period of time enhanced 5-LOX activity up to 250% of the control value. Remarkably, the activity changes were paralleled by changes in the content of 5-LOX protein, which decreased down to 45% of the $1 \times g$ control upon exposure of K562 cells to $0.00049 \times g$, and increased up to 220% after centrifugation of cells at $22 \times g$ (Table 2). Therefore, altered gravity affected cellular 5-LOX activity by modulating gene expression at the translational level. It seems noteworthy that hypo- and hypergravity affected cellular 5-LOX in opposite ways, suggesting a dependence of 5-LOX gene expression on the gravitational field. A similar dependence was found upon lymphocyte activation by mitogens, which was depressed by hypogravity and enhanced by hypergravity [39–41]. Since 5-lipoxygenase plays a critical role in lymphocyte activation [37], it is tempting to suggest that changes in 5-LOX activity might be instrumental in modulating lymphocyte sensitivity to the gravistimulus. This would explain the reduced T lymphocyte activation under microgravity, while enhancement of 5-LOX activity might be instrumental in promoting membrane lipid peroxidation [4,5], which in turn can contribute to cellular sensitivity to hypergravity.

The other branch of the arachidonate cascade in mammals is catalyzed by prostaglandin H synthase (arachidonate, hydrogen-donor:oxygen oxidoreductase, E.C. 1.14.99.1; PHS), which is responsible for the first committed step in the so-called

8

Table 2

Effect of altered gravity on 5-lipoxygenase (5-LOX) activity and expression in human erythroleukemia K562 cells

Force of gravity (g)	5-LOX specific activity (%)	5-LOX protein content (%)
0.00049	35±4	45±5
0.024	55±6	60±6
0.110	80±8	85±9
1	100[a]	100[b]
8	170±17	150±15
14	200±20	180±18
22	250±25	220±22

[a]100% = 0.40±0.05 nmol 5-HPETE per min per mg protein.
[b]100% = 0.45±0.05 absorbance unit at 405 nm.

"cyclooxygenase pathway" of arachidonate metabolism. This pathway leads to the production of prostanoids (prostaglandins, thromboxanes and prostacyclin), potent bioactive molecules [44]. PHS is a dual enzyme, with cyclooxygenase and peroxidase activity on the same protein molecule. Both activities are required for prostanoid biosynthesis [45]. Arachidonate metabolites generated by PHS regulate several aspects of cell physiology [44], including programmed cell death (apoptosis) of human cells [46]. As mentioned before, evidence is being accumulated on the role of programmed cell death in human adaptation to space conditions [35,36,42,47]. In a recent study, we extended the observations on the effects of altered gravity on the "lipoxygenase pathway" in human erythroleukemia K562 cells [43], by assessing the role of gravity in modulating the activity and expression of PHS in the same cells [48].

Exposure of K562 cells to hypogravity or hypergravity under the same conditions described above for 5-LOX [43], significantly affected the cyclooxygenase activity of PHS (Table 3). This activity reached 185% of the $1 \times g$ control when K562 cells were subjected to $0.00049 \times g$ for 12 hours, whereas hypergravity stimulus at $22 \times g$ for the same period of time reduced PHS activity down to 50% of the control value. Peroxidase activity of PHS in K562 cells (1.2 ± 0.1 mU per μg protein) followed the cyclooxygenase activity, increasing up to 2.0 ± 0.2 mU per μg protein at $0.00049 \times g$, and decreasing down to 0.5 ± 0.1 mU per μg protein at $22 \times g$. Remarkably, the activity changes were paralleled by changes in the content of PHS protein, which increased up to 155% of the $1 \times g$ control upon exposure of K562 cells to $0.49 \times 10^{-3} \times g$, and decreased down to 60% after exposure to $22 \times g$ (Table 3). Therefore, altered gravity affected also PHS activity by modulating gene expression at the translational level. Thus hypo- and hypergravity affected PHS of K562 cells in opposite ways, suggesting also in this case a dependence of PHS gene expression on the gravitational field. It seems noteworthy, however, that the effects of gravitational field were specular in the two enzymes under the same conditions (Table 2), suggesting that hypergravity splitted the arachidonate cascade in favour of the lipoxygenase products (leuko-trienes and lipoxins), whereas hypogravity favoured the formation of prostanoids.

Table 3

Effect of altered gravity on prostaglandin H synthase (PHS) cyclooxygenase activity and expression in human erythroleukemia K562 cells

Force of gravity (g)	PHS specific activity (%)	PHS protein content (%)
0.00049	185±20	155±16
0.024	145±15	130±13
0.110	110±12	105±11
1	100[a]	100[b]
8	85±9	90±10
14	75±8	80±8
22	50±6	60±6

[a]100% = 0.50±0.05 nmol O_2 per min per μg protein.
[b]100% = 0.40±0.05 absorbance units at 405 nm.

The possible involvement of this switch in arachidonate metabolism on lymphocyte activation by mitogens, depressed by hypogravity and enhanced by hypergravity, should be investigated. At any rate, these results suggest that modulation of PHS activity and expression might mediate the sensitivity of human cells to the gravi-stimulus, which might be relevant to the control of the immune system, as well as of the hormonal networks in which eicosanoids play a role. The switch of the arachidonate cascade in favour of either the cyclooxygenase or the lipoxygenase pathway, under microgravity or hypergravity, respectively, might also be relevant in determining the choice between cell survival or (programmed) death.

Involvement of lipoxygenases in apoptosis and lymphocyte maturation

Cellular membranes are the primary site of action of several PCD inducers [49,50] and lipid messengers have long been known to act as regulators of apoptosis [51]. Indeed, mobilization of esterified fatty acids from membranes represents a key regulatory step in cellular responses to various stimuli, such as growth factors, cytokines, chemokines and circulating hormones [52]. The influence of fatty acids on both signal transduction through interactions with protein kinases, lipases or G proteins [37], and gene transcription through interactions with nuclear receptors in the peroxisomal proliferator-activated receptor (PPAR) family [53], is increasingly apparent. More recently, attention has been drawn to the role of the lipoxygenase branch of the arachidonate cascade [54] in the execution of mammalian cell apoptosis. In fact lipoxygenases, by introducing oxygen into the fatty acid moieties of (phospho)lipids [55–57], can alter fluidity and permeability of biomembranes [58]. These membrane modifications may play a regulatory role also in the physiological clearance mechanism of apoptotic cells [59]. Moreover, lipid oxidation products can cause protein damage by reacting with lysine, cysteine, and histidine side chains [60]. Besides inducing membrane and protein modifications, LOXs can also generate reactive oxygen species (ROS), which are known to greatly impact PCD [61,62].

Table 4

Pro-apoptotic stimuli which upregulate lipoxygenase activity and cellular targets for lipoxygenase interaction

Stimulus	Ref.	Molecular target	Ref.
TNFα	69	Membrane lipids	70
TGFβ1 ± cisplatin	70	Phospholipase A$_2$	78
Fas/APO-1 ligand	71	Phospholipase C	79
Retinoic acid	38	Mitochondria	80
Hydrogen peroxide	38	Free radicals	81
Thapsigargin	72	Calcium stores	82
Bleomycin	73	Caspases	71
Tamoxifen	74	MAPK phosphatase-1	83
Sodium butyrate	75	P38-MAPK	84
X-ray irradiation	76	Protein kinase C	79
NSAIDs	77	ERK1/2, Ras	79, 85

TNFα, tumour necrosis factor α; TGFβ1, transforming growth factor β1; NSAIDs, nonsteroidal anti-inflammatory drugs; MAPK, mitogen-activated protein kinase; ERK, extracellular regulated kinase.

Furthermore, the hydroperoxides generated by LOXs from the free fatty acids released from membranes by phospholipase activity [63] can act as lipid messengers along different apoptotic pathways. Lipid peroxides have long been considered critical in apoptosis and LOX products have been shown to induce PCD in human T cells [64,65], neutrophils [66], PC12h cells [67] and Jurkat cells [68]. Therefore, it is not surprising that the activation of lipoxygenase pathway has been correlated with programmed death of different cells and tissues, challenged by different, unrelated stimuli (Table 4 and references therein). The molecular targets for lipoxygenase interaction which might contribute to the induction of apoptosis have also been identified (Table 4 and references therein). Altogether, the available experimental evidence supports the scheme shown in Fig. 4. Chemical (drugs, oxidative stress, serum starvation) or physical (UV light) PCD inducers, as well as apoptotic pathways mediated by receptors such as CD28, interleukin receptor or Fas, appear to involve LOX activation, associated to increased intracellular calcium and decreased mito-chondrial potential. These events are paralleled by alterations of: (i) membrane properties (lipoperoxidation, exposure of phosphatidylserine, increased levels of cholesterol and consequent altered Ras expression, generation of free hydroper-oxides); (ii) cytoskeleton (at the level of coactosin-like protein, lamins, Gas 2 and α-fodrin), and (iii) gene transcription (through NF-κB, poly(ADP-ribose) poly-merase and ROS).

It has been shown that the inhibition of 5-LOX favours apoptosis in immune and inflammatory cells [66,86], whereas the activation of 5-LOX causes PCD in various human cancer cells in culture [38]. Some lipoxygenase-related events have been recently identified also in apoptotic human peripheral lymphocytes, forced into PCD by treatment with bleomycin [73]. It should be noted that even the expression of interleukin-2 (IL-2) and the activation of NF-κB in peripheral T lymphocytes require

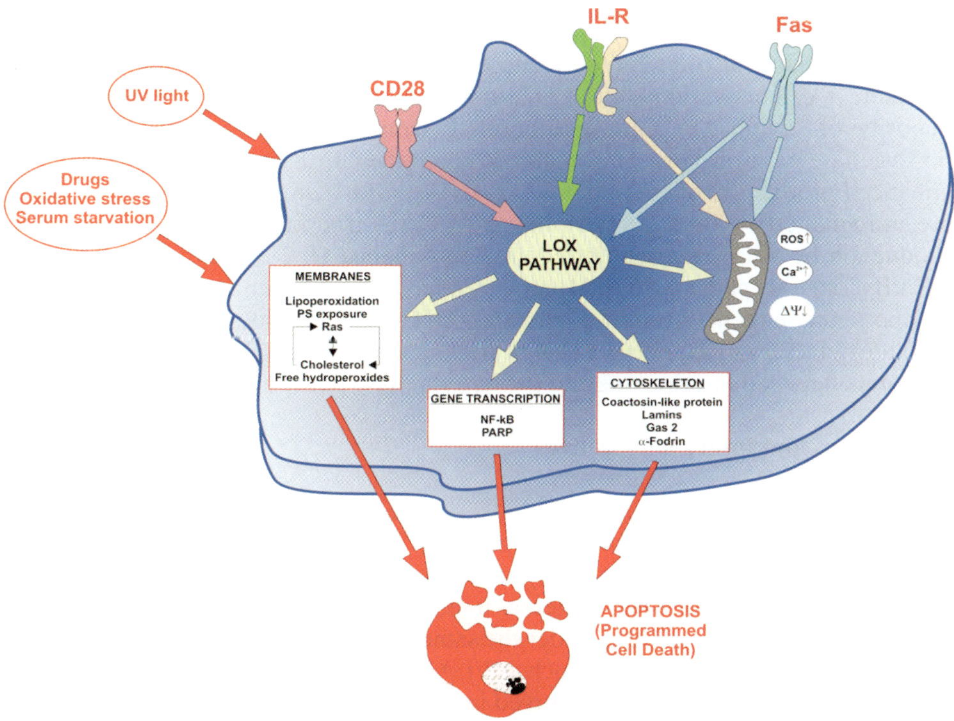

Fig. 4. The lipoxygenase (LOX) pathway can have a central role in apoptosis (programmed cell death), induced by chemical (drugs, oxidative stress, serum starvation) or physical (UV light) agents. Also apoptotic pathways mediated by receptors, such as CD28, interleukin receptor (IL-R) or Fas, can occur through LOX activation, which is associated to increased levels of reactive oxygen species (ROS) and cytosolic calcium (Ca^{2+}), decreased mitochondrial potential ($\Delta\Psi$), and alteration of: (i) membranes, (ii) cytoskeleton, and (iii) gene transcription. NF-κB, nuclear factor-κB; PARP, poly(ADP-ribose) polymerase; PS, phosphatidylserine.

production of ROS by 5-LOX [37]. Noteworthy, ROS, IL-2, NF-κB, as well as c-fos, c-myc and ras, are critical elements for T lymphocyte activation. Over the past two decades lymphocyte activation by concanavalin A has been proved to be extremely sensitive to the gravistimulus, being depressed by hypogravity [39,40] and enhanced by hypergravity [41]. However, the biochemical events responsible for this sensitivity of immune cells to the gravitational field are still largely obscure. The series of events leading to full T lymphocyte activation is complex, and several aspects are still unclear and controversial [87,88]. Briefly, the response of T cells might be essentially of two types: (i) after expression of early genes, a number of cytokines are secreted including interferon-γ and IL-2; (ii) the cells which are in a "resting" G_o phase are induced to enter the cell cycle leading to mitosis. This process takes approximately 70 hours and is characterised by important morphological changes. The expression of IL-2 and of its receptors is a key element of the activation of T cells [89], and has been recently reported to be inhibited by microgravity in mitogen-activated T lymphocytes

12

[24]. On the other hand, IL-2 expression is strongly dependent on the 5-lipoxygenase pathway, which is also inhibited by microgravity [43]. An important costimulus for complete T cell activation is the CD28 surface receptor, which is expressed on the majority of thymocytes and peripheral T lymphocytes [37] and activates the lipoxygenase pathway (Fig. 4). A major effect of the CD28 costimulus consists of the enhanced production of cytokines (IL-2, INF-γ, GM-CSF, TNF and others), which are upregulated by both transcriptional and post-transcriptional processes. ROS production by 5-lipoxygenase is required to enhance IL-2 expression and activate NF-κB along this pathway [37]. Moreover, it has recently been reported that human 15-lipoxygenase gene promoter contains DNA binding sites for IL-13-induced regulatory factors in monocytes [90], a finding which strengthens the concept of a cross-talk between interleukins and the lipoxygenase pathway.

Conclusions and future perspectives

Overall, the available data suggest that lipoxygenase is a "gravity sensor" (i.e., a molecular target for gravity), which might contribute to the cell choice between life and death. Indeed, microgravity depresses LOX in cells (e.g., K562 cells) which remain viable in weightlessness, whereas it might activate LOX in cells (e.g., lympho-cytes) which undergo apoptosis under the same conditions. This effect might require recruitment of ancillary proteins such as FLAP, and might be a critical trigger of the reduced lymphocyte activation observed in microgravity. Work is in progress to test this hypothesis, and to ascertain whether the induction of apoptosis in human lymphocytes plays a role in the immunodepression observed in space. This possibility will be investigated in the framework of a NASA project, during one of the future missions of the Space Shuttle.

Acknowledgements

We wish to thank Mr. Graziano Bonelli for the excellent production of the artwork. This investigation was supported by Agenzia Spaziale Italiana (ASI) under contracts ARS-96-31, ARS-98-156, ARS-99-39 and ARS-00-98.

References

1. Kühn, H. and Borngraber, S. (1999) Mammalian 15-lipoxygenases. Enzymatic properties and biological implications. Adv. Exp. Med. Biol. 447, 5–28.
2. Cathcart, M.K. and Folcik, V.A. (2000) Lipoxygenases and atherosclerosis: protection versus pathogenesis. Free Radic. Biol. Med. 28, 1726–1734.
3. Manev, H., Uz, T., Sugaya, K. and Qu, T. (2000) Putative role of neuronal 5-lipoxygenase in an aging brain. FASEB J. 14, 1464–1469.
4. Maccarrone, M., Bari, M., Corasaniti, M.T., Nisticò, R., Bagetta, G. and Finazzi-Agrò, A. (2000) HIV-1 coat glycoprotein gp120 induces apoptosis in rat brain neocortex by deranging the arachidonate cascade in favor of prostanoids. J. Neurochem. 75, 196–203.

5. Maccarrone, M., Taccone-Gallucci, M., Meloni, C., Cococcetta, N., Manca di Villa-hermosa, S., Casciani, U. and Finazzi-Agrò, A. (1999) Activation of 5-lipoxygenase and related cell membrane lipoperoxidation in hemodialysis patients. J. Am. Soc. Nephrol. 10, 1991–1996.

6. Montero, A. and Badr, K.F. (2000) 15-Lipoxygenase in glomerular inflammation. Exp. Nephrol. 8, 14–19.

7. Heidt, M., Furstenberger, G., Vogel, S., Marks, F. and Krieg, P. (2000) Diversity of mouse lipoxygenases: identification of a subfamily of epidermal isozymes exhibiting a differentiation-dependent mRNA expression pattern. Lipids 35, 701–707.

8. Grechkin, A. (1998) Recent developments in biochemistry of the plant lipoxygenase pathway. Prog. Lipid Res. 37, 317–352.

9. Brash, A.R. (1999) Lipoxygenases: occurrence, functions, catalysis, and acquisition of substrate. J. Biol. Chem. 274, 23679–23682.

10. Boyington, J.C., Gaffney, B.J. and Amzel, L.M. (1993) The three-dimensional structure of an arachidonic acid 15-lipoxygenase. Science 260, 1482–1486.

11. Minor, W., Steczko, J., Stec, B., Otwinowski, Z., Bolin, J.T., Walter, R. and Axelrod, B. (1996) Crystal structure of soybean lipoxygenase L-1 at 1.4 Å resolution. Biochemistry 35, 10687–10701.

12. Sudharshan, E. and Appu Rao, A.G. (1999) Involvement of cysteine residues and domain interactions in the reversible unfolding of lipoxygenase-1. J. Biol. Chem. 274, 35351–35358.

13. Schilstra, M.J., Nieuwenhuizen, W.F., Veldink, G.A. and Vliegenthart, J.F.G. (1996) Mechanism of lipoxygenase inactivation by the linoleic acid analogue octadeca-9,12-diynoic acid. Biochemistry 35, 3396–3401.

14. Glickman, M.H. and Klinman, J.P. (1996) Lipoxygenase reaction mechanism: Demonstration that hydrogen abstraction from substrate precedes dioxygen binding during catalytic turnover. Biochemistry 35, 12882–12892.

15. Clapp, C.H., McKown, J., Xu, H., Grandizio, A.M., Yang, G. and Fayer, J. (2000) The action of soybean lipoxygenase-1 on 12-iodo-cis-9-octadecenoic acid: the importance of C11–H bond breaking. Biochemistry 39, 2603–2611.

16. Shibata, D., Steczko, J., Dixon, J.E., Hermodson, M., Yazdanparast, R. and Axelrod, B. (1987) Primary structure of soybean lipoxygenase-1. J. Biol. Chem. 262, 10080–10085.

17. Began, G., Sudharshan, E. and Appu Rao, A.G. (1999) Change in the positional specificity of lipoxygenase 1 due to insertion of fatty acids into phosphatidylcholine deoxycholate mixed micelles. Biochemistry 38, 13920–13927.

18. Ford-Hutchinson, A.W., Gresser, M., and Young, R.N. (1994) 5-Lipoxygenase. Annu. Rev. Biochem. 63, 383–417.

19. Provost, P., Samuelsson, B. and Radmark, O. (1999) Interaction of 5-lipoxygenase with cellular proteins. Proc. Natl. Acad. Sci. USA 96, 1881–1885.

20. Lam, B.K., Penrose, J.F., Freeman, G.J. and Austen, K.F. (1994) Expression cloning of a cDNA for human leukotriene C4 synthase, an integral membrane protein conjugating reduced glutathione to leukotriene A4. Proc. Natl. Acad. Sci. USA 91, 7663–7667.

21. Jakobsson, P.J., Mancini, J.A., Riendeau, D. and Ford-Hutchinson, A.W. (1997) Identification and characterization of a novel microsomal enzyme with glutathione-dependent transferase and peroxidase activities. J. Biol. Chem. 272, 22934–22939.

22. Lepley, R.A. and Fitzpatrick, F.A. (1998) 5-Lipoxygenase compartmentalization in

granulocytic cells is modulated by an internal bipartite nuclear localizing sequence and nuclear factor kappa B complex formation. Arch. Biochem. Biophys. 356, 71–76.

23. Chen, X.S., Zhang, Y.Y. and Funk, C.D. (1998) Determinants of 5-lipoxygenase nuclear localization using green fluorescent protein/5-lipoxygenase fusion proteins. J. Biol. Chem. 273, 31237–31244.

24. Walther, I. Pippia, P. Meloni, M.A., Turrini, F., Mannu, F. and Cogoli, A. (1998) Simulated microgravity inhibits the genetic expression of interleukin-2 and its receptor in mitogen-activated T lymphocytes. FEBS Lett. 436, 115–118.

25. Hammond, T.G., Lewis, F.C., Goodwin, T.J., Linnehan, R.M., Wolf, D.A., Hire, K.P., Campbell, W.C., Benes, E., O'Reilly, K.C., Globus, R.K. and Kaysen, J.H. (1999) Gene expression in space. Nature Med. 5, 359.

26. Papaseit, C., Pochon, N. and Tabony, J. (2000) Microtubule self-organization is gravity-dependent. Proc. Natl. Acad. Sci. USA 97, 8364–8368.

27. Maccarrone, M., Valensise, H., Bari, M., Lazzarin, N., Romanini, C. and Finazzi-Agrò, A. (2000) Relation between decreased anandamide hydrolase concentrations in human lymphocytes and miscarriage. Lancet 355, 1326–1329.

28. Tacconi, M., Veratti, R., Falciani, P., Giachetti, E., Ranaldi, F. and Vanni, P. (1996) Enzymatic catalysis in low gravity conditions: the E.M.E.C. module. ESA SP 390, 333–338.

29. Maccarrone, M. and Finazzi Agrò, A. (2001) Microgravity increases the affinity of lipoxygenases for free fatty acids. FEBS Lett. 489, 283.

30. Maccarrone, M., Bari, M., Battista, N. and Finazzi-Agrò, A. (2001) The catalytic efficiency of soybean lipoxygenase-1 is enhanced at low gravity. Biophys. Chem. 90, 97–101.

31. Mora, C. and Gharib, T. (1999) Parabolic flight with A300 Zero-G. User's manual. Novespace, Paris.

32. Papaseit, L., Vuillard, C. and Tabony, J. (1999) Reaction–diffusion microtubule concentration patterns occur during biological morphogenesis. Biophys. Chem. 79, 33–39.

33. Maccarrone, M. and Finazzi-Agrò, A. (2001) Enzyme activity in microgravity: a problem of catalysis at the water–lipid interface? FEBS Lett. 504, 80.

34. Giachetti, E., Ranaldi, F. and Vanni P. (2001) Enzyme catalysis in microgravity: an intricate problem to be solved. FEBS Lett. 504, 78–79.

35. Lewis, M.L., Reynolds, J.L., Cubano, L.A., Hatton, J.P., Lawless, B.D. and Piepmeier, E.H. (1998) Spaceflight alters microtubules and increases apoptosis in human lymphocytes (Jurkat). FASEB J. 12, 1007–1018.

36. Sarkar, D., Nagaya, T., Koga, K., Nomura, Y., Gruener, R. and Seo, H. (2000) Culture in vector-averaged gravity under clinostat rotation results in apoptosis of osteoblastic ROS 17/2.8 cells. J. Bone Miner. Res. 15, 489–498.

37. Los, M., Schenk, H., Hexel, K., Baeuerle, P.A., Droge, W. and Schulze-Osthoff, K. (1995) IL-2 gene expression and NF-κB activation through CD28 requires reactive oxygen production by 5-lipoxygenase. EMBO J. 14, 3731–3740.

38. Maccarrone, M., Catani, M.V., Finazzi-Agrò, A. and Melino, G. (1997) Involvement of 5-lipoxygenase in programmed cell death of cancer cells. Cell Death Differ. 4, 396–402.

39. Cogoli, A., Valluchi-Morf, M., Mueller, M. and Briegleb, W. (1980) Effect of hypogravity on human lymphocyte activation. Aviat. Space Environ. Med. 51, 29–34.

40. Cogoli, A., Tschopp, A. and Fuchs-Bslin, P. (1984) Cell sensitivity to gravity. Science 225, 228–230.

41. Lorenzi, G., Fuchs-Bslin, P. and Cogoli, A. (1986) Effects of hypergravity on "whole-blood" cultures of human lymphocytes. Aviat. Space Environ. Med. 57, 1131–1135.

42. Moore, D. and Cogoli, A. (1996) Biological and Medical Research in Space. Springer, Berlin.

43. Maccarrone, M., Putti, S. and Finazzi-Agrò, A. (1998) Altered gravity modulates 5-lipoxygenase in human erythroleukemia K562 cells. J. Gravit. Physiol. 5, 97–98.

44. Eling, T.E., Thompson, D.C., Foureman, G.L., Curtis, J.F. and Hughes, M.F. (1990) Prostaglandin H synthase and xenobiotic oxidation. Annu. Rev. Pharmacol. Toxicol. 30, 1–45.

45. Kulmacz, R.J. and Wang, L.-H. (1995) Comparison of hydroperoxide initiator requirements for the cyclooxygenase activities of prostaglandin H synthase-1 and 2. J. Biol. Chem. 270, 24019–24023.

46. Maccarrone, M., Lorenzon, T., Guerrieri, P. and Finazzi-Agrò, A. (1999) Resveratrol prevents apoptosis in K562 cells by inhibiting lipoxygenase and cyclooxygenase activity. Eur. J. Biochem. 265, 27–34.

47. Wang, E. (1999) Age-dependent atrophy and microgravity travel: what do they have in common? FASEB J. 13, S167–S174.

48. Maccarrone, M., Bari, M., Lorenzon, T. and Finazzi-Agrò, A. (2000) Altered gravity modulates prostaglandin H synthase in human K562 cells. J. Gravit. Physiol. 7, 61–62.

49. Fadok, V.A., Bratton, D.L., Rose, D.M., Pearson, A., Ezekewitz, R.A. and Henson, P.M. (2000) A receptor for phosphatidylserine-specific clearance of apoptotic cells. Nature 405, 85–90.

50. Kagan, V.E., Fabisiak, J.P., Shvedova, A.A., Tyurina, Y.Y., Tyurin, V.A., Schor, N.F. and Kawai, K. (2000) Oxidative signalling pathway for externalization of plasma membrane phosphatidylserine during apoptosis. FEBS Lett. 477, 1–7.

51. McGahon, A.J., Nishioka, W.K., Martin, S.J., Mahboubi, A., Cotter, T.G. and Green, D.R. (1995) Regulation of the Fas apoptotic cell death pathway by Abl. J. Biol. Chem. 270, 22625–22631.

52. Locati, M., Zhou, D., Luini, W., Evangelista, V., Mantovani, A. and Sozzani, S. (1994) Rapid induction of arachidonic acid release by monocyte chemotactic protein-1 and related chemokines. Role of Ca^{2+} influx, synergism with platelet-activating factor and significance for chemotaxis. J. Biol. Chem. 269, 4746–4753.

53. Nolte, R.T., Wisley, G.B., Westin, S., Cobb, J.E., Lambert, M.H., Kurokawa, R., Rosenfeld, M.G., Willson, T.M., Glass, C.K. and Milburn, M.V. (1998) Ligand binding and co-activator assembly of the peroxisome proliferator-activated receptor-γ. Nature 395, 137–143.

54. Ara, G. and Teicher, B.A. (1996) Cyclooxygenase and lipoxygenase inhibitors in cancer therapy. Prostaglandins Leukot. Essential Fatty Acids 54, 3–16.

55. Kühn, H., Belkner, J., Wiesner, R. and Brash, A.R. (1990) Oxygenation of biological membranes by the pure reticulocyte lipoxygenase. J. Biol. Chem. 265, 18351–18361.

56. Feussner, I., Balkenhohl, T.J., Porzel, A., Kühn, H. and Wasternack, C. (1997) Structural elucidation of oxygenated storage lipids in cucumber cotyledons. J. Biol. Chem. 272, 21635–21641.

57. Schnurr, K., Belkner, J., Ursini, F., Schewe, T. and Kühn, H. (1996) The selenoenzyme phospholipid hydroperoxide glutathione peroxidase controls the activity of the 15-lipoxygenase with complex substrates and preserves the specificity of the oxygenation

products. J. Biol. Chem. 271, 4653–4658.
58. Wratten, M.L., Van Ginkel, G., Van't Veld, A.A., Bekker, A., Van Faassen, E.E. and Sevanian, A. (1992) Structural and dynamic effects of oxidatively modified phospholipids in unsaturated lipid membranes. Biochemistry 31, 10901–10907.
59. Godson, C., Mitchell, S., Harvey, K., Petasis, N.A., Hogg, N. and Brady, H.R. (2000) Cutting edge: lipoxins rapidly stimulate nonphlogistic phagocytosis of apoptotic neutrophils by monocyte-derived macrophages. J. Immunol. 164, 1663–1667.
60. Refsgaard, H.H., Tsai, L. and Stadtman, E.R. (2000) Modifications of proteins by polyunsaturated fatty acid peroxidation products. Proc. Natl. Acad. Sci. USA 97, 611–616.
61. Eu, J.P., Liu, L., Zeng, M. and Stamler, J.S. (2000) An apoptotic model for nitrosative stress. Biochemistry 39, 1040–1047.
62. Schaul, Y. (2000) c-Abl: activation and nuclear targets. Cell Death Differ. 7, 10–16.
63. Cummings, B.S., McHowat, J. and Schnellmann, R.G. (2000) Phospholipase A_2s in cell injury and death. J. Pharmacol. Exp. Ther. 294, 793–799.
64. Sandstrom, P.A., Tebbey, P.W., Van Cleave, S. and Buttke, T.M. (1994) Lipid hydroperoxides induce apoptosis in T cells displaying a HIV-associated glutathione peroxidase deficiency. J. Biol. Chem. 269, 798–801.
65. Sandstrom, P.A., Pardi, D., Tebbey, P.W., Dudek, R.W., Terrian, D.M., Folks, T.M. and Buttke, T.M. (1995) Lipid hydroperoxide-induced apoptosis: lack of inhibition by Bcl-2 over-expression. FEBS Lett. 365, 66–70.
66. Hébert, M.-J., Takano, T., Holthöfer, H. and Brady, H.R. (1996) Sequential morphologic events during apoptosis of human neutrophils. Modulation by lipoxygenase-derived eicosanoids. J. Immunol. 157, 3105–3115.
67. Aoshima, H., Satoh, T., Sakai, N., Yamada, M., Enokido, Y., Ikeuchi, T. and Hatanaka, H. (1997) Generation of free radicals during lipid hydroperoxide-triggered apoptosis in PC12h cells. Biochim. Biophys. Acta 1345, 35–42.
68. Liberles, S.D. and Schreiber, S.L. (2000) Apoptosis-inducing natural products found *in utero* during murine pregnancy. Chem. Biol. 7, 365–372.
69. O'Donnell, V.B., Spycher, S. and Azzi, A. (1995) Involvement of oxidants and oxidant-generating enzyme(s) in tumour-necrosis-factor-α-mediated apoptosis: role for lipoxygenase pathway but not mitochondrial respiratory chain. Biochem. J. 310, 133–141.
70. Maccarrone, M., Nieuwenhuizen, W.F., Dullens, H.F.J., Catani, M.V., Melino, G., Veldink, G.A., Vliegenthart, J.F.G. and Finazzi-Agrò, A. (1996) Membrane modifications in human erythroleukemia K562 cells during induction of programmed cell death by transforming growth factor β1 or cisplatin. Eur. J. Biochem. 241, 297–302.
71. Wagenknecht, B., Schulz, J.B., Gulbins, E. and Weller, M. (1998) Crm-A, bcl-2 and NDGA inhibit CD95L-induced apoptosis of malignant glioma cells at the level of caspase 8 processing. Cell Death Differ. 5, 894–900.
72. Zhou, Y.P., Teng, D., Dralyuk, F., Ostrega, D., Roe, M.W., Philipson, L. and Polonsky, K.S. (1998) Apoptosis in insulin-secreting cells. Evidence for the role of intracellular Ca^{2+} stores and arachidonic acid metabolism. J. Clin. Invest. 101, 1623–1632.
73. Vernole, P., Tedeschi, B., Caporossi, D., Maccarrone, M., Melino, G. and Annicchiarico-Petruzzelli, M. (1998) Induction of apoptosis by bleomycin in resting and cycling human lymphocytes. Mutagenesis 13, 209–215.
74. Maccarrone, M., Fantini, C., Ranalli, M., Melino, G. and Finazzi-Agrò, A. (1998)

Activation of nitric oxide synthase is involved in tamoxifen-induced apoptosis of human erythroleukemia K562 cells. FEBS Lett. 434, 421–424.

75. Ikawa, H., Kamitani, H., Calvo, B.F., Foley, J.F. and Eling, T.E. (1999) Expression of 15-lipoxygenase-1 in human colorectal cancer. Cancer Res. 59, 360–366.

76. Matyshevskaia, O.P., Pastukh, V.N. and Solodushko, V.A. (1999) Inhibition of lipoxygenase activity reduces radiation-induced DNA fragmentation in lymphocytes. Radiat. Biol. Radioecol. 39, 282–286.

77. Shureiqi, I., Chen, D., Lee, J.J., Yang, P., Newman, R.A., Brenner, D.E., Lotan, R., Fischer, S.M. and Lippman, S.M. (2000) 15-LOX-1: a novel molecular target of nonsteroidal anti-inflammatory drug-induced apoptosis in colorectal cancer cells. J. Natl. Cancer Inst. 92, 1136–1142.

78. Tang, K., Nie, D. and Honn, K.V. (1999) Role of autocrine motility factor in a 12-lipoxygenase dependent anti-apoptotic pathway. Adv. Exp. Med. Biol. 469, 583–590.

79. Szekeres, C.K., Tang, K., Trikha, M. and Honn, K.V. (2000) Eicosanoid stimulates extracellular regulated kinase1/2 in epidermoid carcinoma cells via multiple signalling pathways. J. Biol. Chem. 275, 38831–38841.

80. Biswal, S.S., Datta, K., Shaw, S.D., Feng, X., Robertson, J.D. and Kehrer, J.P. (2000) Glutathione oxidation and mitochondrial depolarization as mechanisms of nordihydroguaiaretic acid-induced apoptosis in lipoxygenase-deficient FL5.12 cells. Toxicol. Sci. 53, 77–83.

81. Datta, K., Biswal, S.S. and Kehrer, J.P. (1999) The 5-lipoxygenase-activating protein (FLAP) inhibitor, MK886, induces apoptosis independently of FLAP. Biochem. J. 340, 371–375.

82. Buyn, T., Dudeja, P., Harris, J.E., Ou, D., Seed, T., Sawlani, D., Meng, J., Bonomi, P. and Anderson, K.M. (1997) A 5-lipoxygenase inhibitor at micromolar concentration raises intracellular calcium in U937 cells prior to their physiologic cell death. Prostaglandins Leukot. Essent. Fatty Acids 56, 69–77.

83. Metzler, B., Hu, Y., Sturm, G., Wick, G. and Xu, Q. (1998) Induction of mitogen-activated protein kinase phosphatase-1 by arachidonic acid in vascular smooth muscle cells. J. Biol. Chem. 273, 33320–33326.

84. Aoshiba, K., Yasui, S., Nishimura, K. and Nagai, A. (1999) Thiol depletion induces apoptosis in cultured lung fibroblasts. Am. J. Respir. Cell. Mol. Biol. 21, 54–64.

85. Maccarrone, M., Bellincampi, L., Melino, G. and Finazzi-Agrò, A. (1998) Cholesterol, but not its esters, triggers programmed cell death in human erythroleukemia K562 cells. Eur. J. Biochem. 253, 107–113.

86. Tang, D.G., Chen, Y.Q. and Honn, K.V. (1996) Arachidonate lipoxygenases as essential regulators of cell survival and apoptosis. Proc. Natl. Acad. Sci. USA 93, 5241–5246.

87. Berridge, M. (1993) Inositol triphosphate and calcium signalling. Nature 351, 315–325.

88. Rasmussen, H. (1995) Intracellular signalling: from simplicity to complexity. ASGSB Bulletin 8, 7–17.

89. Gaulton, G.N. and Williamson, P. (1994) Interleukin-2 and the interleukin-2 receptor complex. In: Lymphocyte Activation. Chem. Immunol. (Samelson, L.E. ed.), pp. 91–114. Basel, Karger.

90. Kelavkar, U., Wang, S., Montero, A., Murtagh, J., Shah, K. and Badr, K. (1998) Human 15-lipoxygenase gene promoter: analysis and identification of DNA binding sites for IL-13-induced regulatory factors in monocytes. Mol. Biol. Rep. 25, 173–182.

Cell Biology and Biotechnology in Space
A. Cogoli (editor)
© 2002 Elsevier Science B.V. All rights reserved

Microtubule Self-organisation and its Gravity Dependence

James Tabony[1]*, Nicolas Glade[1,2], Cyril Papaseit[1] and
Jacques Demongeot[2]

[1]*Commissariat à l'Energie Atomique, Département de Biologie Moléculaire et
Structurale, Laboratoire de Résonance Magnétique en Biologie Métabolique, D.S.V,
C.E.A. Grenoble, Grenoble, France*
[2]*Institut d'Informatique et Mathématique Appliquées de Grenoble, Laboratoire de
Technique de l'Imagerie, de la Modélisation et de la Cognition, Faculté de Médecine,
Domaine de la Merci, La Tronche, France*

Abstract

The molecular processes by which gravity affects biological systems are poorly, if at
all, understood. Under equilibrium conditions, chemical and biochemical reactions
do not depend upon gravity. It has been proposed that biological systems might
depend on gravity by way of the bifurcation properties of certain types of non-linear
chemical reactions that are far-from-equilibrium. In such reactions, the initially
homogenous solution spontaneously self-organises by way of a combination of
reaction and diffusion. Theoreticians have predicted that the presence or absence of
an external field, such as gravity, at a critical moment early in the self-organising
process may determine the morphology that subsequently develops. We have found
that the formation *in vitro* of microtubules, a major element of the cellular skeleton,
shows this type of behaviour. The microtubule preparations spontaneously self-
organise by way of reaction and diffusion, and the morphology of the state that forms
depends upon gravity at a critical bifurcation time early in the process. Experiments
carried out under low gravity conditions show that the presence of gravity at the
bifurcation time actually triggers the self-organising process. This is an experimental
demonstration of how a very simple biochemical system, containing only two
molecules, can be gravity sensitive. At a microscopic level the behaviour results from
an interaction of gravity with the concentration and density fluctuations that arise
from processes of microtubule shortening and elongation. We have developed a
numerical reaction–diffusion scheme, based on the chemical dynamics of a population of microtubules, that simulate self-organisation. These simulations provide

---

*Corresponding author.

insight into how self–organisation occurs at a microscopic level and how gravity triggers this process. Recent experiments on cell lines cultured in space suggest that microtubule organisation may not occur properly under low gravity conditions. As microtubule organisation is essential to cellular function, it is quite plausible that the type of processes described in this article provide an underlying explanation for the gravity dependence of living systems at a cellular level.

Introduction

Experiments in space furnish evidence that various cellular processes, such as growth rates, signalling pathways and gene expression are modified when cells are grown under conditions of weightlessness [1–3]. At present there is no coherent explanation for these observations, nor is it known what biomolecules are involved. No cellular component has been identified as having a sufficiently large density difference with the surrounding medium that the force exerted on it by gravity is larger than the forces involved with random thermal motions. Chemical and biochemical reactions are mostly thought of as being independent of gravity. One possible mechanism by which gravity may intervene in biochemical mechanisms is by way of the bifurcation properties of certain types of reaction–diffusion processes [4–7]. These processes lead to the progressive appearance of a macroscopically self-organised state from an initially homogenous solution and it has been calculated that the presence of gravity at a critical moment during the early stages of self-organisation can determine the morphology of the state that subsequently develops [8,9].

Since the early 1950s, certain theoreticians [4,5] have proposed that some particular types of chemical or biochemical reactions that are far-from-equilibrium might exhibit non-linear phenomena. Turing and Prigogine predicted that such chemical systems could show macroscopic self-ordering, and that a chemical pattern could spontaneously arise from an initially homogeneous solution. At a molecular level this process involves an appropriate combination of reaction and diffusion, and the patterns that form are comprised of periodic variations in the concentration of some of the reactants. Structures of this type are called reaction–diffusion or Turing structures. They also go under the name of dissipative structures. The latter term was widely used by Prigogine and co-workers because a dissipation of chemical energy through the system is required to drive and maintain the system far from equilibrium. It is this energy dissipation that provides the thermodynamic driving force for the self-ordering process. Turing, Prigogine et al. [4,10–12] and others have proposed that biochemical mechanisms of this type could provide an underlying explanation for biological pattern formation and morphogenesis.

In addition to self-organisation, such systems can also show bifurcation properties [12–14]. At a critical moment early in the development of the self-organised state, the system may bifurcate between dynamic pathways leading to self-organised states of different morphology. Prigogine and co-workers explicitly calculated that in some reaction–diffusion systems, the presence of gravity at the bifurcation point [8,9] could determine the morphology of the state that subsequently forms. Various authors [15,16] have proposed that biochemical systems acting upon such principles could act as a gravity transducer and provide a possible underlying physical–chemical explanation for the dependence of biological processes on gravity.

These concepts, although the subject of interest and debate, have not been adopted by the majority of chemists and biologists. One of the reasons has been that until recently no experimental examples of this general behaviour were known of. For example, it was not until 1990 [17] that a variation of a chemical reaction, initially discovered by Belousov around 1950 [18], was finally recognised as the first example of a Turing structure. Similarly, in biology, there have been no examples of *in-vitro* biochemical reactions showing the self-ordering properties due to these causes. We have found that under appropriate conditions, *in-vitro* preparations of microtubules, a major component of the cytoskeleton, spontaneously self-organise by reaction–diffusion processes [15,19–30]. They form macroscopic patterns whose overall morphology depends upon the gravity direction at a critical moment early in the progressive development of the self-organised state. Experiments under low gravity conditions show that self-organisation is triggered by the presence of gravity at the bifurcation time [26].

Microtubules [31,32] are chemically anisotropic, growing and shrinking along the direction of their long axis. This leads to the formation from their shrinking and growing ends of chemical trails comprised of regions of high and low local tubulin concentration, respectively. These concentration trails are oriented along the direction of the microtubule. Because reaction rates increase with increasing concentration, neighbouring microtubules will preferentially grow into regions where the local concentration of tubulin is highest. When microtubules first form from the tubulin solution they are still in a growing phase and have an isotropic arrangement. However, this isotropic arrangement becomes unstable once significant disassembly from the shrinking end occurs. At this point, if a few microtubules start to take up a preferred orientation then neighbouring microtubules will also grow into the same orientation. Once started, the process mutually reinforces itself with time and leads to self-organisation. At the bifurcation time, when the isotropic arrangement is unstable, any effect that leads to a slight directional bias in the reaction–diffusion process, such as somewhat different rates of molecular transport in the up–down and left–right directions, will trigger self–organisation. Gravity acts by way of its directional interaction with the macroscopic density fluctuations present in the solution arising from microtubule disassembly at the instability.

Conceptual background

Self-organisation by way of non-linear dynamics

One phenomenon shown by most living systems—and one that partially distinguish them from the majority of inert systems—is self-organisation. This phenomenon occurs at numerous levels encompassing different distance scales and types of basic

22

units. Macromolecules self-organise within a cell, cells organise into organisms, organisms self-organise into colonies, etc. It is becoming increasingly apparent that macroscopic self-organisation need not be a consequence of static interactions, but can arise by way of the non-linear properties that describe the dynamics of the system. What is beginning to emerge is that such systems show certain general types of behaviour that are largely independent of the detailed properties of the individual components as such, be they molecules or collections of organisms.

In molecular systems, non-linear properties arise under far-from-equilibrium conditions. Systems at thermodynamic equilibrium are described by mathematical relationships of linear form. On the contrary, systems that are far-from-equilibrium can be described by non-linear mathematical relationships. Energy of some sort is needed to initially drive and maintain the system out-of-equilibrium. Living systems continually consume energy. This maintains them out of thermodynamic equilibrium, and they are thus likely to show a multitude of non-linear phenomena. On the contrary inert or dead systems, are mostly either at, or close to, thermodynamic equilibrium, and they hence in general show a linear behaviour.

The turbulent movement of liquids and gases are physical systems that show a non-linear behaviour and with which we are familiar on a daily basis. If a liquid is vigorously stirred, a vortex appears. When a bathtub is emptied, a stationary spiral pattern appears close to the plug-hole where the liquid is in turbulent movement. The cyclic variations in atmospheric pressure shown on meteorological maps result from the movement of the gases in the atmosphere.

These phenomena illustrate an important property of out-of-equilibrium systems; namely macroscopic self-organisation due to dynamics. A stationary or temporal pattern arises from a homogenous, or close to homogenous, starting point. The thermodynamic driving force for the formation of and maintenance of this macroscopic order, is the flux or dissipation of energy through the system. For this reason, structures of this type are often called dissipative.

Well-studied examples of hydrodynamic self-ordering occur in the Bénard and Taylor instability experiments [12,14,33,34]. In the Bénard experiment, liquid several millimetres deep in a Pétri dish, is heated from below so as to produce convection. Under suitable conditions, the convection currents spontaneously self-organise, and a macroscopic hexagonal pattern forms. In the Taylor experiment a viscous liquid is contained between two cylinders that can be rotated with respect to one another at a chosen speed. At rest, and low rates of rotation, the liquid appears uniform. With increasing rate of rotation, a number of changes occur, and for certain values, equally spaced stationary bands are seen. In the out-of-equilibrium region corresponding to the banded structure, the equations describing the hydrodynamics of the liquid are non-linear. The solution to these non-linear equations leads to periodic variations as a function of position in the parameters that describe the movement of the liquid. A central feature of this type of behaviour is that the pattern arises through the dynamics of the system, and not by way of static interactions between the individual components. We are mostly familiar with the concept that stationary structures arise through static forces between objects. We are much less familiar with the idea that stationary patterns can arise through dynamics. The Bénard and Taylor instability experiments are good illustrations of the latter type of phenomena. Although the patterns are stationary, they are in fact made up of liquid which is in constant movement. At any one point in the stationary pattern, the liquid is being constantly replaced. Moreover, when the movement stops, and no further energy is dissipated through the system, the pattern disappears and the system returns to an homogenous uniform state.

Another example of dynamic self-organisation is the macroscopic patterns generated by certain colonies of bacteria [35]. Over a distance scale of several centimetres, the colony appears as a stationary pattern. Observations at higher magnification show that the pattern is comprised of regions containing differing bacterial densities. At even higher magnification, where the individual bacteria are observed, the bacteria are seen to be undergoing a rapid, seemingly random movement. It is out of this dynamic process involving the collective movement of many bacteria that the stationary pattern arises. The individual bacteria interact indirectly with one another via macroscopic gradients in the concentration of chemicals that they themselves produce. The energy source is the chemicals consumed by the bacteria. When energy runs out, the bacteria stop moving and the pattern disappears.

Another example is the formation of ant colonies and the collective behaviour of the ants [36]. Individual ants leave behind them trails of various chemicals to which other ants react. There are two major types of chemicals emitted by ants; those that attract other ants and those that repel them. The combination of the chemical trails produced by the movement of the animals' result in the complex "social" behaviour of the ant population.

Hydrodynamic and ant colony patterns are examples of macroscopic self-organisation that occur via non-linear dynamics. The overall type of behaviour is a consequence of the mathematical form of the equations describing the non-linear dynamics of the system. Nevertheless, the systems are very different; one is a liquid in motion, the other a population of living organisms. The question obviously arises as to how similar processes might occur in systems that are intermediate between these two extremes. In particular, how could they come about at a molecular level by way of biochemical reactions within biological objects such as a cell, or an egg.

Self-organisation by way of reaction–diffusion processes

D'Arcy Thompson [37] at the beginning of the last century saw the resemblance between the patterns and forms produced by physical processes, such as liquids in turbulent motion, and the patterns and forms that arise in biology. He was convinced that there was a common conceptual origin but was unable to express this in mathematical form. Rashevsky in the 1940s [38] proposed that certain types of chemical reactions, when combined with diffusion might lead to a partial separation of some of the chemicals in a developing embryo. However, it was not until 1952 [4] that this approach took a substantial leap forward with the publication by the mathematician Alan Turing of an article entitled "The chemical basis of

24

morphogenesis". Turing expressed in a concise mathematical form the counter-intuitive idea that a combination of reaction and diffusion, can under certain circumstances, lead to spontaneous macroscopic self-organisation from an initially homogenous solution. At the time of its publication in 1952, Turing's paper was largely ignored. The applicability of the second law of thermodynamics under all circumstances was firmly entrenched. In the 1960s and 70s, Prigogine, Glansdorff, Nicolis and co-workers, substantially developed the approach initially proposed by Turing, and in particular the thermodynamic aspects [5,10–12].

Unfortunately, it has taken almost 50 years for the ideas of Turing and Prigogine to be accepted. Their approach was revolutionary because it implied a change in paradigm. The second law of thermodynamics teaches us that order is progressively and ineluctably lost with time. In particular, at a molecular level, an existing macro-scopic order is gradually attenuated by molecular diffusion. Two miscible liquids, initially separated from one another, will slowly mix by way of diffusion, and the existing order progressively lost. Turing predicted that under some circumstances the contrary might occur, i.e. diffusion could lead to the partial separation of chemicals that are initially homogenous. One of the conditions for this is an auto-catalytic reaction that produces other chemicals that both activate and inhibit the formation of the reaction product. If a local fluctuation gives rise to slight excess of activator, then the reaction accelerates and even more activator is produced at this point. Consider a situation in which the inhibitor diffuses at a much faster rate than the activator. This leads to progressively more inhibitor compared to activator with increasing distance from the source of the fluctuation. This in turn now leads to a decreased production of both activator and inhibitor, and a progressive slowing down of the reaction as a function of distance from the fluctuation. At this point in space, a small fluctuation in activator concentration will once again lead to acceleration in the reaction and to a repetition of the behaviour. Stationary periodic variation in the concentration of the reactants can hence spontaneously emerge from the initially homogenous solution.

This pattern arises not from static interactions but is a consequence of the non-linear dynamics describing the way that rates of chemical reaction are coupled and modified by molecular diffusion. It is important to note that at the moment that the chemical pattern starts to emerge from the homogenous solution, the chemical composition of the solution is unstable to any small fluctuations in the concentration of the reactants. This concept of instability is important. Without it the pattern cannot emerge. In the same manner, in the Bénard and Taylor instability experi-ments, the turbulent patterns can only start to form when the homogenous equili-brium state becomes unstable. In the Bénard convective instability, liquid is heated from below. Its density decreases, and it is subject to a buoyancy force due to the density difference between the bottom and the top of the sample. However, initially this buoyancy force is insufficient to overcome viscous drag, and there is no net movement. When the density difference is sufficiently large, viscous drag is over-come, the equilibrium state becomes unstable, less dense liquid rises from the bottom, and more dense liquid descends from the top. The liquid at the top then cools, whereas that at the bottom is heated. In this way, the process is repeated, and a stationary pattern of convective flow is established. This pattern is formed and maintained by the flow of energy through the system implicit in heating the liquid from below.

There are obvious similarities between this type of process and those proposed by Turing and Prigogine. Although, one process is of a physical nature, whereas the other is chemical, both involve driving a system sufficiently far-from-equilibrium by way of a flux of energy, until the equilibrium and homogenous state become unstable. Beyond this point, the non-linear relationships that describe the dynamics coupling macroscopic transport to composition result in the appearance of a pattern. For the Bénard instability the pattern is comprised of density variations resulting from thermal effects, whereas in the Turing system the patterns is comprised of variations in chemical composition. Gravity participates directly in the Bénard experiment in that convection does not arise in the absence of gravity. It is the interaction of gravity with density variations resulting from thermal expansion that generates the buoyancy forces that both destabilise the equilibrium state and subsequently produce the pattern of convective movement.

Bifurcation properties

In addition to self-organisation, non-linear out-of-equilibrium systems also differ from equilibrium systems by showing bifurcation properties, of for example the bistable type [12–14]. Non-linear systems can exist in multiple stationary states. At the bifurcation point, a field too weak to effect an equilibrium state, can favour the dynamic pathway leading to one of the multiple states over the others, and hence determine the morphology of the self-organised state that forms. Just after the bifurcation, the morphology that will subsequently develop is determined, even though at that time no organisation is as yet visible. Prigogine and co-workers explicitly calculated that in some reaction–diffusion systems, the earth's gravitational field could cause a bifurcation [8,9].

This behaviour (Fig. 1) can be qualitatively understood in the following manner. A linear equation such as $y = mx + c$ has only one solution in x for a given value in y. Systems at or close to equilibrium are described by linear equations. Under given conditions, the solution to these linear equations lead to the thermodynamic state as the only stable state. Far-from-equilibrium systems are described by non-linear relationships. A non-linear equation, such as the quadratic $y = ax^2 + bx + c$, has two solutions in x for a given value in y. Hence multiple states, corresponding to different solutions to the non-linear relationships, can exist under the same experimental conditions. In some cases, these mathematical solutions correspond to self-organised states showing macroscopic patterns of different morphology.

As a system is progressively moved away from equilibrium, a point is reached at which its behaviour is no longer approximated by linear relationships. In this region the equilibrium state becomes unstable and the non-linear multiple states appear as stable solutions. At this point, the system adopts, from the several possible new

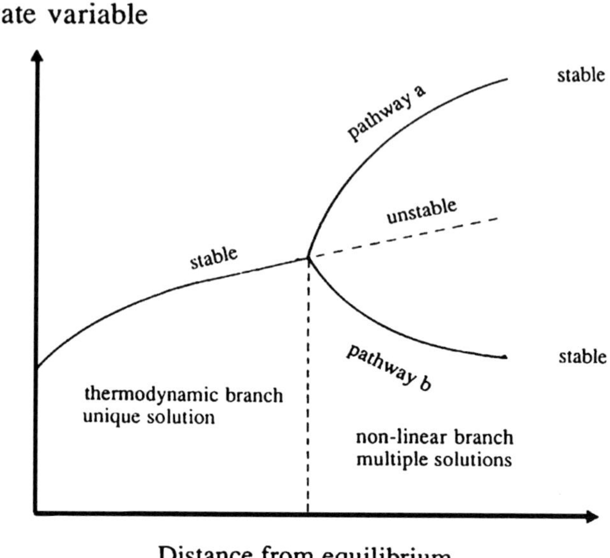

Fig. 1. Schematic representation of a bistable bifurcation in a chemical system. The horizontal axis represents the distance of the system from equilibrium, and the vertical axis a state variable. Near equilibrium, the system has a linear behaviour, and the thermodynamic state is the only permitted stable solution. Far from equilibrium, the system has a non-linear behaviour. At a critical point shown by the dotted line and that corresponds to the bifurcation point, the thermodynamic state becomes unstable, and new stable non-linear states arise. At the bifurcation point, a weak field, such as gravity, can result in a small separation between the different non-linear pathways, and thus favour the formation of one of the possible non-linear states.

non-linear dynamic pathways that are open to it, the pathway that leads to the self-organised morphology that subsequently forms. At the bifurcation point, a field too weak to effect equilibrium states, can determine the dynamic pathway the system takes. Furthermore, the weak field need only be present at the critical moment when the equilibrium state is unstable. Once the bifurcation has occurred, the system evolves progressively along the selected pathway to the pre-determined morphology. It behaves as though it retained a memory of the conditions prevailing at the bifurcation. The bifurcation coincides with the instability in the homogenous equilibrium state. As already mentioned it is this instability which permits the self-organised structures of the Turing and Bénard type to progressively develop from the initially homogenous equilibrium state. The equilibrium state being homogenous, contains no positional information and has perfect symmetry. On the other hand, an organised state contains positional information and is of lower symmetry. For example, in a striped structure, the directions parallel and perpendicular to the stripes are readily differentiated. This is not the case in an homogenous solution where all directions are indistinguishable from one another. The appearance of pattern and form from an homogenous solution is hence associated with symmetry breaking [12,39]. This symmetry breaking coincides with the instability in the uniform state and the bifurcation, and precedes the progressive appearance of the self-organised state. The overall morphology of the self-organised state is determined by conditions at this critical moment. Small differences in conditions at the bifurcation point are sufficient to favour one of the dynamic pathways over the others. Weak fields, and environmental factors, can play a decisive role in selecting out a specific morphology.

A simple analogy to help obtain an understanding for phenomena of this type, is that of a man standing on the edge of a cliff. If his centre of gravity is low, then he is stable and remains stationary. If he progressively raises a large boulder, so that his centre of gravity is raised to the point that he becomes unstable, he will fall. The topology around him is such that he can fall in one of two manners; either to the bottom of the cliff or onto the ground. At the moment of instability, he bifurcates between the two dynamic pathways. A strong gust of wind at this moment will suffice to determine the dynamic pathway taken and hence the final state. Immediately after he has fallen over the cliff he has not yet reached his new state, but it is inevitable. In addition, the determining effect of the wind occurs at the moment of instability. After his fall, it may stop or change direction without any consequence. Imagine now a slightly different situation. The person raises the boulder until he is unstable, then after a time he slowly lowers it again until he is stable once again. During the time that he is unstable, no external effect, such as the wind blowing, occurs that would make him fall. If subsequently, the wind does start to blow, it will not have any effect since he is once again in a stable configuration.

Another situation somewhat closer to the microtubule case is that of ants. For example, when an ant finds a food supply it leaves behind it a trail of a specific chemical. When another ant encounters such a trail it follows it until it arrives at the food source. As it moves away with food, it also leaves a chemical trail. This results in a reinforced trail leading to the food supply. Consider now a situation where there are two identical food sources close to a population of ants [36,40]. One of the food sources is slightly closer to the colony than the other. Two ants from the colony find the different food sources at the same time. As they return to the colony with food they leave behind chemical trails that are then followed by other ants. They in turn deposit chemical indicators that reinforce the original trails. In such a way progressively more and more ants follow the paths to the food sources. However, because the trail from the closer of the two sources is shorter, it takes less time for an ant to return to the colony. This results in a slightly larger number of ants taking the path to the closest food source. This in turn reinforces the strength of the chemical trail of the shorter path at the expense of the longer path. Hence, progressively more and more ants take the shorter path to the closer food supply until they all follow this route. This illustrates how the progressive reinforcement of chemical trails by moving objects that themselves produce the trails, results in self-organisation and a bifurcation. A small effect, in this case the fact that one of the food sources is slightly closer than the other, leads to a progressive reinforcement of one dynamic pathway over the other and thus determines the final state. As we shall see there are many similarities between the principles that govern microtubule self-organisation and the behaviour of ant populations.

Prigogine and Kondepudi [8,9] explicitly calculated that in some types of reaction–diffusion systems, the presence of gravity could strongly effect self-organisation. According to them, the interaction with gravity, of the density fluctuations associated with the macroscopic concentration variations, results in an additional transport process in the vertical direction. This small effect contributes to the reaction–diffusion mechanism and thus effects the final morphology. They predicted that a 1 g field could destabilise the equilibrium state at the bifurcation point, break the symmetry of the system, and thus favour the formation of a macroscopic pattern.

Microtubules

Microtubules are long tubular shaped objects, with inner and outer diameters of about 16 nm and 24 nm, respectively [31,32]. They arise from the self-assembly of a protein, tubulin, by way of reactions involving the hydrolysis of a nucleotide, guanosine triphosphate (GTP), to guanosine diphosphate (GDP). Their length is variable; but often they are several microns long. Tubulin has a monomer molecular weight of about 50 kDa, a diameter of about 4 nm, and occurs as a dimer of the alpha and beta monomer forms. When warmed, from 4 to 35°C, in the presence of GTP, tubulin assembles into microtubules. During this process a series of chemical reactions occurs and GTP is hydrolysed to GDP. Microtubules can be disassembled into tubulin by placing the solution on ice. However, the system is chemically irreversible in that GTP does not reform. Once microtubules are formed, chemical activity continues through processes whereby tubulin is added and lost from opposing ends of microtubules by reactions involving GTP hydrolysis. One such process called "treadmilling" involves already formed microtubules adding and losing tubulin from their growing and shrinking ends at similar rates. In a different process known as "dynamic instability", the microtubules rapidly shrink during a certain time. This is followed by a period of rapid re-growth, followed by shrinking etc. The *in-vitro* formation of microtubules from tubulin generally involves a rapid initial increase in the number of microtubules. In many cases, this is followed by a regime in which the amount of microtubules in co-existence with free tubulin attains a stationary state. In other cases, a stationary state is not reached and the microtubule mass either decreases to a lower level or oscillates.

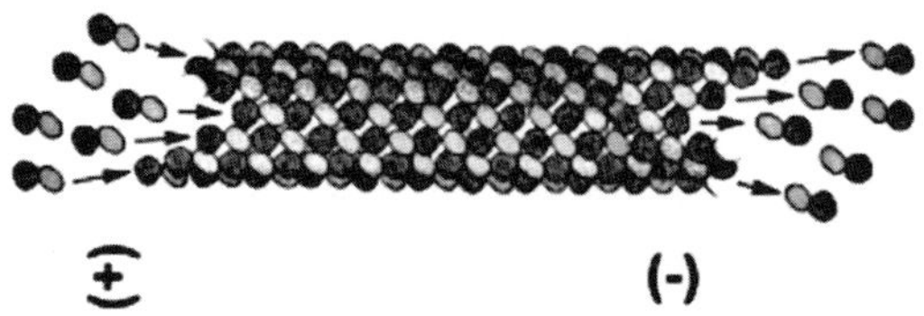

Fig. 2. Schematic illustration of a microtubule undergoing "treadmilling". Tubulin is added to one end of the microtubule (+) and is lost from the other (–) and GTP is hydrolysed to GDP.

The source of chemical energy for the formation and maintenance of microtubules is the hydrolysis of GTP to GDP. As the chemical reaction proceeds, the concentration of GTP falls while that of GDP rises. For many purposes it is preferable to work with a solution of constant GTP concentration, and that also contains a large reservoir of chemical energy. This can be achieved by adding an organic phosphate, together with the corresponding kinase enzyme, to the solution. As GDP is produced, it is re-phosphorylated to GTP from the organic phosphate, by the action of the kinase: the organic phosphate being reduced to inorganic phosphate. The overall reaction is one where the concentration of organic phosphate decreases whilst that of the inorganic phosphate increases. The concentration of GTP remains constant at its initial value until all the organic phosphate is consumed. The organic phosphate acts as a reservoir of chemical energy.

One of the particularities of microtubules is that they frequently found to continually grow from one end and shrink from the other. This is due to differences in reactivity at opposing ends. Since the rates of growth and shrinking are often comparable, individual microtubules change position and appear to move at speeds of several microns per minute. The shrinking end of a microtubule will leave behind it a chemical trail having a high local concentration of tubulin. Likewise the growing end of a microtubule will create a region depleted in tubulin. Neighbouring microtubules will preferentially grow into regions of high local tubulin concentration whilst avoiding the regions of low concentration. The chemical trails produced by individual microtubules, activate and inhibit the formation of their neighbours. Thus neighbouring microtubules will "talk to each other" by depleting and accentuating the local concentration of active chemicals, and the coupling of reaction with diffusion can progressively lead to macroscopic variations in the concentration and orientation of the microtubules. In this respect the behaviour of a population of microtubules shows many analogies with that of a population of ants.

Together with actin filaments, microtubules make up the majority of the cytoskeleton. During the interphase part of the cell cycle, microtubules irradiate across the cell interior. When cell division occurs, the microtubules disassemble and then reassemble in aligned arrays. These constitute the mitotic spindle along which the chromosomes move. Microtubules are believed to control the self-organisation of the cytoskeleton through processes that are not yet understood. They are also involved in many other cellular functions including, the maintenance of shape, motility, signal transmission, and intracellular transport. They play a role in the organisational changes that occur during the early stages of embryogenesis.

Microtubule self-organisation

Stationary structures

Tubulin solutions at concentrations of the order of 10 mg/ml, made up in a suitable buffer [19], are assembled into microtubules by warming the solution from 4 to 35°C in the presence of GTP. Microtubules form rapidly within 2–3 minutes.

30

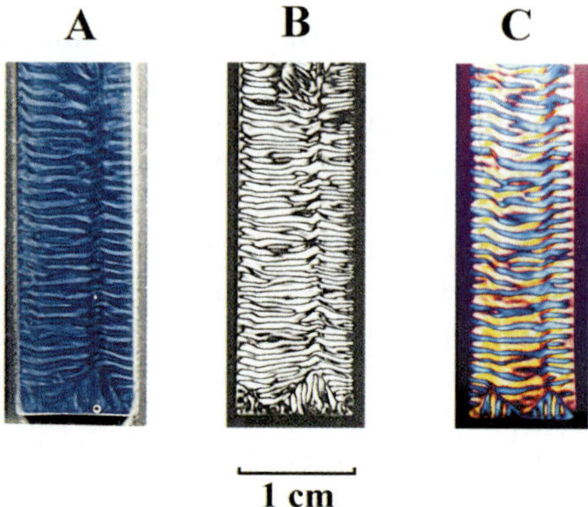

A   **B**   **C**

1 cm

Fig. 3. Striped patterns formed by microtubules at 35°C observed in: (A) reflected light; and (B) through crossed linear polars (0° and 90°); (C) through cross polars (0° and 90°) with a wavelength retardation plate (550 nm) at 45°. The retardation plate produces a uniform mauve background. Microtubule orientations of about 45°, such that their birefringence adds to the birefringence of the wavelength plate, produce a blue wavelength shift whereas orientations at about 135° subtract from the birefringence and result in a yellow interference colour. The alternating blue and yellow stripes arise from periodic variations in microtubule orientation from obtuse to acute. The sample dimensions are 40×10×1 mm. The strong optical birefringence indicates that the microtubules are highly aligned. Microtubules were formed by warming a solution containing 10 mg/ml of phosphocellulose purified tubulin from 4 to 35°C in the presence of GTP. Microtubules form within 2–3 minutes after warming the solution, and the structure shown progressively develops over the next 5–6 hours. The structure once formed is stationary and the solution is stable for about 3 days. The buffer solution was comprised of 100 mM MES (2-N morpholino ethanesulphonic acid),1 mM EGTA (ethylene glycol-bis-(B-aminoethyl) N, N, N^1, N^1 tetra-acetic acid), and 1 mM MgCl$_2$, in D$_2$O at pH 6.75.

Progressively, over a period of about five hours, the initially homogenous solution spontaneously self-organises to form a macroscopic structure. Figure 3 shows the striped arrangement of about 0.5 mm separation that arises when microtubules are formed in spectrophotometer cells 40×10×1 mm. Similar results are obtained in the presence or absence of the GTP regenerating system. Once formed the structure is stationary, and the solution remains stable for about three days. After this time, the reaction runs out of chemicals and the microtubules progressively disassemble and denature.

The preparations have strong birefringence and this indicates that the microtubules are highly aligned. The very high degree of microtubule orientation can be observed by small angle neutron scattering measurements [19,22–25]. In these measurements, a neutron beam passes through the sample and the neutrons scattered at low angles fall onto a two-dimensional detector placed behind the sample and perpendicular to the incident beam. The angular dependence of the intensity of scattered neutrons is related both to the shape, size, and concentration of the scattering particles, and to the manner in which they are disposed in space. When the scattering particles have a uniform isotropic distribution, the distribution of the scattering on the detector is of radial symmetry. Conversely, a net orientation of the

Fig. 4. Electron microscope image of part of the self-organised microtubule preparation showing arrays of highly aligned microtubules.

scattering particles gives rise to scattering concentrated into arcs. The direction of the arcs indicates the direction of orientation, and their spread decreases with increasing particle alignment.

The analysis of the neutron scattering patterns obtained from adjacent stripes in a self-organised microtubule preparation yield tubular shaped objects of inner and outer radii in agreement with the dimensions of microtubules. The scattering is concentrated into arcs, which depending on the stripe examined, are at either approximately 45° or 135° to the horizontal. This demonstrates that in each striped region the microtubules are highly oriented. Neighbouring stripes differ in that the microtubule orientation is either at about 45° or 135°. The striped optical pattern shown in Figs. 3A and B is hence formed of macroscopic regions in which micro-tubules are highly aligned at either 45° or 135°, but in which the orientation periodically alternates between the two orientations every 0.5 mm up the length of the cell. Figure 4 shows an electron microscope image in which the arrays of oriented microtubules are clearly visible.

This pattern of variations in orientation can also be observed by the samples between crossed linear polars with a wavelength retardation plate placed at 45° between the polars (Fig. 3C). The retardation plate produces a uniform mauve background. Microtubule orientations, such that their birefringence adds to the birefringence of the wavelength plate, produce a blue wavelength shift; whereas orientations that subtract, cause a yellow shift. Sample regions made up of micro-tubule orientations that are either acute or obtuse, differ by producing either yellow or blue interference colours. Due to the alternate periodic variations in the micro-tubule orientation, the sample when viewed this way appears as a series of alternating yellow and blue stripes. The method is rapid, inexpensive, and gives an overall image of the pattern of microtubule orientations.

32

The structure is considerably more complicated than it appears at first sight [21,25,26]. The horizontal 0.5 mm stripes also contain within them another series of stripes of about 100 μm separation. These, in their turn, contain another striped structure of about 20 μm separation. At distances below this, there exists another level of organisation of about 5 μm periodicity and, with care, it is possible to observe aligned microtubule bundles of about 1 μm separation. Some of these structures are shown in Fig. 5. Samples made up in larger sample containers also show an additional level of ordering of several mm in separation. These large stripes in turn contain the lower levels of organisation already mentioned. Hence similar types of pattern spontaneously arise over distances ranging from a few microns up to several centimetres. Self-organised structures can also form in sample containers of smaller dimension. Figure 7 shows the type of striped arrangement that arises in glass capillaries of 150 μm radius. The range of dimension over which these microtubule structures occur is typical of those found in many types of higher organisms. Cells are about 10 μm in size, eggs are often about a mm, and a developing mammalian embryo is several centimetres long.

Development of the self-organised structure.

The self-organised structure takes approximately five hours to form [19–25]. The structure, once formed, is stationary and the solution remains stable for about three days. Figure 6 illustrates the kinetics of its formation from the initially homogenous solution. After warming the solution to 36°C, microtubules form within 2–3 minutes. The first major change occurs after six minutes, with the appearance of a longitudinal bar. This corresponds to a breaking of the symmetry of the solution. As we subsequently discuss, this symmetry breaking is triggered by gravity. It coincides both with an instability in the chemical composition of the sample, and a bifurcation between self-organised states of different overall morphology. Just after this symmetry breaking, the sample is divided into two halves on either side of the longitudinal bar. One side of the sample contains microtubules oriented at 45° (blue interference colour) whereas the other half contains microtubules oriented at 135° (yellow interference colour). Approximately two hours later, further breaks in the symmetry occur, and a series of stripes perpendicular to the first longitudinal bar begin to develop. These stripes first appear at the sides and then at the top and bottom of the sample. They spread until they progressively encompass the whole cell. A film of this process, accelerated about 200 times, shows a wave of self-organisation sweeping across the sample. Once this process is completed after about 5 hours, there are no further changes.

Self-organisation results from chemical reactions associated with microtubule formation

An important feature to be established is whether or not self-organisation results from the flux of chemical energy through the system, or if it arises through static interactions related to the liquid crystalline properties of the microtubule solution.

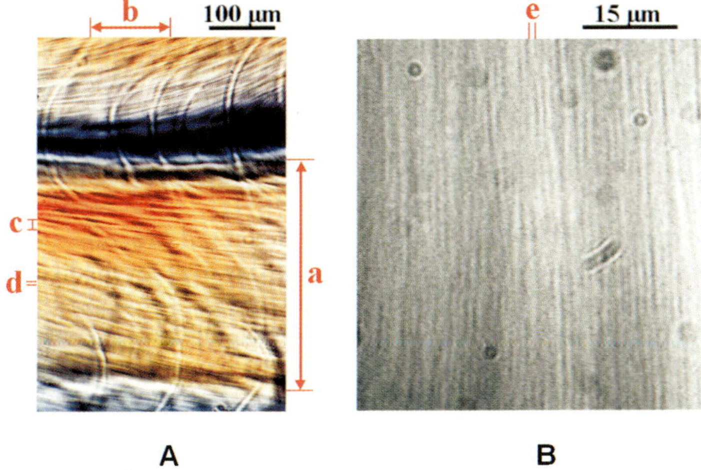

Fig. 5. The striped structure, as shown in Fig. 3, is itself comprised of stripes of smaller spacing. (A) and B) show one of the individual stripes at higher magnification. Regular separations of approximately, $100\,\mu$m, $20\,\mu$m and $5\,\mu$m are clearly visible.

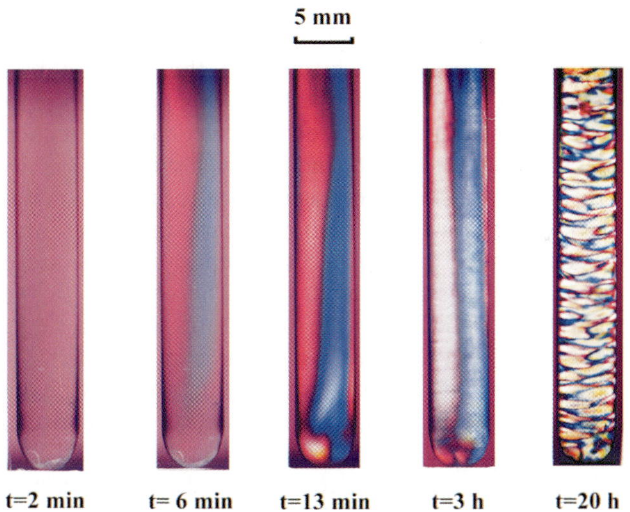

Fig. 6. Development of the self-organised microtubule structure formed in a 5 mm diameter test tube. The photographs show the pattern of birefringence at different times, t, after instigating microtubule assembly, as observed through cross polars (0° and 90°) with a wavelength retardation plate (550 nm) at 45° to the polars. The microtubules form about two minutes after warming the preparation to 35°C. At $t = 6$ minutes, a longitudinal bar develops breaking the symmetry of the sample and dividing it into two halves having microtubule orientations of either 45° or 135°. This symmetry breaking is simultaneous both with an instability in the chemical composition of the sample and with the morphological bifurcation. About two hours later, there is further symmetry breaking, when the horizontal stripes start to form. The structure remains stationary, once it is completely formed after about five hours.

The rate of hydrolysis of GTP to GDP in the preparations can be determined using P^{31} NMR spectroscopy [19–21]. Over a period of 24 hours there is a progressive reduction in the dissipation of chemical energy through the system. The behaviour can be divided into two approximately linear regions. The first period of about six hours corresponds to the time during which the structure forms, whereas the second region at longer times can be attributed to the energy consumption required to maintain the structure.

A simple way to test whether or not the self-organised structure arises from static interactions is to form the striped structure, then destroy it by mixing, and then wait and see whether or not the structure reforms [23,25]. After mixing, the solution contains microtubules at the same concentration and temperature as before mixing. The major difference is that chemical energy consumption is significantly less than when the microtubules were initially formed. If the self-organised structure arises from static interactions, such as occur in some liquid crystals, then the structure should completely reform after mixing. This is not the case. However, some partial organisation does occur. This can be attributed to the chemical activity still present in the preparation after mixing. This activity can be substantially reduced by the addition at the time of mixing, of the microtubule drug, taxol. When the energy consumption is further reduced in this manner, no reorganisation of the preparation occurs after mixing.

If the self-organised microtubule structure is formed then destroyed by mixing; then the microtubules disassembled by cooling the solution to 4°C; and then reformed by warming to 35°C, the striped self-organised structure also reforms [23,25]. These results show that the striped structure arises via chemical processes associated with microtubule formation, and not from static interactions between the microtubules. A further strong argument against static interaction is the dependence of the self-organisation on gravity at a moment early in the self-organising process [26]. Static interactions, such as may occur in liquid crystals, are equally present under conditions of weightlessness, as at 1 g. The absence of microtubule self-organisation under low g conditions at the bifurcation time is a clear demonstration that self-organisation does not arise from such static interactions.

A different possibility is that the pattern might in some way involve a coupling of the reactive process involved in microtubule formation with convective flow due to thermal gradients. The microtubule preparations are gels [41,42] of high viscosity ($\approx$5000 poise). This renders convective motion difficult. In addition, samples prepared in a hot room at 35°C, in which we checked that there was no convective motion, gave identical patterns to samples prepared under conditions where the bottom of the sample was 5°C warmer than the top [21,23,25]. It is not necessary to form microtubules by warming a premixed solution from 4 to 35°C. They can be equally well formed by mixing, and maintaining at 35°C, solutions of tubulin and GTP already pre-warmed to 35°C. The self-organised structure that develops has the same appearance as that formed by mixing tubulin and GTP in the cold and then warming [21,23,25]. Hence thermal convection appears to play no part in the self-organising process.

Reactive contribution to the self-organising process

One of the important variables in a dissipative system is the rate of energy dissipation. In chemical systems this will be related to the various reaction rates and will be strongly dependent upon experimental variables such as concentration and temperature. In a chemically dissipative reaction–diffusion system of the Turing type, the periodicity or wavelength of the structure corresponds to approximately the distance over which groups of molecules diffuse before reacting. The periodicity, L, is related to the reaction rate, R, and diffusion constant, D, by terms involving $L^2 = R/D$ [7,43]. Increasing the reaction rate will shorten the time over which molecules diffuse before reacting, and thus decrease the periodicity, L.

In the microtubule system, P^{31} NMR can measure the rate of energy dissipation. The reaction rates determined during self-organisation for preparations assembled at 30°C and 35°C increases by a factor of 2.1, respectively. Under the same conditions the average distance between stripes decreases by a factor of 1.47 from 0.78 mm to 0.53 mm [21]. The reduction in the spacing is thus close to that predicted for a reaction–diffusion mechanism by taking the square root in the ratio of reaction rates.

Diffusive contribution to the self-organising process

The rate of diffusion is not varied as readily as the reaction rate. One simple approach is to examine the effect on self-organisation of the addition to the initial reaction mixture of small quantities of gelling agents. Increasing quantities of gelling agent will presumably progressively inhibit diffusion. When microtubule assembly is carried out in gels of increasing agarose concentration, the effect is to perturb and eventually inhibit self-organisation [22,23,25]. This observation concords with a diffusive element to the self-organising process. However, there also exists the possibility that the gel might inhibit self-organisation by way of its macroscopic network. Another approach is to diminish diffusion by reducing the accessibility of the different reacting molecules to one another by filling a substantial amount of the sample volume with a bulky inert molecule. Ficoll 400 is a non-ionic synthetic polymer of sucrose, molecular mass 400 KDa, Stokes radius 10 nm, that is well established as being biologically inert. When microtubules are assembled in solutions containing progressively increasing quantities of Ficoll 400, self-organisation is initially perturbed and then inhibited [22,23,25]. The sum of these observations strongly suggests that molecular diffusion contributes strongly to the self-organising process.

Microtubules disassemble and reassemble during self-organisation

During the initial stages of self-organisation, the left and right hand sides of the cell show either yellow or blue birefringent interference colours that correspond to either obtuse or acute microtubule orientations. The stripes subsequently arise by blue zones forming in the yellow region, and yellow zones forming in the blue region. In the zones where there is no colour change the microtubules retain their initial orientation, whereas in the regions the colour change occurs, the microtubule orientation flips from either acute to obtuse or vice versa. In neutron small angle scattering [21–25] from a horizontal band having the approximate dimensions of a stripe, this process is manifested as a change in direction of the microtubular scattering on the detector, from an acute to an obtuse arc. Simultaneous with this re-ordering, the intensity of the microtubule scattering declines, rises and declines again. The microtubule re-ordering, which is itself the stripe-forming process, is thus concurrent with a chemical wave, involving different concentrations of microtubules and free tubulin, crossing the sample area under investigation.

The neutron scattering experiments also show higher counts for the sample areas undergoing orientational re-ordering than for the areas that does not re-order. Since, the sample areas that change their orientation, repeat themselves periodically as the visible stripes, this implies that periodic differences in microtubule concentration arise throughout the sample during the self-organising process.

In other words the stationary pattern arises because microtubules disassemble and reassemble with different orientations and concentrations in alternating parts of the sample. This neutron scattering experiment takes us to the heart of the self-organising process. It clearly shows that self-organisation result from the reaction dynamics of microtubules.

Microtubule concentration patterns

The central prediction of reaction–diffusion theories is the formation of macroscopic concentration patterns from an initially homogenous solution. The images described above reflect variations in the orientation of the microtubules, and not their concentration as such. It is not always easy to demonstrate, as in the present case, the existence of concentration variations in preparations in which a pattern of orientational changes is superimposed on a concentration pattern. To unambiguously prove the presence of microtubule concentration patterns in these self-organised preparations, we used two independent methods—small angle neutron scattering and fluorescent imaging [24,25].

As already mentioned, neutron small angle scattering is dependent upon the shape, dimensions, concentration, and orientation of the scattering particles. Shape and size information is contained in the variation, in the plane perpendicular to the incident beam, of the scattered intensity as a function of the angle from the centre. If, as in the present case, the solution is dilute the polar angular average of the scattered intensity is proportional to the concentration of scattering objects. Orientational information is in the polar angular distribution of the scattered intensity on the two-dimensional detector. Small angle neutron scattering measurements using a neutron beam 1×0.15 mm were carried out on self-organised microtubule preparations. Information as a function of position in the sample was determined by measuring the microtubule scattering, in steps of 0.2 mm, along the longitudinal axis perpendicular to the stripes. The changes in microtubule orientation, detected in the neutron measurements as changes in the polar angular distribution coincide with those seen optically. The intensity of the scattering taken over all polar angles varied periodically with position that correspond to variations in microtubule concentration are observed; the concentration is approximately constant in neighbouring stripes, but drops by about 25% when the orientation changes from acute to obtuse. To be sure that the intensity variation did not contain any orientational effects, scattering curves were measured at fixed sample positions for various microtubule orientations produced by suitable rotations about sample axis. The scattered intensity, after correction for geometrical factors, was independent of orientation to within 4%. Microtubule preparations that do not form self-organised structures did not show any variation in scattered intensity as a function of position.

The fluorophore, 4′,6-diamidino-2-phenylindole (DAPI), is strongly fluorescent when bound to microtubules, but only weakly fluorescent when associated with free tubulin or in buffer solution (44). Self-organised microtubule structures were prepared containing $5\,\mu$M DAPI (emission, 450 nm). The striped fluorescent pattern observed is interpreted as arising from periodic variations in microtubule concentration. To check that this pattern does not arise from other optical properties of the solution, samples also contained another fluorophore, rhodamine chloride, whose fluorescence occurs at higher wavelength (610 nm) and is independent of microtubule formation. As expected, the rhodamine image, obtained from the same sample by changing the optical filters, is almost uniform. Dividing the DAPI image by the rhodamine image corrects it both for this effect and for variations over the sample of incident light intensity and camera response. Figure 7 shows such a

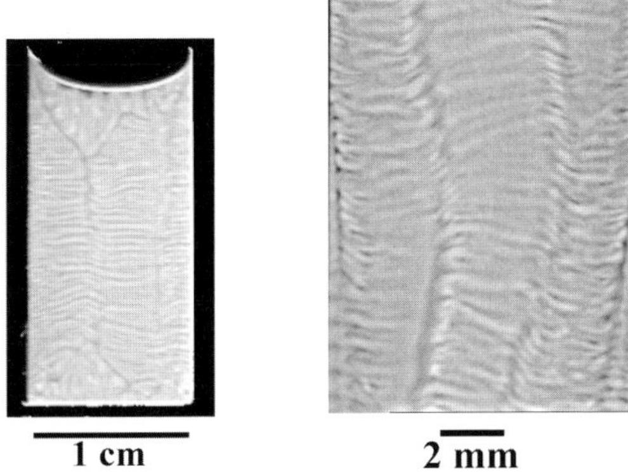

Fig. 7. Microtubule concentration patterns as shown by fluorescent imaging. The fluorophore DAPI (emission, 450 nm) is strongly fluorescent when associated with microtubules but only weakly fluorescent when associated with free tubulin in buffer. Rhodamine chloride is a fluorophore (emission, 610 nm) that does not bind to microtubules. Microtubule samples were prepared containing 10 mg/ml of tubulin, $5\,\mu$M DAPI and 95 μM rhodamine chloride. The DAPI image before assembly of the microtubules was uniformly dark. The rhodamine image is close to uniform, thus confirming that the DAPI image arises from periodic variations in microtubule concentration. (A) shows the DAPI image corrected for variations in incident light intensity and camera response by dividing it by the rhodamine image. (B) shows an image of the DAPI fluorescence at higher magnification demonstrating that the smaller 100 μm stripes also correspond to microtubule concentration variations.

38

corrected image. Periodic changes of the order of 25% of the mean occur. In agreement with the neutron scattering observations, variations in microtubule concentration occur when the orientation flips from acute to obtuse. Figure 7 also shows a detail at higher magnification in which the smaller (100 μm) stripes are also visible. These smaller stripes hence also correspond to periodic microtubule concentration variations.

Effect of gravity on microtubule self-organisation

Effect of the gravity direction

The microtubule solutions discussed above were all prepared by assembling tubulin in sample containers that were vertical, and remained vertical, during structure formation. When samples are assembled in the same spectrophotometer cells, but positioned horizontally, face down, a different morphology forms (Fig. 8A) [15, 19–27]. Both microtubule orientation and concentrations vary in a circular manner. The pattern of birefringence very frequently shows the type of cross, shown in Figs. 8, 9 and 11.

The fact that different types of pattern arise when preparations self-organise in optical cells that are either horizontal or vertical, shows that gravity contributes to the self-organising process. To check this, a series of samples was prepared with the

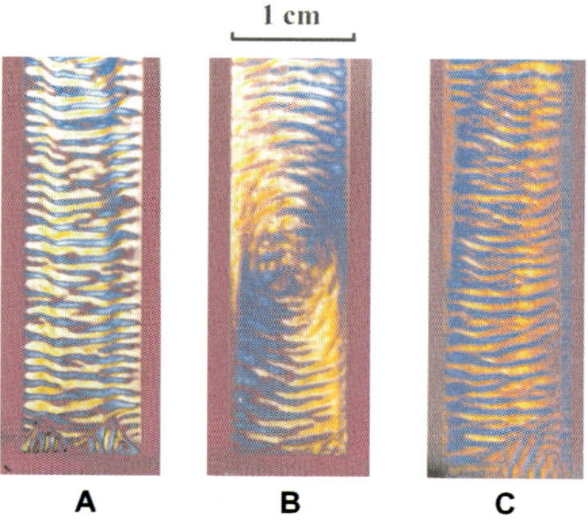

Fig. 8. Different stationary morphologies form depending upon whether the sample cells are horizontal or vertical. In (A) and (B) the microtubules were assembled in a 1 mm optical pathlength cell positioned respectively vertical or horizontal during the entire period of structure formation. (B) Shows the structure when the microtubules were assembled with the cell placed face down on the turntable of a record player with its long axis along the direction of the centrifugal field (0.14 g). The striped morphology resembles those that arise in vertical containers, and the stripes are perpendicular to the direction of the centrifugal field. Samples were photographed through linear cross polars with a wavelength retardation plate at 45°. The structures, once formed are independent of the orientation of the cell with respect to gravity.

spectrophotometer cells placed flat down on the turntable of a record player, such that the long axis of the sample cells coincided with the direction of the centrifugal force [15]. As shown in Fig. 8B, this resulted in stripes perpendicular to the direction of the applied field. The difference between the "horizontal" and "vertical" morphologies is hence attributed to the direction of the sample with respect to gravity.

Morphological bifurcation

The sample morphology once formed after five hours, is independent of whether the samples are horizontal or vertical. To determine at what moment in time gravity acts upon the samples, the following experiments were carried out [21]. Twenty samples made from the same tubulin preparation were assembled simultaneously in identical optical cell positioned upright. At one-minute intervals, consecutive samples were rotated from vertical to horizontal. The samples were then left undisturbed for about 12 hours and examined through crossed polars (Fig. 9). As the samples were hori-

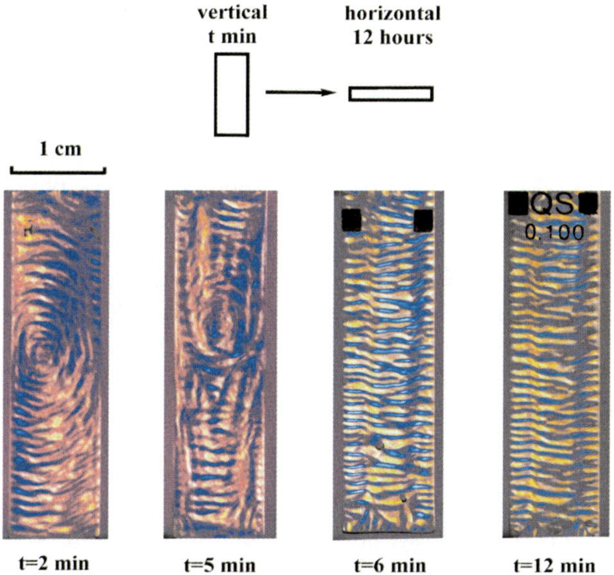

Fig. 9. Bifurcation properties of microtubule solutions. Preparations containing tubulin and GTP were placed in twenty spectrophotometer cells. Microtubules formation was instigated by warming to 35°C at time, $t = 0$, with the cells vertical. Consecutive cells, were turned from vertical to horizontal at intervals of one minute, then left 12 hours for the structures to form and observed with polarisation optics. The photographs show the final stationary morphologies for samples rotated from upright to horizontal at different times, t, during the first 20 minutes. Samples that remained vertical for six minutes or more formed striped structures as though they had remained vertical throughout the entire period of structure formation. The final morphology depends upon the orientation of the sample with respect to gravity at a critical moment prior to the formation of the self-organised state. The samples show an elementary form of memory together with transduction of the gravitational signal. The gravity direction has a decisive effect on the final sample morphology at the moment of the chemical instability.

zontal for all except the first few minutes of the self-organising process, one might expect the horizontal morphology would form. This is the case for the samples that were turned from vertical to horizontal during the first four minutes after instigating microtubule assembly. On the contrary, samples that remained vertical for six minutes or longer formed striped vertical patterns. Although they were vertical for only a brief initial period, they behaved as though they were vertical for the entire period of the pattern forming process.

The critical moment at which the morphology is determined is 6 ± 1 min after instigating assembly, and is prior to the formation of the stripes. At this moment, the system bifurcates between two different pathways that subsequently lead to different morphologies. The position of the sample with respect to gravity at the bifurcation time determines which morphology forms. The samples behave as though they retained a memory of their conditions at this critical moment. The direction of the weak field at a critical moment results in the formation of permanently different macroscopic pattern from that which would have arisen for a different field direction.

Chemical instability in the microtubule solution

During the self-organising processes that occur in far-from-equilibrium systems, bifurcations are associated with the presence of an instability in the initially homogenous state. For a self-organising process based on physical mechanisms, such as in the Bénard and Taylor experiments, the instability involves physical parameters. Likewise when self-organisation is based on chemical processes, then reactive mechanisms will be involved. Hence, establishing whether an instability arises from physical or chemical mechanisms provides evidence as to whether self-organisation is due to either physical or chemical factors. In a chemically dissipative system, the instability will involve changes in the concentration of the principal reactants. In the present case, we would expect a chemical instability to occur at the bifurcation time that involves the relative concentrations of microtubules and free tubulin.

Microtubules are several microns long and scatter light strongly. The optical density at a given wavelength, often around 350 nm, is taken as being proportional to their concentration [32]. The kinetics of microtubule self-assembly is routinely measured in this way. Frequently, after an initial increase due to the formation of microtubules caused by warming the tubulin solution, the value of the optical density remains approximately constant. In general, microtubule solutions showing this type of behaviour do not show any macroscopic self-organisation. In the present case, the optical density does not show this type of kinetics [21–23,25,26]. After an initial rapid increase, it decreases to a value about 20% lower (Fig. 14). The optical density maximum occurs approximately six minutes after warming. As expected, this coincides with the bifurcation time and with the first symmetry breaking in the system with the appearance of the longitudinal bar. There is hence a chemical instability in the microtubule concentration at the bifurcation time. This also corresponds to the first appearance of the longitudinal bar already described. This is the first symmetry-breaking event to occur and at this point in time the preparation is no longer

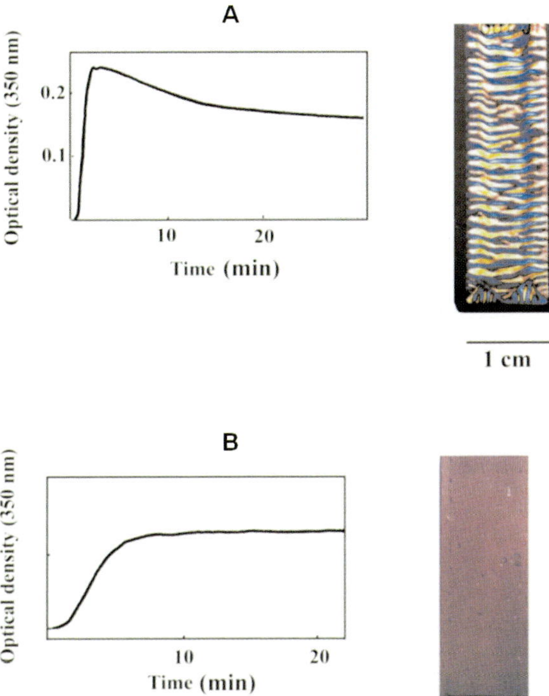

Fig. 10. A chemical instability at the bifurcation time is associated with self-organisation. The left hand side shows the kinetics of microtubule assembly, as measured by the optical density at 350 nm, for preparations that (right-hand side): (A) self-organise, (B) do not self-organise. Preparations that self-organise show a maximum in the optical density, demonstrating the existence of an instability in the chemical composition (relative proportions of microtubules and tubulin) of the sample. This instability coincides with the morphological bifurcation and symmetry breaking in the initially homogenous solution. Preparations that do not show this instability do not self-organise.

homogeneous. Just prior to the appearance of this bar the initially homogeneous solution is unstable. As we shall see since this bar does not arise when samples are assembled under low gravity conditions then in the present case its appearance is triggered by gravity.

Microtubules assembled under low gravity conditions do not self-organise

Sounding rocket experiments.

So as to establish whether the self-organising process is directly dependent on gravity, microtubules were assembled under conditions of weightlessness [26]. The experiment was carried out during the flight of a MAXUS sounding rocket of the European Space Agency. This provides approximately 13 minutes of low gravity $(2 \cdot 10^{-4}$ g) before the payload falls back to earth and is recovered. During re-entry, samples are subjected to accelerations from $10\,g$ to $50\,g$ for about three minutes. Since on the ground, the sample morphology is determined by the orientation with respect to gravity six minutes after instigating microtubule assembly, 13 minutes of low gravity should suffice to investigate the effect of weightlessness on the self-organising process. The possible effect of re-entry was simulated on the ground by assembling microtubules in sample cells either vertical or horizontal for 13 minutes, centrifuging them at $50\,g$ for approximately three minutes, and then leaving them until the structures formed about five hours later. The stationary morphologies that arose were the same as for samples left all the time in the initial sample orientation.

Flight samples were contained in an experimental module built by the Swedish Space Corporation and Ferrari and divided into two compartments. One, the "low-gravity" compartment contained eight rectangular cells upright and 11 rectangular cells flat. In the other compartment samples were placed on a rotating circular plate that produced a $1\,g$ centrifugal field either along, or perpendicular to, the long axis of the sample cells. Samples were disposed in the same way in the "$1\,g$" and "low-gravity" compartments. Hence, the samples formed under low gravity conditions could be compared with those formed at $1\,g$ under otherwise identical conditions.

A ground experiment, the same as the flight experiment and using the same tubulin preparation was carried out several days before the launch. Microtubules assembled in the "low-gravity" part of the module with the cells either vertical or horizontal, formed the striped and circular morphologies that normally form in the laboratory. Preparations lying flat on the centrifuge, and spun with the centrifugal field $(1\,g)$ along the long axis of the cell for the first 13 minutes only, formed stripes.

For the flight experiment, sample preparations were installed in the payload 14 hours before lift-off and maintained at 7°C. Once low gravity conditions were obtained, the sample temperature was raised to 36°C, thus instigating microtubule formation, and the $1\,g$ centrifuge switched on. Just prior to re-entry the centrifuge was stopped. Hence, the only difference between the two sets of samples is that one set was subject to weightlessness during the first 13 minutes of the self-organising process. The payload was recovered, and the experimental module reinstalled in the laboratory at the launch site within two hours from lift-off. The samples were then left for another four hours before the module was opened and the samples examined. The sample temperature was maintained at 36±0.5°C.

Figure 15 shows the results of the flight experiment. The samples formed in the "centrifuge" part of the module formed stripes when the centrifugal field was parallel to the long axis of the cell, and the circular morphology when it was perpendicular. For the samples perpendicular to the centrifugal field, the pattern is centred at the bottom rather than in the middle of the sample. This arises on the ground when microtubules are prepared in samples tilted by 1–2° from the horizontal. The comparison of the $1\,g$ flight morphologies with those that arise under ground conditions shows that as expected the self-organising process is unaffected by payload re-entry and recovery. In contrast, the samples formed in the "low-gravity" part of the module show practically no self-organisation. To check that microtubule assembly occurred during the flight experiment, the optical density of the samples at 350 nm was measured. Both the "low-gravity" and "1 *g*" samples had very similar values close to those measured for samples formed under ground conditions. So as to be able to observe images of microtubule concentration some samples contained 5 μM of the fluorophore DAPI. Fluorescence images for the "microgravity "samples show an homogenous microtubule distribution, whereas the "1 *g*" samples show periodic variations in microtubule concentration identical to the pattern in orientation. Contrary to the "1 *g*" samples, the "microgravity" preparations possess only very weak birefringence, demonstrating that the microtubules do not have any preferred orientation. Observations down to a distance of ~1 μm taken under a polarising microscope showed similarly weak birefringence. Hence, individual microtubules are relatively disordered with respect to one another. This contrasts with the "1 *g*" preparations which show strong optical birefringence and in which many micro-tubules are highly oriented along the same direction.

Experimental problems frequently occur in space experiments due to air bubbles. In our case, although care was taken to prevent it, in some samples small air bubbles formed in the neck of some of the sample containers. During re-entry, when the sample is subject to high centrifugal fields, the air bubble is propelled through the sample. In one sample this process was filmed. A line of strong birefringence formed along the trajectory of the air bubble showing that the bubble oriented the micro-tubules along its trajectory [28,29]. Subsequently, striped regions limited in extent, developed perpendicular to this trajectory (Fig. 11D). Hence, orienting some micro-tubules at an early stage in the process can also trigger self-organisation. We were expecting that self-organisation would not occur in the absence of gravity, and we found the effect of the air bubbles both intriguing and instructive.

In addition to the rectangular cells described, we also assembled microtubules in 2.5 mm radius tubes that mimic the shape of developing drosophila fruit fly eggs [24]. Figures 12A and 12B, show respectively photographs of such samples five hours after being assembled on the "on-board 1 *g*" centrifuge and at low gravity. The samples prepared under low gravity differ from those assembled in the rectangular cells in that some self-organisation has developed. In particular, the longitudinal bar as described above is present. As discussed later, there are strong reasons to believe that the difference in sample shape compared to the rectangular cells is responsible for this effect. It is a further illustration of just how sensitive the microtubule system is and demonstrates how easily self-organisation can be triggered by a variety of different causes.

Rotating clinostat experiments

One of the barriers to studying the effect of gravity on various physical and biological processes is the experimental difficulty in achieving low gravity levels. Very low gravity levels of sufficient duration are obtained in spacecraft under free fall condi-tions. However, the cost and complexity of the apparatus limit the number of experiments that can be carried out. In addition, for practical reasons, experimental procedures are often restricted to relatively simple actions and observations.

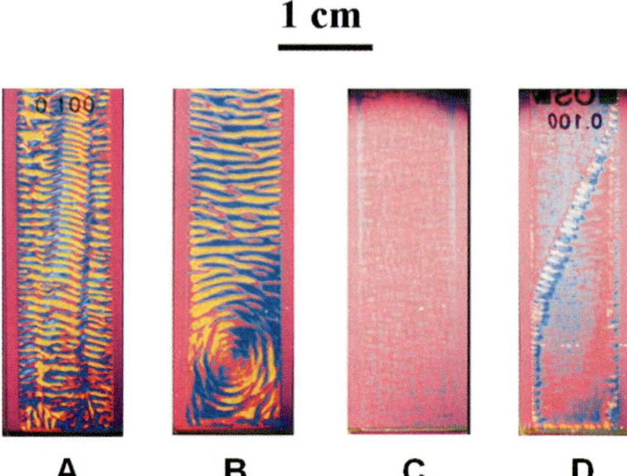

Fig. 11. Microtubule structures as formed during space flight. Microtubules were assembled once micro-gravity conditions were obtained. Photographs (A) and (B) show the self-organised morphologies that arise for samples placed on the on-board "1 g" centrifuge, with the centrifugal field along and perpendic-ular to the long axis of the sample cell. The centrifuge was stopped after 13 minutes, immediately prior to re-entry, and the samples left under 1 g conditions for a further five hours while the structures developed. (C) shows that almost no self-organisation occurs for samples subject to weightlessness during the first 13 minutes. (D) is a photograph of a sample in which an air bubble in the neck of the cell crossed the sample during re-entry. Microtubules along the trajectory of the bubble, were oriented, thus triggering partial self-organisation perpendicular to the trajectory.

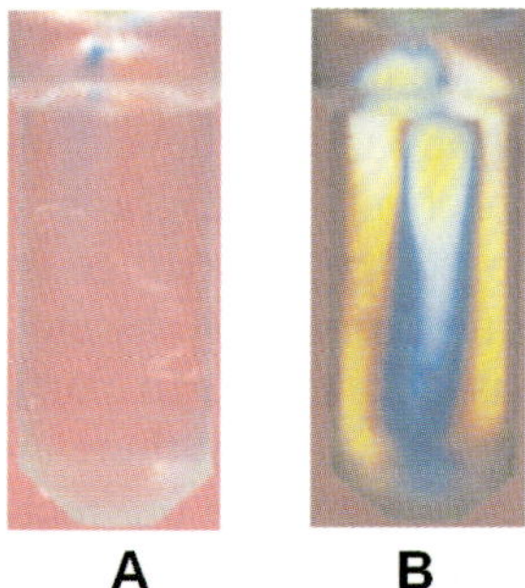

Fig. 12. Microtubule structures as formed in 2.5 mm radius tubes. Microtubules were assembled and photographed one hour after instigating microtubule assembly: (A) on a rotating clinostat revolving at 360°/s; (B) is a reference 1 g sample. Unfortunately, for practical reasons, no photographs corresponding to this stage were taken during the sounding rocket experiment. They would show a close resemblance to (A). Sample (A) eventually self-organised after 24 hours.

Another severe limitation for some biological systems is the difficulty of storing the sample in a functional state on the spacecraft until the experiment can be carried out.

On the other hand, various ground-based methods have been developed, which, although they do not produce the low gravity levels produced in space, nevertheless substantially reduce or modify the gravity vector. These methods are inexpensive and allow normal laboratory procedures and observations.

The simplest type of apparatus, the rotating clinostat, is comprised of a tube rotating about the horizontal axis. A sample placed at the centre of rotation does not change position. However, the direction of "fall" changes all the time; being "down" at one moment and "up" the next, thus reducing the average gravity level along the vertical direction. For a point object at the centre and rotating at a speed of 360°/s, the residual gravity value along one direction, obtained by integrating the gravity vector over time, is less than 10^{-2} g. Although these gravity levels are higher than those obtained in space experiments (10^{-5}–10^{-3} g), they are nevertheless sufficient to produce substantial effects. The major disadvantage with the method is that the gravity levels are not uniform over the sample. A point object at the centre of the axis of rotation will experience the residual gravity level; however positions away from the centre will also be subject to a centrifugal force due to rotation of the sample. For example, a point 2.5 mm from the axis rotating at 360°/s will be subject to a gravity vector of 10^{-2} g whereas a point 10 mm from the axis will be at 4 10^{-2} g.

We carried out experiments [30] using the rotating clinostat installed at the Biology Space Laboratory at the ETH Zurich [2]. So as to keep "gravity" levels at the edges of the sample to reasonably low values, we assembled the microtubules in cylindrical tubes of 2.5 mm internal radius. The clinostat experiment was carried out in the same way as the space experiment, and it was thus possible to make a meaningful comparison between the two. In the clinostat experiment the micro-tubule preparations all self-organised. However, self-organisation was very slow compared with that at 1 g. The rate of self-organisation depended on the rate of rotation, being slowest for samples rotated at 360°/s. Under 1 g conditions self-organisation is completed after five hours, whereas for samples assembled on the clinostat it took 24 hours or more to occur. After five hours, the clinostat samples show little self-organisation. At this point in time they compare favourably with the results of the space experiment.

Molecular basis of self-organisation

The reaction scheme for microtubule formation is approximately the following. Tubulin is added to microtubules in the form of a tubulin–GTP complex. This occurs either by addition to the growing end of a microtubule, or by nucleation of a new microtubule that subsequently grows. Hydrolysis occurs and the tubulin is incorp-orated into the microtubule as tubulin–GDP. Likewise, tubulin is lost from the shrinking end of a microtubule as the complex tubulin–GDP. When a microtubule shrinks or collapses it leaves in its place a very high local concentration of tubulin–GDP. As a microtubule grows from one end while shrinking from the other, it appears to move at a rate of several microns per minute and leaves behind a trail of tubulin–GDP of high local concentration. Similarly, as it grows from the front by incorporating tubulin–GTP, it will create zones that are locally depleted in tubulin–

46

GTP. So straightaway, starting from an initially homogenous solution, the micro-tubule dynamics create regions of high local variations in concentration. Since microtubules are several microns in length we can estimate the distance scale of these concentration fluctuations as about the same. At this point, as soon as microtubule disassembly becomes significant, the solution is already very different from an homogenous solution in that it contains substantial concentration fluctuations. The size and number of these concentration fluctuations then evolve as a function of the chemical dynamics of the preparation. Concentration differences also correspond to differences in density. The macroscopic density fluctuations in the solution will interact with gravity. This interaction produces a directional bias in the reaction–diffusion process that triggers self-organisation

The tubulin–GTP complex promotes the formation of microtubules whereas the tubulin–GDP complex inhibits their formation. Once liberated, the inactive tubulin (tubulin–GDP) is progressively regenerated into the active form (tubulin–GTP) with a certain time constant. During this time, the tubulin is free to diffuse, after which it is once again available to be incorporated into microtubules. If there were no tubulin diffusion whatsoever, then the tubulin would be found along exactly the same trail as the disassembled tail of the microtubule from which it came. Hence, the micro-tubules which grow or reform with this tubulin will do so along exactly the same trail as before. Under these conditions, since the microtubule solution that initially forms from the tubulin solution is macroscopically homogenous, then the preparation will remain so and no self-organisation will occur. Likewise, if tubulin diffusion is very fast, then by the time the tubulin–GDP becomes tubulin–GTP, the solution is once again homogenous in tubulin–GTP. Microtubules will reform or grow in a random manner and once again there will be no self-organisation. If, however, the tubulin diffusion has an intermediate value, then self-organisation can occur. Tubulin will progressively diffuse out from the trails left by shrinking microtubules. Since rates of chemical reaction are proportional to concentration, growing microtubules will progressively grow along the direction and in the zone where these trails exist. Likewise, microtubule nucleation will be greater in a zone of high tubulin–GTP concentration. These microtubules in their turn will leave behind new trails of high local concentration in tubulin–GDP and so on. Obviously, the zones depleted in tubulin–GTP formed by the growing end of microtubules will behave in a comple-mentary manner. Once regions of increased microtubule concentration and of preferred orientation start to form then the feedback mechanism outlined above will lead to a progressive reinforcement of this process. In such a way the microtubules "talk to each other" and progressively form a self-organised structure comprised of periodic variation in microtubule concentration and orientation.

When the microtubule solution first forms from the tubulin preparation it is isotropic. However, this isotropic arrangement is unstable. Once a few microtubules have taken up a preferred orientation, then neighbouring microtubules will start to grow in the same direction. Orientational order then spreads to its neighbours and so on. Hence any small external effect that will preferentially orient just a few micro-tubules will trigger progressive self-organisation. As outlined above, the passage of

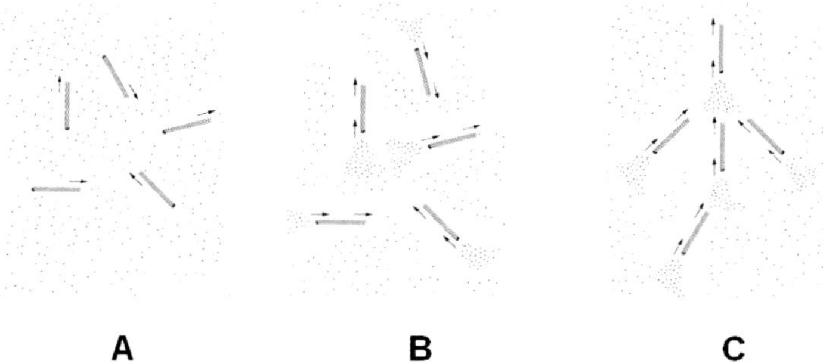

Fig. 13. A possible mechanism for the formation of the self-organised structure. Microtubules are chemically anisotropic, growing and shrinking along the direction of their long axis. This leads to the formation of chemical trails, comprised of regions of high and low local tubulin concentration from their shrinking and growing ends respectively. These concentration trails (density fluctuations) are oriented along the direction of the microtubule. Neighbouring microtubules will preferentially grow into regions where the local concentration of tubulin is highest. (A) Microtubules have just formed from the tubulin solution; they are still in a growing phase and have an isotropic arrangement. (B) Microtubule disassembly has started to occur at the bifurcation time; this produces trails of high tubulin concentration from the shrinking ends of the microtubules. (C) Microtubules are growing and forming preferentially into these tubulin trails. The isotropic arrangement shown in (B) is unstable. Once a few microtubules start to take up a preferred orientation, then neighbouring microtubules will also grow into the same orientation. Once started, the process mutually reinforces itself with time and leads to self-organisation. Any small effect that leads to a slight directional bias, such as slightly different rates of molecular transport in the up-down and left-right directions, will trigger self–organisation. Gravity acts by way of its directional interaction with the macroscopic density fluctuations present in the solution.

an air bubble through the solution orients the microtubules along its path and triggers self-organisation in its wake.

Using the same argument, we believe that magnetic fields would also trigger self-organisation. Sample geometry sets the boundary conditions operating in the reaction–diffusion system. A strongly anisotropic shape can break the symmetry of the reactive processes and hence trigger self-organisation. Gravity acts by way of its directional interaction with the macroscopic density fluctuations present in the solution. This could result in "diffusion" that is faster in the up–down direction, thus causing a slight directional bias for the formation of microtubules that triggers self-organisation. According to Kondepudi and Prigogine [8,9], the presence of a $1\,g$ field, in a suitable reaction–diffusion system can destabilise the equilibrium state at the bifurcation point and thus favour the formation of a macroscopic pattern.

Obviously we wanted to confirm the plausibility of this hypothesis. One approach is experimental verification. However it is difficult to devise experiments that would unambiguously demonstrate the presence and role of the chemical trails outlined above. Another approach is to numerically simulate the behaviour of a population of growing and shrinking microtubules using experimentally determined values of reaction rates and diffusion constants, and then to compare this behaviour with experimental observations.

Numerical simulations of microtubule self-organisation

Numerical simulations [28,29] were carried out based upon the following premises. Microtubules can grow and shrink from their ends. Tubulin molecules are added as the complex tubulin–GTP at a rate dependent on the local concentration and the reaction rate. Tubulin–GDP is lost from the opposing end. The liberated tubulin–GDP diffuses into the surrounding medium with a given diffusion constant. Due to the large excess of GTP, it is converted to tubulin–GTP at a given rate. This tubulin–GTP is available to be incorporated into the growing end of neighbouring microtubules, or of nucleating to form new microtubules.

The following equations represent the reaction scheme; despite their simplified appearance all three are of non-linear form.

Microtubule assembly:
$$n(\text{GTP·Tubulin}) + \text{Microtubule}_m \rightarrow \text{Microtubule}_{m+n}$$

Microtubule disassembly:
$$\text{Microtubule}_{m+n} \rightarrow \text{Microtubule}_m + n(\text{GDP·Tubulin}) + n\text{Pi}$$

Tubulin regeneration:
$$(\text{GDP·Tubulin}) + \text{GTP} \rightarrow (\text{GTP·Tubulin}) + \text{GDP}$$

The microtubule length is not fixed; microtubules are free to grow and shrink according to the reaction scheme outlined above. These simulations used rates of reaction and diffusion within the range of the experimentally determined values found in the literature [28] and they reproduced the kinetics of microtubule assembly as observed in the self-organising process. To limit calculations to a manageable size we restricted dimensions to two, even though the experimental samples obviously have three. However, reducing the dimensions to two does not seem to cause any major changes in the calculated self-organising process.

Initially, simulations were carried out on a population containing a small number of microtubules and then extended to larger populations. Obviously, one of the things that we wanted to know was whether the chemical trails hypothesised above might be predicted by such numerical simulations. To test this hypothesis, simulations were carried out with five microtubules contained on a two-dimensional surface, 100×60 μm. Two microtubules of 10 μm length (a, b) were placed close together, parallel to one another, and with their shrinking ends pointing in the same direction. Behind these microtubules were placed the remaining microtubules (c, d, e). These microtubules were also parallel to one another, and to the first two microtubules, but were significantly further apart from each one another than were microtubules a and b (Fig. 14). The concentration of tubulin–GTP and tubulin–GDP is initially uniform. The simulation was then started. Figure 14 shows the position of the microtubules and the concentration profiles of tubulin–GTP and tubulin–GDP, at different times. Due to tubulin–GTP being added at the growing end, and tubulin–GDP being liberated at the shrinking end, all the microtubules progressively change position at a rate of about 5 μm per minute. During this process the microtubules form regions, both at the growing end and down the sides, that are depleted in tubulin–GTP. Because of this, the two forward microtubules (a, b) that are close together, grow apart. This is because it is more favourable for them to grow into regions rich in tubulin–GTP than into the regions depleted in tubulin–GTP caused by the growth of the nearby neighbour. The shrinking ends of the microtubules liberate tubulin–GDP that is progressively converted to tubulin–GTP. Hence, just behind each microtubule is a small region of high tubulin–GDP concentration. This is followed by a long diffuse trail of tubulin–GTP. The growing ends of the three backward microtubules (c, d, e) grow into the tubulin–GTP trails formed by the forward microtubules (a, b). Figure 14 illustrates the manner by which, in the simulations, an individual microtubule directs itself towards regions of high tubulin–GTP concentration.

We then extended the simulations, to a population of about thirty thousand microtubules contained on a surface 100×100 μm. In experiments, microtubules initially form from an homogenous solution of tubulin and GTP. So as to imitate these conditions, we started the simulation with a uniform distribution of tubulin–GTP of given concentration, no tubulin–GDP, and no microtubules. Nucleation of tubulin–GTP into microtubule germs was taken as being exponentially dependent on tubulin–GTP and tubulin–GDP concentrations, as proposed by Marx and Mandelkow [45]. The small microtubule fragments generated this way are initially homogenous and have no preferred orientation. These numerous small fragments are free, by the mechanism already described, either to grow into microtubules, or shrink and disappear. The microtubules that develop attain an average length of about 10 μm comparable with the experimentally determined value in the self-organised preparation. Likewise in agreement with experiment, the microtubules that first form have no preferred orientation and their distribution is uniform. Subsequently, a self-organised structure comprised of regular bands of about 5 μm separation progressively develop (Fig. 15A). Within these bands the microtubules are highly oriented with respect to one another and the microtubule concentration drops by about 25% in the region between the bands. The structure is hence comparable with the type of structure observed experimentally. In the simulation, regular variations in microtubule concentration arise because of the progressive build up of regions that are depleted in tubulin–GTP. Over macroscopic distances, high concentrations of microtubules occur where there are high concentrations of free tubulin, and vice versa. Self-organisation takes about 10^6 iterations and corresponds to about two hours in real time. In experiments, self-organisation takes about five hours. Because of the large number of calculations involved it is not possible to extend the reaction space to a surface of several square centimetres such as was used in the experiments. Nevertheless, it is possible to compare the scale of the periodicity in the simulations with those arising in experiment over comparable distances. As described above, the self-organised structure contains levels of organisation down to distances of about 1 μm. Figure 15 shows a strong similarity between the simulated self-organisation and experimental observations over the same distance scale.

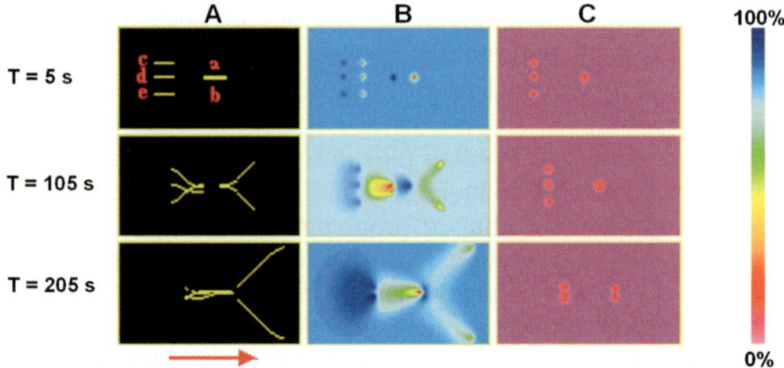

Fig. 14. Numerical simulation illustrating the formation of chemical trails. (A) The microtubules; (B) and (C) the concentration profiles of tubulin·GTP and tubulin·GDP, respectively. Initially, five microtubules were positioned as shown and the simulation started. The growing ends of the microtubules form regions that are depleted in tubulin–GTP, whereas the shrinking ends form trails rich in tubulin–GTP. Microtubules a and b move apart, away from the region of low tubulin–GTP concentration produced by their neighbour. Conversely, microtubules c, d and e, grow into the path of the chemical trails of tubulin–GTP produced by the shrinking ends of microtubules a and b.

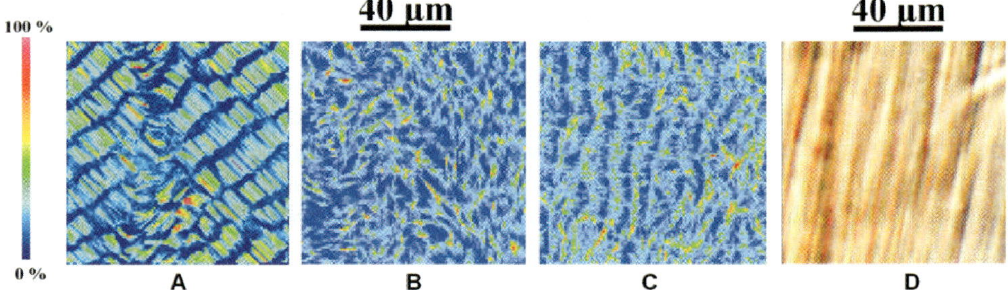

Fig. 15. Reaction–diffusion simulations for a population of microtubules. (A) A simulation on a reaction space, 100×100 μm, containing 5×10^4 microtubules. The diagonal stripes are triggered by a small asymmetry in the part of the algorithm describing diffusion. (B) A simulation in which this asymmetry is eliminated. The reaction space is 100×100 μm and contains the same number of microtubules. Although concentration inhomogeneities are present there is no macroscopic self-organisation. In (C) the simulation is identical to (B) except that diffusion is now twice as fast along the y-axis as along the x-axis. In this case, self-organised stripes develop in which the microtubules are perpendicular to the direction of the stripes. (D) Experimentally observed self-organised structure over the same distance scale.

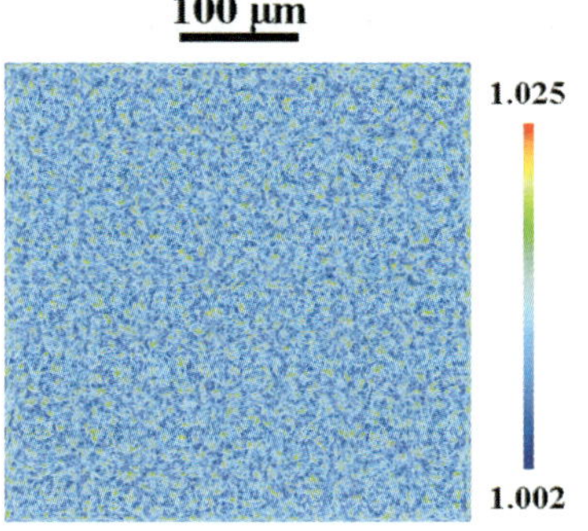

Fig. 16. Reaction–diffusion simulation of density fluctuations arising in a population of microtubules. The reaction space and number of microtubules is the same as in Fig. 14A and as described in the text. The simulation shown corresponds to 36000 iterations after instigating microtubule assembly. This corresponds to a "real" time close to the bifurcation time (six minutes).

One of the aspects of these simulations is that regions of high microtubule concentrations go hand in hand with zones of high tubulin concentration. This inevitably means that substantial density differences are present in the preparation. During the simulation, the density at any one point is readily calculated from the microtubule, tubulin–GTP and tubulin–GDP concentrations. Figure 16 shows the calculated density variations across the sample at different times during self-organisation. Density differences initially appear when net partial disassembly of microtubules first starts to occur. At this time, the length of the density fluctuation is about 5 μm, and their number density is about 5×10^4 mm^{-2}. The difference between maximum and minimum densities is about 3%. The strength of interaction of such a density fluctuation with a 1 g field is about 30 kT. This is considerably greater than kT, thus demonstrating that the interaction with gravity is sufficient to overcome any thermal averaging in the preparation.

As self-organisation develops, the density fluctuation progressively increase in size, until they make up the macroscopic microtubule bands described. At the same time they decrease in number. The net result is that the overall interaction between the sum of these zones and a 1 g field progressively decreases.

One of the features of these simulations is that the stripes always formed along the diagonal of the square reaction space. This led us to suspect that an asymmetry favouring this orientation was somehow unwittingly built into the simulation. This turned out to be the case: a very slight asymmetry occurred in the way that diffusion was digitised as a square wave front, and this small effect sufficed to favour the formation of self-organised stripes along the diagonal. This type of asymmetry is well known in numerical simulations involving diffusion. When we eliminated it, the simulations show concentration and orientation inhomogeneities of about 5 μm, but which are uncorrelated and disorganised (Fig. 15B). When an asymmetry was once again introduced into the simulation by using slightly different rates of diffusion in the x and y directions, a macroscopic self-organisation once again produced (Fig. 15C).

These numerical simulations demonstrate that a relatively simple scheme, based on experimental rates of reaction and diffusion, predict the overall experimental behaviour. They thus permit a link between the microscopic and macroscopic, and between the molecular and phenomenological. The simulations confirm the hypothesis that self-organisation arises by the formation of anisotropic chemical trails produced as microtubules grow and shrink from opposing ends. Ordering arises not from static interactions, but is due to the formation by reactive processes of concentration gradients in chemicals that activate and inhibit the formation of neighbouring microtubules. In this way, microtubules communicate with one another and behave as a macroscopic collective ensemble.

In this system there are at least two significant differences from the type of reaction–diffusion scheme originally proposed by Turing. In the Turing system the molecules communicate with one another across the sample by way of diffusion (fast diffusion of the inhibitor and slow diffusion of the activator). In the microtubule system, the molecules communicate by way of the reactivity of the microtubules. The

52

fact that the microtubules are growing from one end at the rate of the order of several μm per minute whilst shrinking by approximately the same, results in a displacement of the microtubule at the same speed. At the same time, they leave behind a trail having a high concentration in inactive tubulin and create in front of them a furrow depleted in active tubulin. These chemical trails determine the pathways taken by surrounding microtubules. It is a reaction–diffusion system, since as already explained, without tubulin diffusion at the appropriate rate, self-organisation would not occur. In terms of a pure Turing system, active tubulin (tubulin–GTP) plays the role of the activator whereas inactive tubulin (tubulin–GDP) is the inhibitor. However, in this case they both have the same diffusion constant. Self–organisation occurs because displacement of the microtubules by reaction is faster than diffusion of tubulin.

The second major difference with a normal Turing system is the reactive anisotropy of the microtubule system. In a normal reaction–diffusion scheme, the reaction has no inherent anisotropy. This is not the case with microtubules, which can obviously only react at their ends. A microtubule can move in only one direction and this will equally lead to anisotropic reactive trails in active and inactive tubulin along the same direction. The system has an in-built propensity for symmetry breaking under the effect of a weak field or some other external influence.

The simulations also predict another major experimental observation: how self-organisation is dependent upon weak external fields. Any small effect that creates a slight preference for some microtubules to grow along one direction will trigger macroscopic self-organisation. The simulations thus provide a molecular basis for explaining how gravity, magnetic fields, and shearing can act on the system. Magnetic fields, shearing, and sample shape would act by slightly favouring some microtubule orientations, whereas gravity could also act by introducing an asymmetry into molecular diffusion and transport.

Proposed mechanism for how gravity triggers self-organisation

The experimental results clearly demonstrate that the *in-vitro* microtubule preparations self-organise by way of the non-linear dynamics of reaction and diffusion. The molecular basis of this process is the elongation and contraction of individual microtubules by the addition of tubulin–GTP and loss of tubulin–GDP, respectively. Microtubule contraction leads to the formation of chemical trails of free tubulin–GTP that can be incorporated into other microtubules by growth or nucleation, and microtubule elongation produces furrows depleted in tubulin and unfavourable to the growth or formation of other microtubules. Both of these processes give rise to heterogeneities in the chemical composition and density of the sample. In their turn, the elongated regions, both of tubulin depletion and concentration, activate and inhibit the formation of neighbouring microtubules. Thus neighbouring microtubules "talk to each other" by depleting and accentuating the local concentration of active chemicals. The coupling of reaction with diffusion progressively leads to macroscopic variations in the concentration and orientation of the microtubules.

In the present case self-organisation is triggered by the presence of gravity in the early stages of the process. In other words, the system bifurcates due to the effect of gravity. The bifurcation time coincides with the appearance of the vertical bar described above. It is therefore slightly preceded by an instability in the homogeneous state. The breaking of the symmetry of the homogeneous state at the instability leads to self-organisation. This symmetry breaking does not occur under low gravity conditions and hence it involves an interaction of the system with gravity. The concept of instability and symmetry breaking lie at the heart of the gravity dependence of these preparations.

The bifurcation time also corresponds to an instability in the relative composition of the sample in terms of free tubulin and microtubules. From these observations we can state that the gravity acts on the microtubule preparation at the moment that the initially homogeneous state is unstable. Since this instability is a chemical instability involving reactive processes, and since self-organisation arises from a combination of reaction and diffusion, then the gravity interaction must in some way make a contribution to the reaction–diffusion process at the instability that modifies the course of reaction–diffusion and leads to self-organisation. In the present case of microtubule self-organisation, the parameters describing the dynamics are related to the macroscopic concentration and density fluctuations involved in the self-organising process. It is by way of its interaction with these density fluctuations at the bifurcation time that gravity triggers self-organisation.

The observation that self-organisation can also be triggered by the passage of an air bubble through the preparation, strongly suggests that the orientation of a few microtubules at the moment that the isotopic arrangement is unstable, is an essential feature of understanding the role of gravity in this process. The numerical simulations demonstrate how the presence of any small effect in the reactive and transport processes, that leads to a slight directional bias in the distribution of microtubule orientations, will be enough to trigger self-organisation.

When the microtubules first form from the tubulin preparation within the first 3–4 min following the instigation of assembly, they are all in a growing phase with no significant disassembly from the shrinking end. The reaction is autocatalytic, and tubulin–GTP is consumed, where the microtubules grow they generate furrows depleted in tubulin–GTP. There are more and more growing microtubules competing for less and less tubulin–GTP. Under the present conditions this process is non-linear and a large number of microtubules find they lack the tubulin–GTP necessary to sustain growth. At this point in time (after about 4 min), slightly before the bifurcation time, partial disassembly of microtubules starts to occur. This leads to chemical trails of tubulin–GDP. This tubulin–GDP is converted into tubulin–GTP by the excess of GTP in solution with a certain time constant. During this time the tubulin is free to diffuse, after which it is once again available to be incorporated into microtubules, thus progressively stabilising the quantity of microtubules in the preparation. This process of partial microtubule disassembly followed by stabilisation constitutes the "overshoot" in the kinetics of microtubule assembly described above. It lies at the heart of the gravity dependence of these preparations.

54

When partial disassembly occurs at the instability it does not happen uniformly across the sample. Because the microtubule length at this time is between 5 and 10 μm, and as disassembly can only occur at the ends of microtubules, one could expect that close to the instability, fluctuations in composition and density might develop that are about the length of microtubules. Both experimental observations and numerical simulations show that this is indeed the case: density and concentration fluctuations of about this size appear at the instability. Their interaction with gravity, at the moment that the isotropic arrangement is unstable, triggers self-organisation by way of a mechanism that produces a slight bias in the orientation of growing microtubules. Once a few microtubules have taken up a preferred orientation, then neighbouring microtubules will start to grow in the same direction. Orientational order then spreads to its neighbours and so on. Because high concentration of microtubules arise where there are high concentrations of free tubulin, and as high concentrations of microtubules in their turn produce high levels of tubulin, then the self-organisation that occurs also contains macroscopic variations in composition and density.

After the instability, the sample progressively returns to a stable stationary state. Our results show that self-organisation does not occur when gravity is absent during the period of this chemical instability. If self-organisation is not triggered by gravity during the first 13 minutes, then by the time it is applied, the homogenous state has returned to a sufficiently stable condition that it remains unaffected by gravity; and hence little or no self-organisation subsequently occurs.

What we do not yet know is the exact details of how the interaction of gravity with density fluctuations modifies the reaction diffusion process. For example, the interaction with gravity could result in transport of tubulin that is faster in the vertical than in the horizontal direction. We know from simulations that this will trigger self-organisation. On the other hand gravity could also act by introducing in regions of high microtubule concentration, and which also correspond to regions of high tubulin concentration, a relative displacement of free tubulin as compared with microtubules. Another possibility is that the hydrodynamics in the preparation permitted the gravity field to produce a slight directional bias in microtubule orientation. Neither do we fully understand exactly how gravity contributes to the formation of the longitudinal bar at the bifurcation time.

Conclusions

The results described show how a very simple biological system, initially comprised of only tubulin and GTP, is capable of behaving as a gravi-receptor. In the present case, gravity triggers the self-organising process. The gravity direction breaks the symmetry of the initially homogenous state and leads to the emergence of form and pattern. Such processes may have played a role in the development of life on earth. Other external factors, such as magnetic and electric fields, or shearing, could have the same effect. Processes of this type could form a general class of mechanism by which weak environmental factors are transduced by biological systems. The results presented demonstrate that gravity substantially modifies microtubule self-organisation by way of its participation in a reaction–diffusion process. Gravity can thus intervene in a fundamental cellular process and will indirectly affect other cellular processes that are in their turn dependent upon microtubule self-organisation.

In this article we have attempted to elucidate how gravity affects microtubule organisation in terms of what the individual molecules are actually doing. One of the reasons for this effort is that most chemists and biologists reason in these terms and are unfamiliar with the conceptual basis of non-linear dynamics and bifurcation phenomena. Nevertheless, it is worth pointing out that once the tenets of non-linear chemical dynamics are accepted, then it is no longer necessary to understand the details of what all the different molecules are doing all the time. A general understanding can be obtained based on knowing how "emergent" phenomena arise in a "complex" system, and in particular how reaction–diffusion processes lead to chemical instabilities, bifurcations and self-organisation.

At a practical level, one of the striking features of this work is that very simple and inexpensive apparatus and methods, such as turning the sample from vertical to horizontal, spinning on a record player, or rotating about the horizontal axis at 60 rpm, can lead to substantial effects. In the present case, the understanding derived from these simple experiments is comparable with those obtained from experiments carried out in space, and without the associated difficulties and expense.

In humans, weightlessness depresses the immune system, reduces bone mass. These, and other effects, are thought to arise at a cellular level and many experiments point to an involvement of the cytoskeleton [1–3]. Workers have observed modifications in the organisation of the cytoskeleton [46–49] and results on human lymphocyte cells cultured in space show a disorganised microtubule network compared to ground control experiments [48]. More recently, Vassy and co-workers have examined microtubule organisation in epithelial cells cultured in space [49]. For many cells they found a marked degree of disorganisation of the microtubules. These observations are consistent with our results, and raise the possibility that reaction–diffusion processes form an underlying mechanism for the dependence of cellular function on gravity. However, at the moment the evidence that this is the case is suggestive but not conclusive. Further experiments need to be designed and carried out that demonstrate the effect of gravity on microtubule organisation in cells arise from the reaction–diffusion processes described in this article. If this were the case it would also mean that microtubule reaction–diffusion processes occur in living cells, and might lead to new insights into the physical chemical processes controlling the organisation of the cytoskeleton.

Acknowledgements

We thank the European Space Agency for providing the MAXUS flight together with all the flight hardware. We are indebted to the Centre National d'Etudes Spatiales for financial support. We would like to express our warm thanks to Drs. A. Cogoli and I. Walther for use of the rotating clinostat and their hospitality.

References

1. Cell and molecular biology in space (1999) Faseb J., Supplement 13.
2. Cogoli, A. and Gmünder, F. (1991) Gravity effects on single cells: techniques, findings, and theory. Adv. Space Bio. Med. 1, 183–248.
3. Hammond, T.G. et al (1999) Gene expression in space. Nature Medicine 5, 359.
4. Turing, A.M. (1952) The chemical basis of morphogenesis. Phil. Trans. Roy. Soc. 237, 37–72.
5. Glandsdorff, P. and Prigogine I. (1971) Thermodynamic Theory of Structure, Stability and Fluctuations. Wiley, New York.
6. Meinhardt, H. (1982) Models of Biological Pattern Formation. Academic Press, London.
7. Harrison, L.G. (1993) Kinetic Theory of Living Pattern. Cambridge University Press, Cambridge.
8. Kondepudi, D. and Prigogine, I. (1981) Sensitivity of non-equilibrium systems. Physica 107A, 1–24.
9. Kondepudi, D. (1982) Sensitivity of chemical dissipative structures to external fields: formation of propagating bands. Physica 115A, 552–566.
10. Prigogine I. and Stengers I. (1984) Order out of Chaos. Heinemann, London.
11. Nicolis, G. and Prigogine, I. (1977) Self-organisation in Non-equilibrium Systems. Wiley, New York.
12. Nicolis, G. and Prigogine, I. (1989) Exploring Complexity. Freeman, New York.
13. Abraham, R. and Shaw, C. (1983) Dynamics. Ariel Press, Santa Cruz.
14 Thompson, J.M.T. (1982) Instabilities and Catastrophes in Science and Engineering. Wiley, New York.
15. Tabony, J. and Job D. (1992) Gravitational symmetry breaking in microtubular dissipative structures. Proc. Natl. Acad. Sci. USA 89, 6948–6952.
16. Mesland, D. (1992) Mechanisms of gravity effects on cells: are there gravity-sensitive windows? Adv. Space Bio. Med. 2, 211–228.
17. Castets, V., Dulos, E., Boissonade, J. and de Kepper, P. (1990) Experimental evidence of a sustained standing Turing-type non-equilibrium chemical pattern. Phys. Rev. Lett. 64, 2953–2956.
18. Belousov, B.P. (1951) Reprinted in: Oscillations and Travelling Waves in Chemical Systems (Field R. and Burger M. eds.). Wiley, New York, 1985, pp. 605–613.
19. Tabony, J. and Job, D. (1990) Spatial structures in microtubular solutions requiring a sustained energy source. Nature 346, 448–450.
20. Tabony, J. and Job, D. (1992) Microtubular dissipative structures in biological auto-organisation and pattern formation. Nanobiology 1, 131–147.
21. Tabony, J. (1994) Morphological bifurcations involving reaction–diffusion processes during microtubule formation. Science 264, 245–248.
22. Tabony, J. (1996) Self-organisation in a simple biological system through chemically dissipative processes. Nanobiology 4, 117–137.
23. Tabony, J. and Papaseit, C. (1998) Microtubule self-organisation as an example of a biological Turing structure. Adv. Struct. Biol. 5, 43–83.
24. Papaseit, C., Vuillard, L. and Tabony, J. (1999) Reaction–diffusion microtubule concentration patterns occur during biological morphogenesis. Biophys. Chem. 79, 33–39.
25. Tabony, J., Vuillard, L. and Papaseit, C. (2000) Biological self-organisation and pattern

formation by way of microtubule reaction–diffusion processes. Adv. Complex Systems 2, 221–276.

26. Papaseit, C., Pochon, N. and Tabony, J. (2000) Microtubule self-organisation is gravity-dependent. Proc. Nat. Acad. Sci. USA 97, 8364–8368.

27. Tabony J., Pochon N. and Papaseit, C. (2001) Microtubule self-organisation depends upon gravity. Adv. Space Res. 28, 529–535.

28. Glade, N., Demongeot, J. and Tabony, J. (2002) Comparison of reaction–diffusion simulations with experiment in self-organised microtubule preparations. C.R. Acad. Sci. Fr. (in press).

29. Tabony, J., Glade N., Papaseit C. and Demongeot. J. (2001) The effect of gravity on microtubule self-organisation. J. Phys IV France 11, 239–246.

30. Glade, N. and Tabony, J. (2001) The effect on microtubule self-organisation of reduced gravity levels produced in ground based experiments; comparison with space flight experiments. J. Phys IV France 11, 255–260.

31. Alberts, B., Bray, D., Lewis, J., Raff, M., Roberts, K. and Watson, J. (1983) Molecular Biology of the Cell. Garland, New York.

32. Dustin, P. (1984) Microtubules. Springer, Berlin.

33. Babloyantz, A. (1986) Molecules, Dynamics and Life. Wiley, New York.

34. Coveney, P. and Highfield, R. (1995) Frontiers of Complexity. Random House, New York.

35. Ben-Jacob, E., Schochet, O., Tenenbaum, A., Cohen, I., Czirok, A. and Vicsek, T. (1994) Generic modelling of cooperative growth patterns in bacterial colonies. Nature 368, 46–49.

36. Hölldobler, B. and Wilson, E.O. (1991) The Ants. Springer-Verlag. Berlin.

37. D'Arcy Thompson, W. (1917) On Growth and Form. Cambridge University Press, Cambridge.

38. Rashevsky, N. (1940) An approach to the mathematical biophysics of biological self-regulation and of cell polarity. Bull. Math. Biophys. 2, 15–23.

39. Prigogine, I., Lefever, R., Goldbeter, A. and Herschkowitz-Kaufman, M. (1969) Symmetry breaking instabilities in biological systems. Nature 223, 913–916.

40. Nicolis, S.C. and Deneubourg, J.L. (1999) Emerging patterns and food recruitment in ants: an analytical study. J. Theor. Biol. 198, 575–592

41. Buxbaum, R.E., Dennerll, T., Weiss, S. and Heidemann, S. (1987) F-actin and micro-tubule suspensions as indeterminate fluids. Science 235, 1511–1514.

42. Sato, M., Schwartz, W., Selden, S.C. and Pollard, T.D. (1988) Mechanical properties of brain tubulin and microtubules. J. Cell Biol. 106, 1205–1211.

43. Harrison, L.J. (1987) What is the status of reaction–diffusion theory thirty-four years after turing?. J. Theor. Biol. 125, 369–384.

44. Bonne, D., Heusèle, C., Simon, C. and Pantaloni, D. (1985) 4′,6-Diamidino-2-phenyl-indole, a fluorescent probe for tubulin and microtubules. J. Biol. Chem. 260, 2819–2825.

45. Marx, A. and Mandelkow, E. (1994) A model of microtubule oscillations. Eur. Biophys. J. 22, 405–421

46. Hughes-Fulford, M. and Lewis, M.L. (1996) Effects of microgravity on osteoblast growth activation. Exp. Cell Res. 224, 103–109.

47. Cogoli-Greuter, M., Spano, A., Sciola, L., Pippia, P. and Cogoli, A. (1998) Influence of microgravity on mitogen binding, motility and cytoskeleton patterns of T lymphocytes

and Jurkat cells. Experiments on sounding rockets. J.J. Aerospace Env. Med. 35, 27–39.

48. Lewis, M.L., Reynolds, J.L., Cubano, L.A., Hatton, J.P., Lawless, B. and Piepmeier, E.H. (1998) Spaceflight alters microtubules and increases apoptosis in human lymphocytes (Jurkat). Faseb J. 12, 1007–1018.

49. Vassy, J., Portet, S., Beil, M., Millot, G., Fauvel-Lafeve, F., Karniguian, A., Gasset, G., Irinopoulou, T., Calvo, F., Rigaut, J.P. and Schoevaert, D. (2001) The effect of weightlessness on cytoskeleton architecture and proliferation of human breast cancer cell line MCF-7. Faseb J., 15, 1104–1106.

Cell Biology and Biotechnology in Space
A. Cogoli (editor)
© 2002 Elsevier Science B.V. All rights reserved

Gravity-related Behaviour in Ciliates and Flagellates

Ruth Hemmersbach and Richard Bräucker

DLR Institute of Aerospace Medicine, Cologne, Germany

Unicellular organisms, like protists, use gravity as a most reliable environmental parameter for spatial orientation. Thus, gravisensitive protists offer ideal model systems contributing to answering the question whether gravisensation is a common cellular capacity. During the last decades, the gravity related behaviour in several protozoa, mainly ciliates and flagellates, has been investigated. Though the species differ in cell sizes and living conditions, similar mechanisms of graviperception based on mechanosensitive principles seem to exist.

Gravireponses under 1 *g* conditions—ecological consequences

Microorganisms develop movement organelles, such as cilia and flagella, which are triggered by the input of sensors for various stimuli. This enables them to actively reach and remain in environments which offer optimal living conditions with respect to light, oxygen and food and thus for growth, reproduction and safety against excessive solar radiation [1]. Though protozoan cells are heavier than water, most of the species studied so far prefer to swim against the direction of the gravity vector (negative gravitaxis) which guides them to the surface. Well-studied examples are the negative gravitaxis of the heterotrophic ciliates *Paramecium* and *Tetrahymena*, which need oxygen-saturated layers and feed on aerobe bacteria [2,3]. Predatory species, like *Didinium* and *Bursaria*, show a negative gravitaxis depending on their feeding state [4,5].

The green algae *Euglena* and *Chlamydomonas* [6–8] demonstrate negative gravitaxis in the absence of light or at low irradiations, while positive gravitaxis is induced by higher irradiation levels. At a certain depth of the water column, negative phototaxis in algae is balanced by negative gravitaxis, demonstrating that these antagonistic responses protect phytoplankton organisms against extreme and toxic irradiation levels [9–14]. In cases where the stimuli light and gravity are applied perpendicular to each other, flagellates respond on the resultant vector [15]. Under special circumstances—heavy metal ions (mercury, copper and lead) and culture age—the negative gravitaxis of *Euglena* might be inverted to a positive gravitaxis, while the swimming velocity remains unaffected [16]. *Paramecium* cells, which were grown under long-term low temperature conditions, no longer showed negative gravitaxis [17].

60

In the case of the microaerophilic karyorelictan ciliate *Loxodes,* the direction of gravitaxis is triggered by the oxygen concentration of the medium. *Loxodes*, which belongs to the benthos of lakes, leaves the sediment and migrates in the water column at low O_2 concentrations. The populations show a positive gravitaxis at $\geq 40\%$ air saturation. At < 5–10% atm O_2 the cells display either a negative gravitaxis or bimodal (bipolar) distributions [18–22]. It is proposed that in *Loxodes* the mechanism that modulates the mode of gravitaxis is based on a sensor for oxygen [23]. Whether the oxygen concentration actually influences negative gravitaxis in *Paramecium* is the subject of controversy [24,25].

Several species do not only show a tactic response to gravity but also an active regulation of swimming velocities, a phenomenon called gravikinesis [26]. By a polarised kinetic response—speeding up during upward swimming and deceleration during downward swimming—cells become able to compensate part of their sedimentation rate (*Paramecium*) [27] or even overcompensate sedimentation (*Tetrahymena*) [3] (see [26] and references therein) (Fig. 1). The amount of this compensation is obviously not dependent on the cell size but on the preferred living conditions. Gravikinesis is calculated comparing the upward and downward swimming velocities of a cell (population) with its sedimentation velocity:

$$\text{gravikinesis} = \frac{\text{downward swimming rate} - \text{upward swimming rate}}{2} - \text{sedimentation rate}$$

It has been described for all (seven) ciliate species examined so far [28].

According to Vogel et al. [29], the flagellate *Euglena* shows no gravikinesis, but accelerates during reorientation [30,31]. Machemer-Röhnisch et al. [32], however, do describe a gravikinesis in *Euglena*.

The ecological advantage of gravireactions becomes obvious, when taking into account that even giant cells such as *Bursaria truncatella* (cell volume $30 \times 10^6 \ \mu\text{m}^3$, sedimentation rate 923 μm/s), would rapidly sediment to the bottom of a pond if no compensating speed regulating mechanism exists [5].

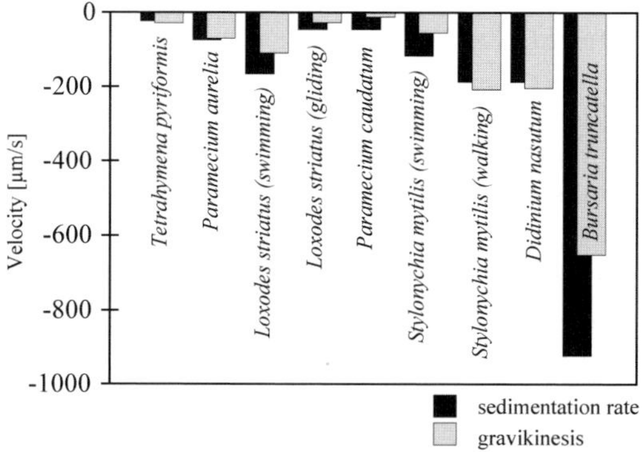

Fig. 1. Comparison of sedimentation rates and gravikinetic values in seven ciliate species. All species tend to counterbalance sedimentation by an orientation-dependent active change in swimming rate (gravikinesis).

Behaviour is the result of environmental inputs from different receptors. In order to learn about the gravireceptor mechanism it is necessary to perform experiments under controlled stimulus conditions. We may then modulate the intensity of the gravity stimulus by either increasing (hypergravity) or by switching it off (microgravity).

Hypotheses of graviperception—historical overview

Due to the fact that gravitaxis of swimming protists can be already observed with simple optical equipments in vertical tubes, the phenomenon has been known since the end of the 19th century [33]. Different hypotheses have been established to explain gravitaxis (for review see [6,26,34–36]). A pure physical model is based on static principles, assuming that a cell reacts as a buoy due to an unequal mass distribution inside the cytoplasm [37–39]. A major argument against this hypothesis is the fact that immobilized *Paramecium* cells sediment in variable orientations of their longitudinal axes [40]. Alternatively, hydrodynamic processes were taken as the reason for negative gravitaxis: the shape of the cell was proposed to cause an upward orientation due to different sedimentation velocities of the posterior and anterior hemisphere [41,42]. Some authors presumed the difference between the centre of gravity of the cell and the centre of propulsion to be the reason for spatial orientation [43].

In contrast, physiological models are based on the assumption that a protozoan cell perceives the direction of the gravity vector and uses this information for an active regulation of the motor response. It has been proposed that *Paramecium* either registers the hydrostatic pressure difference between its upper and lower hemisphere [44] or swims into the direction of increased resistance (that means upwards) by measuring its energy consumption [45]. Both assumptions presume storage and temporal evaluation of measured values, which is unlikely in a unicellular organism.

Alternatively, other physiological models assume that heavy compartments of the cell trigger a selective stimulation of the locomotor apparatus. Due to the lack of a morphologically distinct statocyst-like organelles in most protists, it was proposed that the mass of the whole cytoplasm causes a mechanical load on the lower cell membrane thus stimulating gravisensory ion channels ("statocyst hypothesis") [27, 46–48]. In addition to such a membrane-bound (peripheral) mechanism, representatives of Karyorelictida (ciliates), *Loxodes* and *Remanella*, bear highly specialized statocyst-like organelles [49–51]. A variable number of these so-called Müller vesicles is always arranged along the aboral surface of the cell. A Müller organelle consists of a vacuole containing a heavy mineral body of bariumsulfate fixed to a microtubular stick (Fig. 2). It is proposed that gravity is perceived by deformation of a sensitive area near the cilium. This should result in a shift of the membrane potential, which modulates the activity of the cilia and thus the swimming velocity [19,52].

Gravitaxis research acquired new inputs after the electrophysiological characterisation of mechanoreceptor channels in ciliates [53] and the discovery of the

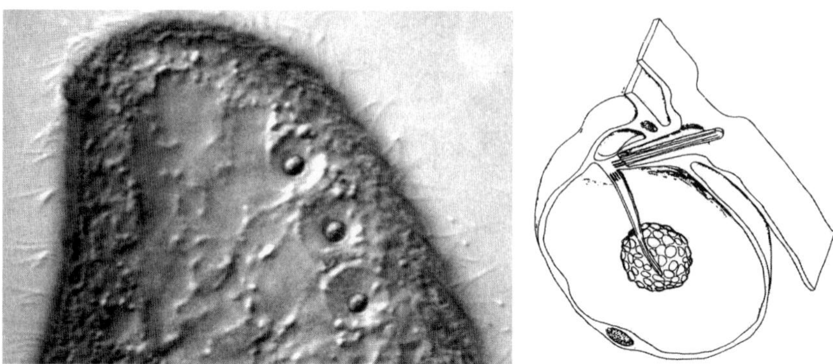

Fig. 2. Light micrograph of the rostral region and schematic drawing of a Müller vesicle in *Loxodes striatus.* The vacuole (7 μm diameter) bears a mineralized granulum fixed to a microtubular stick (from Rieder, personal communication). During spatial movements of the protist, the Müller body causes a mechanical shear to the microtubular stick due to altered direction of the gravity vector, thus providing the stimulus for graviperception.

electromotor coupling describing the relationship between membrane potential and swimming behaviour [54]. Furthermore, new experimental possibilities (e.g., experimentation in microgravity) were a further fruitful step towards identifying gravity-related signal transduction chains.

Distinct mechanosensitivity—a prerequisite for gravisensation

From electrophysiological studies in different ciliated species it is known that mechanosensitive ion channels are distributed in a bipolar manner in the plasma membrane. In *Paramecium,* mechanosensitive K-channels prevail posteriorly, mechanosensitive Ca-channels are mainly found anteriorly (Fig. 3). Stimulation of these channels leads to either hyperpolarization (K-channels) or depolarization (Ca-channels) of the membrane potential which in turn increases (hyperpolarization) or decreases (depolarization) the swimming rates, or even induces backward swimming (for a review see [55,56]). This gradient-type distribution of mechanosensitive ion channels and the density difference between the cytoplasm and the medium offers ideal preconditions for gravisensation in ciliates: depending on the orientation of the cell with respect to gravity a distinct stimulation of mechanosensitive ion channels occurs. It can be assumed that during upward swimming the mass of the cytoplasm leads to an outward deformation of the cell membrane of the posterior cell hemisphere, thus stimulating K-channels. The resulting hyperpolarization of the membrane potential induces an increased swimming rate. Alternatively, stimulation of the anterior cell pole in a downward swimming cell leads to a decreased swimming velocity. The arrangement of the ion channels predicts that de- and hyperpolarizing sensory inputs are compensated in horizontally swimming *Paramecium* cells. Electrophysiological studies of *Loxodes* also suggest a bipolar mechanosensitivity [57] (Fig. 3). The bipolar mechanosensitivity seems to be a characteristic of many

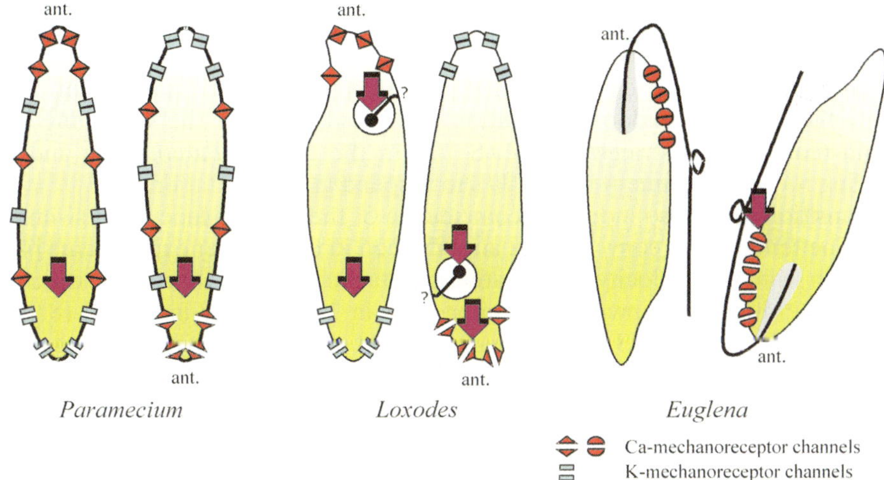

Fig. 3. Models of graviperception in three protist species (ant. = anterior cell pole). Ca- and K-mechano-receptor channels are incorporated in the cell membrane. They are activated by the mechanical load of the cytoplasm (forces symbolised by arrows; see text for details). Additionally, *Loxodes* bears specialised gravireceptors, the Müller vesicles (not to scale). A connection of the Müller vesicle to the cell membrane is uncertain so far. The activation of the receptor channels influence the membrane potential of the cells which triggers the activity of motion organelles: cilia (not shown) in *Paramecium* and *Loxodes*; flagellum in *Euglena*.

ciliates. *Didinium*, as an exception, lacks hyperpolarizing mechanoreceptors [58], thus offering the possibility to prove the statocyst hypothesis and to characterise the role of the Ca-mechanoreceptors in gravisensation [59].

The model of graviperception in the flagellate *Euglena* is also based on membrane-bound gravireceptors. Electrophysiological data are not available so far from *Euglena*, as intracellular potential measurements were not successful. It is postulated that mechanosensitive ion channels are concentrated at the front end of the *Euglena* cell beneath the flagellum [60]. If a cell is not aligned with the gravity vector, these channels are supposed to be activated by the pressure difference between cell and medium. The resulting depolarization modulates the activity of the flagellum until *Euglena* swims upwards again. In contrast, Machemer-Röhnisch et al. [32] assume a K-gravireceptor in *Euglena* which should be located at the posterior cell pole.

Experiments with ciliates and flagellates under changed gravitational stimulation and in density-adjusted media proved the different hypotheses: gravisensation via membrane-bound and/or intracellular gravireceptors.

Experiments in actual and simulated microgravity

The important role of gravity as a cue for cellular orientation was confirmed by experiments under microgravity conditions (μg). In preparation for these experiments some species were also studied on a fast-rotating clinostat microscope [61–67]. This method is based on the assumption that in fast rotated biological systems the gravity-induced relative movement of masses is neutralized and thus the direction of the *g*-vector no longer perceived.

When gravitactically oriented ciliates are exposed to the conditions of microgravity they show a random distribution after a short time. The swimming paths remain more or less straight [29,64,68–70]. In the case of *Paramecium* and in a sounding rocket experiment it took 80 s to suspend graviorientation [64].

Transition to μg also induces a coincidence of the gravity-dependent swimming rates. The tendency to converge was already seen in drop tower experiments [4,22]. The mean swimming velocity of a negative gravitactic *Paramecium* culture, which was mainly determined by upward swimming cells in 1 *g*, initially increased (8.9% in TEXUS 27; 3% in TEXUS 38) in sounding rocket experiments before they declined within 3 min to the level of the former horizontal swimming velocity [65]. This proves the hypothesis that the swimming velocity of *Paramecium* in horizontal direction mimics the gravity-unrelated situation (as in μg) (see also [59]). Further support for the hypothesis of distinct channel activation comes from *Didinium* due to its lack of hyperpolarizing mechanoreceptors [58]. As a consequence, its horizontal swimming velocity under 1 *g* is determined by a stimulation of depolarizing mechanoreceptors, thus leading to an increased swimming velocity in μg [71].

Cultivation of *Euglena, Paramecium* and *Loxodes* for up to 12 days in space did not affect the reactivity of the cells to changing accelerations; thus, no long-term influence of μg has been stated so far [72,73].

On a fast-rotating clinostat—as in microgravity—*Paramecium* and *Euglena* cultures show a random distribution pattern within two (*Paramecium*) to three (*Euglena*) minutes, which may indicate that the threshold for gravitaxis is higher than the maximal residual acceleration under the chosen clinostat conditions (4×10^{-2} g at the border of the observation chamber) [29,64,69]. Though paramecia showed no significant difference between upward, downward and horizontal swimming velocities in the clinostat their mean swimming velocity remained increased (8.9%) (compared to the former horizontal swimming velocity) in contrast to the rocket experiment (see above), thus indicating a persisting mechanostimulation of the cells [65].

Experiments under hypergravity conditions

Hypergravity is a useful tool to achieve more pronounced gravireactions. In ciliates, centrifuge experiments have been performed using various devices [74–78]. A slow rotating centrifuge microscope (NIZEMI) was used in ground-based studies (1–5 *g*) and in flight for threshold studies (μg to 1.5 *g*, IML-2 mission). The NIZEMI consists of a horizontally mounted microscope and a recording video system. Further centrifuge microscopes have been flown on the sounding rocket MAXUS. Ground-based studies revealed that with rising gravity the orientation of a gravitactic cell population becomes more pronounced, indicating that the response at 1 *g* is not saturated [21,28,30,69,73,79–82]. Even at twofold natural gravity, the giant cell *Bursaria* swims against the acceleration vector [5], while *Paramecium* manages up to 9 *g* (own observation). Hypergravity also potentiates the gravikinetic response of

Paramecium, *Didinium*, and *Bursaria*. With rising acceleration the downward swimming velocities increase, while the upward swimming velocities show a minor increase or remain virtually constant documenting an active regulation [5,28].

Hypergravity revealed that gravisensitive ciliates respond within seconds to an increase in gravitational force. Even during repetitive stimulations between 1 g and 5 g, *Paramecium*-cells maintained their reactivity, showing no change in the intensity of the responses, thus indicating that the cells do not adapt to altered gravitational stimulation. The behavioural responses to increasing or decreasing acceleration profiles did not differ significantly [28,72,73,82].

Hypergravity has also been used as a tool to determine sedimentation rates of various protists as a function of acceleration, as knowledge of these data is essential for the isolation of the graviresponses. Obviously, Stoke's law is not fully applicable for objects the size of ciliates, demanding an experimental determination of the sedimentation rate [78].

Another possibility to enhance the density difference between cell and medium is to feed *Paramecium*-cells with iron particles. As a consequence the sedimentation rate increased by 13% and overall gravikinesis by 46% compared to the control cells [83]. As predicted, the gravitaxis was much more precise in iron fed cells than in control cultures. This became evident especially at increased acceleration levels (e.g., 4 g). The results support the statocyst hypothesis.

Centrifuge experiments in space

In order to determine the minimum acceleration necessary to induce a gravi-response, protists were exposed to increasing acceleration steps from μg to 1.5 g or *vice versa* on a centrifuge in space. Using this method threshold values for gravitaxis have been determined: *Euglena* $\leq$0.16 g [72,84] and 0.12 g [30]; *Paramecium* 0.3 g; *Loxodes* $\leq$0.15 g [73,82,85]. The measurement of a threshold implies that gravitaxis is the result of a physiological signal transduction chain. The sigmoidal response curve of *Euglena* for the dependence of the precision of orientation on acceleration (μg to 1.5 g) has been interpreted to indicate the physiological nature of the response [84]. Machemer-Röhnisch et al. [32] used a custom-made centrifuge (CECILIA) on ground for experiments in *Euglena*. They found linear relationships between gravi-responses and acceleration, which intersect at 0 g, when extrapolated. These authors deny the existence of a threshold for gravikinesis in *Euglena*. The data obtained so far allow no discrimination between a first sign of orientation and the first sign of graviperception. The question of a threshold of graviresponses has to be settled under real hypo-gravity conditions, so further experimental approaches seem to be necessary.

Manipulations of the gravisensors

If buoyancy or the density-difference between cell and medium triggers the gravi-responses, they should disappear in neutral-buoyancy experiments, while an

66

intracellular gravisensor should work without restrictions under these conditions. Increasing density of the medium (Ficoll or Percoll) impaired gravitaxis of *Euglena* and *Paramecium* as well as gravikinesis in *Paramecium*. Orientation was completely disturbed at density values which correspond to the mean cell density (in the range 1.04–1.05 g/cm^3, depending on cell age and culture conditions) [85,86]. Higher densities reversed the direction of movement [85–88]. Based on other indications for a physiologically guided mechanism, the results support the hypothesis that gravi-perception in *Paramecium* and *Euglena* takes place at the plasma membrane.

In contrast, the ciliate *Loxodes* maintained its gravitaxis under isodensity conditions, thus demonstrating that these responses are controlled by an intracellular gravireceptor [85].

In order to prove the role of the Müller vesicles as gravisensors, they were destroyed by laser beams. As a consequence, *Loxodes* lost its capacity for orientation though vitality and swimming velocity were undisturbed. The swimming patterns of cells with destroyed Müller organelles were similar to the behaviour in microgravity. As only isolated cells could be manipulated by laser beam, the impact on gravikinesis, which can only be separated in larger cell populations, remains open [85]. The general morphology of the intracellular gravisensors of *Loxodes* did not change during a 12-day cultivation in space, though there were indications for less minerali-zation in *Loxodes* cells under μg, however, without impact on the graviresponse after landing [73,89].

Neugebauer et al. [90] showed that gravikinesis in *Loxodes* was slightly reduced under isodensity conditions. From this it is concluded that in *Loxodes* an intracellular gravisensing mechanism is supported by a membrane-located one. It may be specul-ated that membrane-located gravisensors, which are found in several protozoan species, are early inventions of evolution, while the Müller bodies of *Loxodes* are later acquisitions due to adaptation to special living conditions.

Identification of a gravireceptor potential

Great effort is undertaken in order to directly measure a gravireceptor potential by means of intracellular electrophysiological recording. First evidences of a gravi-receptor potential in *Paramecium* have been given [91]. Depending on the cell's orientation with respect to the gravity vector, a hyperpolarization or a depolarization was registered after a 180° turn.

So far, direct electrical potential measurements have not been successful in *Euglena*. However, manipulation of the membrane potential by means of the iono-phore calcimycin (A23187) or the lipophilic cation TPMP$^+$ (triphenyl-methylphos-phonium) resulted in a loss of gravitaxis, thus confirming that changes in the membrane potential are involved in graviperception [92]. Recent investigations focus on the use of voltage-sensitive dyes such as Oxonol in order to demonstrate changes in membrane potential depending on cellular orientation [93]. Gentle mixing of *Euglena* cells induced potential changes for 200s, initiated by a short period of hyperpolarization, followed by a depolarization. The experimental set-up did not allow a registration of the orientation of the Oxonol-labelled cells. From comparison with behavioural studies of controls, Richter et al. [93] concluded a close relationship between precision of orientation and membrane potential.

Channel blockers

Different channel blockers have been used to identify gravisensory channels, but with controversial results. Gadolinium blocks the gravitaxis in *Euglena*, leading to the hypothesis that stretch-activated channels function as gravisensory ion channels [87,94]. However, behavioural and electrophysiological studies in *Paramecium* did not confirm these results indicating that gadolinium is not specific for mechano-sensitive channels in *Paramecium* and affects different channel types, K-channels as well as Ca-channels [95]. Unfortunately, the search for suitable blockers for gravi-sensitive ion channels in ciliates has so far been without success [96].

Limits of gravisensation?

Gravikinesis and gravitaxis have been demonstrated in several protists of various sizes and cell volumes ranging from *Euglena gracilis* (2.6×10^3 μm^3), *Tetrahymena* ($22 \times 10^3 \mu m^3$), *Paramecium caudatum* (262×10^3 μm^3) to *Bursaria truncatella* (30×10^6 μm^3). Using the specific density of the cytoplasm (1.03–1.05 g/cm^3), the cell shape and volume, and the gating distance of a mechanoreceptor channel (here assumed to be 10 nm) we may estimate the gating energy for gravireceptor channels. Calculations show that in larger species channel gating energy is clearly above the thermal noise level (2×10^{-21} J). For species, as *Tetrahymena* and *Euglena*, or even smaller cell systems such as lymphocytes the signal to noise ratio may be critical.

The energetic considerations as well as the threshold values in the range 0.1–0.3 g demand the involvement of supporting structures, such as the microfilament system in graviperception. This assumption is supported by results from electric-field experiments [97] which have documented the influence of weak external voltage gradients on the gravikinesis of *Paramecium*. It is assumed that the voltage gradient causes a contraction of the filament system via a Ca^{2+} influx. The changed tension of the filament system would then influence the graviresponse.

Second messengers

The role of second messengers (such as Ca^{2+}, cAMP, cGMP, calmodulin) has been demonstrated in ciliary and flagellary activity and thus swimming behaviour [98–100]. Therefore, second messengers are assumed to play a role in the gravi-transduction chain. Different inhibitors and antagonists of the cAMP pathway were tested with respect to their impact on gravitaxis in *Euglena*. Increasing cAMP levels by means of the phosphodiesterase-inhibitors IBMX, caffeine, and theophyllin or by application of 8-bromo-cyclo-AMP induced an increase in the precision of negative gravitaxis in *Euglena*, while decreasing cAMP inhibited the response [87,94,101].

Chemical assays of cells, which had been exposed to varied accelerations, are currently under investigation. cAMP levels of *Euglena* cells which had been exposed to different acceleration steps in a sounding rocket experiment of 7 min duration revealed acceleration-dependent changes: transition from above threshold ($= 0.16\,g$) to μg does not significantly change the cAMP content, while acceleration from μg to above threshold results in a significant increase in cAMP, with a subsequent decline to initial values within 40 s. As a saturation of the graviresponse was not observed under 1 g conditions, hypergravity experiments may further support the hypothesis that an activation of mechanosensitive ion channels is coupled to cAMP production. In the case of *Paramecium*, exposure to 5 g for up to 10 min led to a significant increase in cAMP concentration [96]. The results indicate an involvement of cAMP in the gravity signal transduction chain, though the exact mechanism needs to be clarified.

A promising approach is the visualisation of second messenger changes in living cells. However, autofluorescence or photophobic responses of cells induced by laser light make experimental problems in protists. A suitable system is the chlorophyll-free flagellate *Astasia longa*. Its cytosolic Ca-changes during reorientation and acceleration changes have been investigated. Therefore, cells were loaded with calcium Crimson dextran (Molecular Probes; 10,000 MW) by means of electroporation and excited with 538 nm laser light. A 180° turn of a negative gravitactic *Astasia* culture increased the Ca signal with a maximum after 30 s, followed by a decrease back to the original intensity level within the following 10 s. These Ca kinetics were correlated to the reorientation of the cells [102].

Conclusions—models for gravisensation

The results obtained so far are in favour of an active gravisensation in protists. Graviperception is obviously based on mechanosensitive ion channels in the cell membrane which are stimulated by the mechanical load of the cytoplasm. Figure 3 shows models for graviperception in three protist species:

(1) *Paramecium* (based on the modified statocyst hypothesis) [27]. Two species of mechanoreceptor channels are incorporated in the cell membrane in a gradient-like distribution. At the anterior hemisphere, Ca-mechanoreceptors prevail, while in the posterior hemisphere K-mechanoreceptors dominate. Since the mechanical load of the cytoplasm stimulates the lower region of the membrane, the resulting change in membrane potential depends on the orientation of the cell with respect to the gravity vector. In an upward swimming cell (left) K-mechanoreceptors open, which results in a hyperpolarization and an increase in swimming rate. Correspondingly, a downward swimming *Paramecium* is depolarized by an increase in Ca-conductance. The gravity related changes in swimming rates (gravikinesis) have been documented (see above). Whether gravitaxis of *Paramecium* is controlled by a similar mechanism is under discussion (see below).

(2) *Loxodes*. First electrophysiological tests [57] indicate that the mechanosensitivity in *Loxodes* is similar to *Paramecium*, at least regarding the receptor

channels at the anterior and posterior hemispheres. Hence, *Loxodes* should show a similar gravikinesis. However, *Loxodes* additionally bears the Müller organelles, which are obviously involved in gravisensation (one organell and not drawn to scale in Fig. 3). So far, it remains ambiguous whether the gravitational pull of the heavy body is mediated directly to ion channels in the cell membrane or if second messengers are involved. As stated above, the statocystoid of *Loxodes* may be a specialisation due to the living conditions. As the cells often creep on surfaces of detritus or in narrow gaps of the benthos, it may be speculated, that the mechano-receptors in the cell membrane are often stimulated by external contacts. This would impede gravisensation and favour the development of an intracellular gravisensor.

(3) *Euglena* (after an unpublished model by Häder, personal communication). Stretch-activated gravireceptor channels are located asymmetrically in the anterior cell membrane. In an upward swimming cell the channels are closed, they open if the longitudinal axis deviates from the vertical or if the cell swims downwards [31,60, 103,104]. The resulting depolarization induces an intracellular signal chain and reorientation of the cells is started.

Likewise, a physiological explanation of gravitaxis in *Paramecium* has been proposed. According to the hypothesis of Baba [105] different receptor populations are activated in the course of the helical swimming path of *Paramecium*: in the downward swimming phase of the helix the cell is depolarized by stimulation of the anterior gravireceptors, resulting in a decrease in the swimming velocity. During the following upward phase, the swimming velocity of the cell is increased due to a stimulation of the posterior, hyperpolarizing gravireceptors. Integration of these changes in swimming rate over several helical turns leads to upward curved swimming paths, which have been documented in graviorienting *Paramecium* cells (e.g. [42,106,107]). The model implies that the resulting curvature is smaller the more the swimming track is parallel to the gravity vector, thus resulting in negative gravitaxis.

Perspectives

Perception of gravity is obviously linked to the known principles of mechano-sensation. An evolutional question is, whether the distribution and function of mechano-(gravi)receptors in single cells depend on the presence of gravity. For short-term periods (up to 12 days) in μg no adaptation was found, though increased proliferation rates in ciliates [108] might be an indication of an accelerated cell ageing in microgravity. Long-term studies in microgravity, now possible on the International Space Station, require stable cell cultures under controlled conditions. Recent investigations concentrate on the development of minimalised ecosystems, e.g., the AQUARACK-system [109] or the MICROPOND-system [110].

Acknowledgement

The work is supported by the German Aerospace Centre DLR (grant 50WB9923).

References

1. Häder, D.-P. and Griebenow, K. (1988) Orientation of the green flagellate, *Euglena gracilis* in a vertical column of water. FEMS Microbiol. Ecol. 47, 159–167.
2. Hemmersbach-Krause, R. and Häder, D.-P. (1990) Negative gravitaxis (geotaxis) of *Paramecium* demonstrated by image analysis. Appl. Micrograv. Techn. 4, 221–224.
3. Kowalewski, U., Bräucker R. and Machemer, H. (1998) Responses of *Tetrahymena pyriformis* to the natural gravity vector. Micrograv. Sci. Technol. XI/4, 167–172.
4. Machemer, H., Bräucker, R., Takahashi, K. and Murakami, A. (1992) Short-term microgravity to isolate graviperception in cells. Micrograv. Sci. Techn. V/3, 119–123.
5. Krause, M. (1999) Elektrophysiologie, Mechanosensitivität und Schwerkraftbeantwortung von *Bursaria truncatella*. Diploma thesis, Faculty of Biology, Ruhr-Universität Bochum, Germany.
6. Bean, B. (1984) Microbial geotaxis. In: G. Colombetti and F. Lenci (eds.), Membranes and Sensory Transduction. Plenum Press, New York, pp. 163–198.
7. Kessler, J.O. (1986) The external dynamics of swimming microorganisms. In: F.E. Round and D.J. Chapman (eds.), Progress in Physiological Research. Biopress, Oxford, pp. 258–307.
8. Sineshchekov, O., Lebert, M. and Häder, D.-P. (2000) Effects of light on gravitaxis and velocity in *Chlamydomonas reinhardtii*, J. Plant Phys. 157, 247–254.
9. Häder, D.-P. (1987) Polarotaxis, gravitaxis and vertical phototaxis in the green flagellate, *Euglena gracilis*. Arch. Microbiol. 147, 179–183.
10. Häder, D.-P. (1991) Strategy of orientation in flagellates. In: E. Riklis (ed.), Photobiology. The Science and its Applications, Plenum Press, New York, pp. 497–510.
11. Häder, D.-P. (1991) Phototaxis and gravitaxis in *Euglena gracilis*. In: F. Lenci, F. Ghetti, G. Colombetti, D.-P. Häder, and P.-S. Song (eds.), Biophysics of Photoreceptors and Photomovements in Microorganisms. Plenum Press New York and London, pp. 203–221.
12. Eggersdorfer, B. and Häder, D.-P. (1991) Phototaxis, gravitaxis and vertical migrations in the marine dinoflagellate *Prorocentrum micans*. FEBS Letters MS Microbiol. Ecol. 85, 319–326.
13. Häder, D.-P. and Liu, S.-M. (1990) Effects of artificial and solar UV-B radiation on the gravitactic orientation of the dinoflagellate, *Peridinium gatunense*. FEMS Microbiol. Ecol. 73, 331–338.
14. Häder, D.-P. and Liu, S.-M. (1990) Motility and gravitactic orientation of the flagellate, *Euglena gracilis*, impaired by artificial and solar UV-B radiation. Curr. Microbiol. 21, 161–168.
15. Kessler, J.O., Hill, N.A. and Häder, D.-P. (1992) Orientation of swimming flagellates by simultaneously acting external factors. J. Phycol. 28, 816–822.
16. Stallwitz, E. and Häder, D.-P. (1994) Effects of heavy metals on motility and gravitactic orientation of the flagellate, *Euglena gracilis*. Europ. J. Protistol. 30, 18–24.
17. Block, I., Freiberger, N., Gavrilova, O. and Hemmersbach, R. (1999) Putative graviperception mechanisms of protists. Adv. Space Res. 24(6), 877–882.
18. Fenchel, T. and Finlay, B.J. (1984) Geotaxis in the ciliated protozoon *Loxodes*. J. Exp. Biol. 110, 17–33.
19. Fenchel, T. and Finlay, B.J. (1986) The structure and function of Müller Vesicles in

loxodid ciliates. J. Protozool. 33, 69–76.

20. Hemmersbach-Krause, R. and Briegleb, W. (1992) Gravitational effects on *Paramecium* and *Loxodes*. Physiologist 35, 23–26.

21. Machemer-Röhnisch, S., Bräucker, R. and Machemer, H. (1993) Neutral gravitaxis of gliding *Loxodes* exposed to normal and raised gravity. J. Comp. Phys. A 171, 779–790.

22. Machemer-Röhnisch, S., Machemer, H. and Bräucker, R. (1998) Graviresponses of gliding and swimming *Loxodes* using step transition to weightlessness. Euk. Microbiol. 45(4), 411–418.

23. Finlay, B.J., Fenchel, T. and Gardener, S. (1986) Oxygen perception and O_2 toxicity in the freshwater ciliated protozoon *Loxodes*. J. Protozool. 33, 157–165.

24. Hemmersbach-Krause, R., Briegleb, W. and Häder, D.-P. (1991) Dependence of gravitaxis in *Paramecium* on oxygen. Eur. J. Protistol. 27, 278–282.

25. Machemer, H., Machemer-Röhnisch, S. and Bräucker, R. (1993) Velocity and graviresponses in *Paramecium* during adaptation and varied oxygen concentrations. Arch. Protistenk. 143, 285–296.

26. Machemer, H. and Bräucker, R. (1992) Graviperception and graviresponses in ciliates. Acta Protozool. 31, 185–214.

27. Machemer, H., Machemer-Röhnisch, S., Bräucker, R. and Takahashi, K. (1991) Gravikinesis in *Paramecium*: theory and isolation of a physiological response to the natural gravity vector. J. Comp. Physiol. A. 168, 1–12.

28. Bräucker, R., Machemer-Röhnisch, S. and Machemer, H. (1994) Graviresponses in *Paramecium caudatum* and *Didinium nasutum* examined under hypergravity conditions. J. Exp. Biol. 197, 271–294.

29. Vogel, K., Hemmersbach-Krause, R., Kühnel, C. and Häder, D.-P. (1993) Swimming behaviour of the unicellular flagellate, *Euglena gracilis*, in simulated and real microgravity. Microgravity Sci. Technol. 4, 232–237.

30. Häder, D.-P, Porst, M., Tahedl, H., Richter, P. and Lebert, M. (1997) Gravitactic orientation in the flagellate *Euglena gracilis*. Micrograv. Sci. Techn. 10, 53–57.

31. Kamphuis, A. (1998) Digitale Pfadanalyse am Beispiel der Schwerkraftausrichtung von *Euglena gracilis* in Flachküvetten. Faculty of Biology, Doctorial Thesis, Rheinische-Friedrich-Wilhelms-Universität Bonn, Germany.

32. Machemer-Röhnisch, S., Nagel, U. and Machemer, H. (1999) A gravity-induced regulation of swimming speed in *Euglena gracilis*. J. Comp. Physiol. A 185(6), 517–527.

33. Jennings, H.S. (1906) Behaviour of the lower organisms. Indiana University Press, London, Bloomington.

34. Häder, D.-P. and Hemmersbach, R. (1997) Graviperception and graviorientation in flagellates. Planta 203, S7–S10.

35. Hemmersbach, R., Volkmann, D. and Häder, D.-P. (1999) Graviorientation in protists and plants. J. Plant Physiol. 154, 1–15.

36. Hemmersbach, R. and Häder, D.-P. (1999) Graviresponses of certain ciliates and flagellates. FASEB J. 13, S69–S75.

37. Verworn, M. (1889) Psychophysiologische Protistenstudien, pp. 1–219, Fischer Verlag, Jena.

38. Dembowski, J. (1929) Vertikalbewegungen von *Paramecium caudatum* I. Lage des Gleichgewichtszentrums im Körper des Infusors. Arch. Protistenk. 66, 104–132.

39. Dembowski, J. (1929) Vertikalbewegungen von *Paramecium caudatum*. II. Einfluß

einiger Außenfaktoren. Arch. Protistenk. 66, 215–260.

40. Kuznicki, L. (1968) Behaviour of *Paramecium* in the gravity field. I. Sinking of immobilized specimens. Acta Protozool. 6, 109–117.

41. Roberts, A.M. (1970) Geotaxis in motile microorganisms. J. Exp. Biol. 53, 687–699.

42. Fukui, K. and Asai, H. (1985) Negative geotactic behaviour of *Paramecium caudatum* is completely described by the mechanism of buoyancy-oriented upward swimming. Biophys. J. 47, 479–483.

43. Winet, H. and Jahn, T.L. (1974) Geotaxis in protozoa. I. A propulsion-gravity model for *Tetrahymena* Ciliata. J. Theor. Biol. 46, 449–465.

44. Jensen, P. (1893) Über den Geotropismus niederer Organismen. Pflüger's Arch. ges. Physiol. 53, 428–480.

45. Davenport, C.B. (1897) Experimental Morphology. Macmillan, New York, 509 pp.

46. Koehler, O. (1922) Über die Geotaxis von *Paramecium*. Arch. Protistenk. 45, 1–94.

47. Koehler, O. (1930) Über die Geotaxis von *Paramecium*. II. Arch. Protistenk. 70, 279–307.

48. Lyon, E.P. (1905) On the theory of geotropism in *Paramecium*. Am. J. Physiol. 14, 421–432.

49. Penard, E. (1917) Le genre *Loxodes*. Rev. Suisse de Zool. 25, 453–489.

50. Bedini, C., Lanfranchi, A. and Nobili, R. (1973) The ultrastructure of the Müller body in *Remanella*. J. Protozool. 24, 394–401.

51. Rieder, N. (1977) Die Müllerschen Körperchen von *Loxodes magnus* (Ciliata, Holotricha): Ihr Bau und ihre mögliche Funktion als Schwererezeptor. Verh. Dt. Zool. Ges., 254.

52. Neugebauer, D.-C. and Machemer, H. (1997) Is there an orientation-dependent excursion of the Müller body in the "statocystoid" of *Loxodes*? Cell Tissue Res. 287, 577–582.

53. Machemer, H. and Deitmer, J.W. (1985) Mechanoreception in ciliates. Progr. Sensory Physiol. 5, 81–118.

54. Machemer, H. and Ogura, A. (1979) Ionic conductances of membranes in ciliated and deciliated *Paramecium*. J. Physiol. 296, 49–60.

55. Machemer, H. (1988) Electrophysiology. In: H.-D. Görtz (ed.), *Paramecium*. Springer Verlag, Berlin, pp. 185–211..

56. Naitoh, Y. (1984) Mechanosensory transduction in protozoa. In: G. Colombetti and F. Lenci (eds.), Membranes and Sensory Transduction. Plenum Press, New York, pp. 113–134.

57. Nagel, U. (1993) Elektrische Eigenschaften und elektromotorische Kopplung im Verhalten von *Loxodes striatus*. Diploma thesis, Faculty of Biology, Ruhr-University Bochum, Germany.

58. Hara, R. and Asai, H. (1980) Electrophysiological responses of *Didinium nasutum* to *Paramecium* capture and mechanical stimulation. Nature 283, 255–266.

59. Bräucker, R. (1994) Behavioural changes of *Paramecium* and *Didinium* exposed to short-term microgravity. Adv. Space Res. 16 (7), 231–234.

60. Häder, D.-P., Lebert, M. and Richter, P. (1999) Gravitaxis and graviperception in flagellates and ciliates. ESA-SP-437, 479–486.

61. Briegleb, W. (1988) Ground-borne methods and results in gravitational cell biology. Physiologist 31, 44S–47S.

62. Briegleb, W. (1992) Some qualitative and quantitative aspects of the fast-rotating clinostat as a research tool. ASGSB Bull. 5(2), 23–30.

63. Cogoli, M. (1992) The fast-rotating clinostat: a history of its use in gravitational biology and a comparison of ground-based and flight experiment results. ASGSB Bull. 5 (2), 59–68.

64. Hemmersbach-Krause, R., Briegleb, W., Häder, D.-P., Vogel, K., Grothe, D. and Meyer, I. (1993) Orientation of *Paramecium* under the conditions of weightlessness. J. Euk. Microbiol. 404, 439–446.

65. Hemmersbach-Krause, R., Briegleb, W., Vogel., K., Häder, D.-P. (1993) Swimming velocity of *Paramecium* under the condition of weightlessness. Acta Protozool. 32, 229–236.

66. Hemmersbach-Krause, R., Briegleb W., Häder, D.-P., Klein, S. and Mulisch, M. (1994) Protozoa as model systems for the study of cellular responses to altered gravity conditions. Adv. Space Res. 14, (8)49–(8)60.

67. Klaus, D.M., Todd, P. and Schatz, A. (1998) Functional weightlessness during clino-rotation of cell suspensions. Adv. Space Res., 21, 1315–1318.

68. Mogami, Y., Kimura, T., Okuno, M., Yamashita, M. and Baba, S. A. (1988) Free fall experiments on swimming behaviour of ciliates. Proc. Int. Symp. Space Techn. Sci. 6, 2351–2354.

69. Häder, D.-P. (1994) Gravitaxis in the flagellate *Euglena gracilis*—results from Nizemi, clinostat and sounding rocket flights. J. Gravitat. Physiol. 1, P82–P84.

70. Häder, D.-P., Vogel, K. and Schäfer, J. (1990) Responses of the photosynthetic flagel-late, *Euglena gracilis*, to microgravity. Appl. Micrograv. Techn. 2, 110–116.

71. Machemer, H., Bräucker, R., Murakami, A. and Yoshimura, K. (1993) Graviperception in unicellular organisms—a comparative study under short-term microgravity. Micrograv. Sci. Technol. V/4, 221–231.

72. Häder, D.-P, Rosum, A., Schäfer, J. and Hemmersbach, R. (1996) Graviperception in the flagellate *Euglena gracilis* during a shuttle space flight. J. Biotech. 47, 261–269.

73. Hemmersbach, R., Voormanns, R., Briegleb, W., Rieder, N. and Häder, D.-P. (1996) Influence of accelerations on the spatial orientation of *Loxodes* and *Paramecium*. J. Biotech. 47, 271–278.

74. Hiramoto, Y. and Kamitsubo, E. (1995) Centrifuge microscope as a tool in the study of cell motility. Int. Rev. Cytol. 157, 99–128.

75. Kuroda, K., Kamiya, N., Yoshimoto, Y. and Hiramoto, Y. (1986) *Paramecium* behaviour during video centrifuge-microscopy. Proc. Japan Acad 62, Ser. B, 117–121.

76. Hemmersbach-Krause, R., Briegleb, W. and Häder, D.-P. (1992) Swimming behaviour of *Paramecium* - first results with the low-speed centrifuge microscope NIZEMI. Adv. Space Research 1, 113–116.

77. Friedrich, U.D., Joop, O., Pütz, Ch. and Willich, G. (1996) The slow rotating centrifuge microscope NIZEMI—a versatile instrument for terrestrial hypergravity and space microgravity research in biology in materials science. J. Biotech. 47, 225–238.

78. Nagel, U., Watzke, D., Neugebauer, D.-Ch., Machemer-Röhnisch, S., Bräucker, R. and Machemer, H. (1997) Analysis of sedimentation of immobilized cells under normal and hypergravity. Micrograv. Sci. Technol X/1, 41–52.

79. Häder, D.-P., Liu, S.-M. and Kreuzberg, K. (1991) Orientation of the photosynthetic flagellate, *Peridinium gatunense*, in hypergravity. Curr. Microbiol. 22, 165–172.

80. Häder, D.-P., Reinecke, E., Vogel, K. and Kreuzberg, K. (1991) Responses of the photosynthetic flagellate, *Euglena gracilis,* to hypergravity. Eur. Biophys. J. 20, 101–107.

81. Lebert, M. and Häder, D.-P. (1997) Effects of hypergravity on the photosynthetic flagellate, *Euglena gracilis*. J. Plant Physiol. 150, 153–159.
82. Hemmersbach, R., Voormanns, R. and Häder, D.-P. (1996) Graviresponses in *Paramecium* under different accelerations—studies on the ground and in space. J. Exp. Biol. 199, 2199–2205.
83. Watzke, D., Bräucker, R. and Machemer, H. (1998) Graviresponses of iron-fed *Paramecium* under hypergravity. Eur. J. Protistol. 34, 82–92.
84. Häder, D.-P., Rosum, A., Schäfer, J. and Hemmersbach, R. (1995) Gravitaxis in the flagellate *Euglena gracilis* is controlled by an active gravireceptor. J. Plant Physiol. 146, 474–480.
85. Hemmersbach, R., Voormanns, R., Bromeis, B., Schmidt, N., Rabien, H. and Ivanova, K. (1997) Comparative studies of the graviresponses of *Paramecium* and *Loxodes*. Adv. Space Res. 21, 1285–1289.
86. Lebert, M. and Häder, D.-P. (1996) How *Euglena* tells up from down. Nature 379, 590.
87. Lebert, M., Richter, P. and Häder, D.-P. (1997) Signal perception and transduction of gravitaxis in the flagellate *Euglena gracilis*. J. Plant Physiol. 150, 685–690.
88. Ooya, M., Mogami, Y., Izumi-Kurotani, A. and Baba, S.A. (1992) Gravity-induced changes in propulsion of *Paramecium caudatum*: A possible role of graviperception in protozoan behaviour. J. Exp. Biol. 163, 153–167.
89. Hemmersbach, R., Bromeis, B., Wilczek, M., Neubert, J. Tairbekov M, Gavrilova, O., Rieder, N., Send, W. and Mulisch, M. (1999) Morphology and physiology of *Loxodes* after cultivation in space. ESASP-1222, Biorack on Spacehab, 119–126.
90. Neugebauer, D.-C., Machemer-Röhnisch, S., Nagel, U., Bräucker, R. and Machemer, H. (1998) Evidence of central and peripheral graviperception in the ciliate *Loxodes striatus*. J. Comp. Physiol. A 183, 303–311.
91. Gebauer, M., Watzke, D. and Machemer, H. (1999) The gravikinetic response of *Paramecium* is based on orientation-dependent mechanotransduction. Naturwiss. 86, 352–356.
92. Häder, D.-P., Lebert, M. and Richter, P. (1998) Gravitaxis and graviperception in *Euglena gracilis*, Adv. Space Res. 21, 1277–1284.
93. Richter, P., Lebert, M., Korn, R. and Häder, D.-P. (2001) Possible involvement of the membrane potential in the gravitactic orientation in *Euglena gracilis*. J. Plant Physiol. 158, 35–39.
94. Tahedl, H., Richter, P., Lebert, M. and Häder, D.-P. (1998) cAMP is involved in gravitaxis signal transduction of *Euglena gracilis*. Micrograv. Sci. Technol. XI/4, 173–178.
95. Nagel, U. and Machemer, H. (2000) Effects of gadolinium on electrical membrane properties and behaviour in *Paramecium tetraurelia*. Eur. J. Protistol. 36, 161–168.
96. Hemmersbach, R., Bromeis, B., Block, I., Bräucker, R., Krause, M., Freiberger, N., Stieber, C. and Wilczek, M. (2001) *Paramecium* - a model system for studying cellular graviperception. Adv. Space Res. 27 (5), 893–898.
97. Machemer-Röhnisch, S., Machemer, H. and Bräucker, R. (1996) Electric-field effects on gravikinesis in *Paramecium*. J. Comp. Physiol. A 179 (2), 213–226.
98. Bonini, N.M., Gustin, M.C. and Nelson, D.L. (1986) Regulation of the ciliary motility by membrane potential in *Paramecium*: a role of cyclic AMP. Cell Motil. Cytoskel. 6, 256–272.
99. Schultz, J.E., Pohl, T. and Klumpp, S. (1986) Voltage-gated Ca^{2+} entry into *Paramecium*

linked to intraciliary increase in cyclic GMP. Nature 322, 271–273.

100. Schultz, J.E., Guo, Y.L. Kleefeld, G. and Völk, H. (1997) Hyperpolarization and depolarization-activated Ca^{2+}-currents in *Paramecium* trigger behavioural changes and cGMP formation independently. J. Membrane Biol. 156, 251–259.

101. Richter, P. (2000) Untersuchungen zur Gravitaxis von *Euglena gracilis*. Faculty of Biology, Doctorial Thesis, Friedrich-Alexander-Universität Erlangen-Nürnberg, Germany.

102. Richter, P. and Häder, D.-P. (2000) Calcium imaging in living cells. In: D.-P. Häder (ed.), Image Analysis: Methods and Applications. CRC Press, Boca Raton, FL, pp. 373–389.

103. Lebert, M. and Häder, D.-P. (1999) Negative gravitactic behaviour of *Euglena gracilis* can not be described by the mechanism of buoyancy-oriented upward swimming. Adv. Space Res. 24, 851–860.

104. Lebert, M., Porst, M. and Häder, D.-P. (1999) Physical characterization of gravitaxis in *Euglena gracilis*. J. Plant Physiol. 155, 338–343.

105. Baba, S.A., Tatematsu, R. and Mogami, Y. (1991) A new hypothesis concerning graviperception of single cells, and supporting simulated experiments. Biol. Sci. Space 5, 290–291.

106. Taneda, K. (1987) Geotactic behaviour in *Paramecium caudatum*. I. Geotaxis assay in individual specimen. Zool. Sci. 4, 781–788.

107. Taneda, K. and Miyata, S. (1995) Analysis of motile tracks of *Paramecium* under gravity field. Comp. Biochem. Physiol. 111A, 673–680.

108. Planel, H., Richoilley, G., Tixador, R., Templier, J. and Gasset, G. (1988) Effects of microgravity and hypergravity on the cell: Investigations on *Paramecium tetraurelia*. Acta Astronautica 17, 147–150.

109. Lebert, M. and Häder, D.-P. (1998) AQUARACK: long-term growth facility for "professional" gravisensing cells. ESA-SP 433, 533–537.

110. Bräucker, R., Machemer, H. and Fahn, C. (1999) A closed artificial ecosystem for ciliates. Proceedings of the 2nd Symposium on the Utilisation of the International Space Station, ESTEC, Noordwijk, 511–515.

Cell Biology and Biotechnology in Space
A. Cogoli (editor)
© 2002 Elsevier Science B.V. All rights reserved

The Cytoskeleton, Apoptosis, and Gene Expression in T Lymphocytes and Other Mammalian Cells Exposed to Altered Gravity

Marian L. Lewis

Department of Biological Sciences, University of Alabama in Huntsville, Huntsville, AL 35899, USA

Introduction

Lymphocytes exposed to spaceflight or altered gravity exhibit significant growth attenuation and changes in metabolism, cytoskeletal structure, and gene expression. Since 1973 when Konstantinova et al. [1] reported reduced activation response in peripheral T lymphocytes, others have consistently found that mitogen stimulation of T cells is blunted during, and for up to seven days after spaceflight [2]. Numerous experiments flown on aircraft in parabolic flight (20–30 s), sounding rockets (6–13 min) and the space shuttle (days) have failed to discover conclusive reasons for T cell growth arrest in microgravity. Comprehensive reviews of this research, beginning with experiments on the Russian Zond 5 and 6 spacecraft flown in the 1960s, have been published by Cogoli et al. [3–5].

The historic report in 1984 by Cogoli et al. that human T lymphocytes *in vitro* are 90% blunted in response to mitogenic challenge in microgravity [6] has motivated space research at the single cell level for almost two decades. Recent studies describe altered PKC translocation [7,8], altered gene expression [9,10], and changes in integrin production [11] in space-flown lymphocytes. A significant body of evidence is accumulating to implicate the cytoskeleton in the atypical response of lymphocytes in microgravity.

This chapter reviews the effects of spaceflight and altered gravity on cytoskeletal morphology and expression of cytoskeletal genes in T lymphocytes and established T cell lines. Other mammalian cell types are included for comparison. The effects of simulated shuttle launch vibration are discussed based on research showing that vibration during launch may disrupt the cytoskeleton and alter gene expression. Ground-based vibration studies provide baseline information for distinguishing effects of vibration from effects of the microgravity environment *per se*. Since cytoskeletal disruption is one of several factors known to exacerbate apoptosis, and because of the apparent increase in programmed cell death in space-flown cells [12,13], apoptosis is also discussed.

Leukemic cell lines can provide information on cytoskeletal changes, apoptosis and gene expression and although they may not exactly model normal peripheral T lymphocytes, their responses to stress may be similar. Cell lines eliminate variability characteristic of fresh lymphocytes and provide a common cell source within a close range of passage levels for repeated flight experiments. Cells of the T lymphocyte leukemic line, Jurkat, are easily and consistently growth-stimulated by increasing serum from 2 to 10% and raising culture temperature from 20 to 37°C. This permits consistent initiation of experiments in microgravity. A significant response database for space-flown Jurkat cells is available [7,9,10,12–16].

The terms "microgravity" or "microgravity environment", as applied in this chapter, refer to variables characteristic of the orbital phase of spaceflight. Among these are microgravity *per se*, electromagnetic fields, pressure changes, vibrations from motors and crew activities, radiation and absence of sedimentation-induced convective mixing. The term "spaceflight" includes these variables plus space shuttle launch-induced vibration and acceleration (approximately 3 *g*). To define the true effect of gravity or microgravity *per se*, each of these variables, considered separately in ground-based tests where appropriate as well as in combination with microgravity during spaceflight, should be characterized. Achieving this goal can be possible when a staffed orbiting laboratory, capable of accommodating routine cell culture, becomes fully operable on the International Space Station (ISS).

Mechanisms of growth activation in T lymphocytes are complex and not completely understood. Exactly why these cells (even leukemic T cells) fail to grow in microgravity, whereas proliferation of other cell types may be retarded but not arrested, remains a mystery. Some characteristic of T lymphocytes renders them uniquely sensitive to weightlessness or to other attributes of the microgravity environment. Investigators have long suspected that cause and effect may lie with the cytoskeleton.

A number of the observed responses in space-flown cells, including reduced secretory processes (growth hormone) [17,18] blunted T lymphocyte activation [3–6,19] T lymphocyte (Jurkat) growth arrest, [12] and altered PKC translocation [7,8] could result from anomalies in cytoskeletal structure and function. The cause and effect relationship between the cytoskeleton and anomalous cellular responses in microgravity is not confirmed. However, the recent discovery by Papaseit et al. [20] that gravity is required for the *in vitro* self-assembly of microtubules, provides new evidence that unusual cellular responses in the absence of gravity may be linked to changes in cytoskeletal dynamics, structure, and function.

The cytoskeleton

The cytoskeleton maintains cell shape, provides mechanical support, coordinates and directs movement of cells, maintains proper spatial position of organelles and cellular proteins with respect to each other, acts as a conduit within the cytoplasm for organelles and proteins, and segregates chromosomes during mitosis. These functions depend on complex assemblies of proteins binding in cooperative groups to cytoskeletal filaments. Disruption of filaments or filament association with related proteins impacts signal transduction, cell growth and metabolism and can ultimately lead to apoptosis.

The eukaryotic cytoskeleton is composed of three types of interconnected filaments with functions coordinated by hundreds of associated cytoskeletal accessory proteins. Tables 1–4 provide an overview of the cytoskeletal elements and a quick reference for the role of some of the cytoskeleton-associated proteins. The binding of these proteins in cooperative groups to cytoskeletal filaments, the dynamic and rapid polymerization and depolymerization of filaments, and the fact that cytoskeletal functions occur without producing measurable chemical changes, challenge the assessment of the cytoskeleton's role in affecting cellular responses in microgravity.

Spaceflight-induced cytoskeletal changes include altered actin and tubulin polymerization; morphological anomalies in actin stress fibers, microtubules, and intermediate filament (vimentin) networks; coalesced and shortened microtubules; loss of microtubule contact at the cell membrane; and loosening of the perinuclear cytokeratin network. Reports also describe microtubule organizing center (MTOC) disruption and anomalies in the centrosome complex during cell division. In addition to cellular responses, the *in vitro* self-assembly of tubulin into microtubules in cell-free solutions is essentially non-existent during sounding rocket flight [20]. A summary of cytoskeletal changes in lymphocytes and other cell types cultured in clinostats and rotating bioreactors, sounding rockets, the shuttle, and orbital spacecraft is presented in Tables 5–7.

Methods commonly used to evaluate the cytoskeleton in space-flown cells entail fixing cells with formalin, or combinations of formalin and glutaraldehyde, in microgravity and viewing post-flight by immunofluorescence microscopy sometimes coupled with confocal or electron microscopy. While rhodamine phalloidin directly binds to actin, visualization of the microtubule and IF cytoskeleton and the cytoskeletal proteins generally involves incubating permeabilized cells with desired protein-specific antibody. This is followed by incubation with a second antibody tagged with a fluorescent dye such as fluorescein isothiocyanate (FITC). The complexity and dynamic nature of the cytoskeleton demands more sophisticated analytical methodology to ultimately unravel the potential role of the cytoskeleton in cell growth and differentiation responses in space-flown cells.

Vector-averaged studies in clinostats

Effects of altered gravity on cells in ground-based vector averaged studies in clinostats are shown in Table 5. Based on disorganization, condensation, and irregular pattern arrangement of microtubules in *Xenopus* myocytes grown in a slow rotating clinostat, Hoeger and Gruener [21] concluded that the cytoskeleton is highly sensitive to changes in gravity. Rijken et al. [22] reported enhanced EGF-induced, cytoskeleton-related cell rounding in A 431 cells exposed to simulated low-*g* in a fast-rotating clinostat. In osteoblast-like ROS 17/2.8 cells cultured in a fast-rotating clinostat, the actin cytoskeleton disorganized, cell surface integrin beta 1 distribution changed, and cells detached from the substrate and became apoptotic [23]. These

80

Table 1

Cytoskeletal filaments and associated proteins

Filament type	Distribution	Function	Associated proteins
Actin filaments			
Also called micro-filaments. Consist of two-stranded helical polymers of actin protein.	Distributed throughout cell but concentrated just below plasma membrane.	Sub-plasma membrane actin network functions with myosins to control cell surface movements including chemotaxis.	Myosin I and II α-Actinin Thymosin Profilin Actin-depolymerizing factor
Diameter 5–9 nm Filaments organized in linear bundles, two-dimensional networks and three-dimensional gels. Filaments function as networks. Several isoforms, each encoded by different genes.	Location and orientation in cell controlled by actin nucleation sites. Alpha actins found mostly in muscle; beta and gamma actins found in non-muscle cells.	At focal contacts actin relays signals from extracellular matrix to cytoskeleton. Signal transduction: actin and protein kinases in focal contacts regulate cell survival, growth, morphology, movement and differentiation in response to environment. Actin stress fibers anchor cell to substrate or maintain cell–cell focal contacts.	Spectrin (fodrin) Ankyrin bridges Fimbrin Filamin Gelsolin Adseverin/Scinderin Cap-G Tensin Tropomyosin Ezrin/Radixin/Moesin Plectin Tropomodulin Calponin Ankyrin
Microtubules (MT)			
Long straight hollow cylinders composed of tubulin subunits; much more rigid than actin filaments. Outer diameter ~25 nm Filaments function individually. At least six forms of alpha and six forms of beta tubulin, each form encoded by a different gene.	Slow-growing (minus) end embedded in a single microtubule organizing center or centrosome; plus end grows toward cell membrane. Several hundred MTs radiate from one centrosome. MTs highly dynamic, appear and disappear, constantly polymerize and depolymerize.	Polarize T-cells during antigen binding to membrane. Determine location of membrane bound organelles and other molecules. Serve as "highways" to transport organelles and molecules through cytoplasm from cell membrane to nucleus. Major function during mitosis.	Families of motor proteins: Kinesins Dyneins MAP 1,2 tau proteins

continued

Table 1 (continuation)

Filament type	Distribution	Function	Associated proteins
Intermediate filaments (IF)			
Rope-like fibers with diameter approx. 10 nm Large and hetero-geneous family made up of: nuclear lamins; vimentin-like proteins; keratins; neuronal IFs	IFs are arrayed throughout cytoplasm. Nuclear lamina lie just beneath inner nuclear membrane. IFs span cytoplasm and extend from one cell–cell junction to another.	IFs depolymerize at mitosis and repolymerize after mitosis via phosphorylation and dephosphorylation of several serine residues on the lamins. Provide mechanical stability; function to maintain tension in the cell. IFs are not easily deformed when stretched. Lamina associate with chromatin and inner nuclear membrane.	Sub-unit proteins: Lamins A, B, C Vimentin Desmin Glial fibrillary acidic protein Peripherin Keratin type I acidic type II neutral/basic; Neurofilament proteins

From: Molecular Biology of the Cell (3rd Edition).

Table 2

Actin binding proteins

Protein	Function
Myosin	A group of motor proteins, each encoded by a different gene, found in muscle and non-muscle cells. Provide contractility to actin stress fibers. Actin and myosin II form contractile ring during mitosis to separate daughter cells during cytokinesis. Stress fibers consist of actin and myosin-II (myofibrils), span cytoplasm from focal contact on plasma membrane to intermediate filaments surrounding cell nucleus. Span cytoplasm from one cell–cell junction to another; generate tension along lines of a particular stress. Stress fibers disappear rapidly when tension is released by de-taching at focal contact. All eukaryotic cells have simple force-generated mechanisms; example is myosin-II in nonmuscle cells. The phosphorylation of myosin light chains controls myosin/actin interaction to cause contraction, rounding and shape changes in lymphocytes and other cells.
α-actinin	Concentrated in stress fibers; loosely cross-linking actin filaments in contractile bun-dles; functions to anchor ends of stress fibers at focal contacts in plasma membrane. Primarily a transmembrane linker protein. The actin-binding proteins of focal con-tacts belong to the *integrin family*. An integrin external domain binds to an extra-cellular matrix component and the cytoplasmic domain binds to actin filament stress fibers. This linkage is indirect and mediated by multiple attachment proteins. The cytoplasmic domain of an integrin binds to *talin* which binds to *vinculin* associated with α-*actinin* linking all of these proteins to an actin filament.

continued

82

Table 2 (continuation)

Protein	Function
Thymosin	Blocks actin polymerization.
Profilin	Promotes regulated polymerization of actin during cell movement. Present in all cells.
Actin-depoly-merizing factor	Inhibits assembly of actin into filaments.
Spectrin (fodrin)	Found in various cell types; spectrin is connected to short actin filaments and linked to transmembrane carrier protein through *ankyrin* bridges.
Ankyrin bridges	Proteins that link spectrin connected to short actin filaments to transmembrane proteins.
Fimbrin	Cross-links two actin filaments into three-dimensional networks with gel-like properties to form tight association of parallel actin filaments at leading edges of cells.
Filamin	Widely distributed gel-forming actin-associated protein.
Gelsolin	A ubiquitous protein activated by binding Ca^{++}; binds to side of long actin filaments and severs actin filaments; caps the plus end of actin and breaks up the cross-linked actin network making it more fluid [128]. May function in cell crawling over substratum.
Adseverin/ scinderin	Close relatives of gelsolin; found in platelets and possibly lymphocytic cell lines. Similar to gelsolin, actin-severing activity [129].
Cap-G	A capping protein; nucleates actin assembly but does not sever filaments [130].
Tensin	Found at variety of cell–cell and cell–substrate junctions [131]. May regulate actin membrane attachments at integrin-mediated adhesions [132,133].
Tropomyosin	Binds along the length of actin filaments to stabilize and stiffen actin filaments. It inhibits filamin binding to actin and increases binding of myosin-II to actin. Found in a wide variety of cells [123].
Ezrin/ Radixin/ Moesin	Family of proteins; probably bind along sides of actin filaments and are important in connection of actin filaments to the plasma membrane.
Plectin	Ubiquitous. Connects intermediate filaments, microtubules and actin to each other and to cell membrane; cross-links vimentin filaments and binds MAP1 and 2, spectrin, lamin, B [114] anchors intermediate filaments to desmosomes and sub-membrane skeleton in polarized cells [134], regulates actin filament dynamics [112].
Tropomodulin	A tropomyosin-binding protein that regulated tropomyosin–actin interactions [123].
Calponin	Links actin and vinculin to anchor F-actin to cell membrane [124]; associated with signal transduction from membrane receptors to contractile proteins. Binds actin, tropomyosin, tubulin, [125] myosin light-chain protein [126].
Ankyrin	Peripheral membrane protein; interconnects to spectrin-based membrane cytoskeleton to link cytoplasmic structural protein to an integral membrane [127].

From Molecular Biology of the Cell (3rd Edition) and referenced sources.

Table 3

Microtubule binding proteins

Protein	Function
Kinesins	Move along MTs toward plus end; transport organelles during mitosis and meiosis; transport synaptic vesicles along axons.
Dyneins	Move toward MT minus end; are involved in mitosis and organelle transport.
MAP 1 and 2 tau proteins	Speed nucleation step in polymerization and link MTs to other cellular components; act as microtubule motors.
Plectin	Ubiquitous. Connects intermediate filaments, microtubules and actin to each other and to cell membrane; cross-links vimentin filaments and binds MAP1 and 2, spectrin, lamin B [114] anchors intermediate filaments to desmosomes and sub-membrane skeleton in polarized cells [134], regulates actin filament dynamics [112].

From Molecular Biology of the Cell (3rd Edition) and referenced sources.

Table 4

Intermediate filament associated proteins

Protein	Function
Lamins A,B,C	Form network of nuclear lamina in eukaryotic cells.
Vimentin	Found in leukocytes and many adult cell types; cells of mesenchymal origin. Often expressed transiently during development.
Desmin	Found predominantly in muscle cells.
Glial fibrillary acidic protein	Characteristically forms glial filaments in astrocytes in the central nervous system and Schwann cells in peripheral nerves.
Peripherin	Found in neurons
Keratin: type I acidic; type II neutral/basic	Attached to specialized cell junctions (desmosomes) to bind cells to other cells and hemidesmosomes which anchor cells to basal lamina.
Neurofilament proteins	Most abundant protein found in neurons; form the primary cytoskeletal component in mature nerve cells.
Plectin	Ubiquitous. Connects intermediate filaments, microtubules and actin to each other and to cell membrane; cross-links vimentin filaments and binds MAP1 and 2, spectrin, lamin, B [114] anchors intermediate filaments to desmosomes and sub-membrane skeleton in polarized cells [134], regulates actin filament dynamics [112].

From Molecular Biology of the Cell (3rd Edition) and referenced sources.

Table 5

Cytoskeletal changes: vector-averaged studies in clinostats

Cell type	Effect	Altered g source	Ref.
A-431 epidermoid carcinoma cells	Enhanced cytoskeleton-related cell rounding in presence of EGF.	Fast rotating clinostat	Rijken et al. (1989) [22]
Xenopus myocytes	Microtubules disorganized, condensed, arranged in an irregular pattern.	Slow rotating clinostat	Hoeger and Gruener (1990) [21]
Osteoblast-like ROS 17/2.8 cells	At 50 RPM for 2 h, 35% of cells detached from substrate, became apoptotic, associated with perinuclear distribution of cell surface integrin beta 1 and disorganized actin cytoskeleton. Suggests that vector-averaged gravity causes apoptosis by altering cytoskeletal organization.	Clinostat (50 rpm)	Sarkar et al. (2000) [23]

data suggest that altered gravity triggered apoptosis by modifying the organization of the cytoskeleton.

Parabolic flight and sounding rocket experiments

Studies conducted in sounding rockets and parabola-flying aircraft, such as the KC-135, are summarized in Table 6. In 1988, Moos et al. [24] published an abstract describing their preliminary study with suspensions of purified tubulin in which they identified a significant difference between microtubules assembled in the 30-s low-g phase and the $2\,g$ phase of parabolic flight on the KC-135 aircraft. This preliminary early result showing that microtubule polymerization is affected by gravity change has been substantiated by recent sounding rocket [20] and shuttle flight [25] experiments that confirm altered cytoskeletal polymerization in absence of gravity. Guignandon et al. [26] found that transitioning through several seconds each of $1\,g$, $2\,g$, and micro-g during 15–30 parabolas caused a decrease in cell area and focal contact plaque reorganization in osteosarcoma ROS 17/2.8 cells. This study confirmed cell sensitivity to gravity transition in anchorage-dependent cells. Adaptation of these cells to changes in gravity appears to depend on microtubule function. This research was continued on Russian Foton space flights (see Table 7).

Research beginning with experiments in clinostats and centrifuges [22] and advancing to sounding rockets showed that the F-actin content is increased in A431 epidermoid carcinoma cells in microgravity. These studies, summarized by Boonstra [27], show that EGF-induced cytoskeletal changes and the actin microfilament system are sensitive to changes in gravity and suggest that remodeling of actin microfilaments may affect signal transduction.

Table 6

Cytoskeletal changes: parabolic flight and sounding rockets

Cell type	Effect	Space-flight	Ref.
Cell-free solution of purified tubulin	Differences in microtubule assembly between low-g and $2g$ phases of parabolic flight.	Parabolic flight (KC-135 aircraft)	Moos et al. (1988) [24]
Osteoblast-like ROS 17/2.8 cells	Effect of gravity variation during 15 and 30 parabolic flights caused significant cell shape changes, decreased cell area and focal contact plaque reorganization. Centrifugation at $2g$ or clinorotation had no major effect on focal adhesions. Concluded that these cells are sensitive to gravity switches and adaptation is at least dependent on microtubule function.	Parabolic flight	Guignandon et al. (1997) [26]
Primary human skin fibroblast	Six minutes of microgravity did not affect electrofusion of cells. Launch and microgravity did not have visible effect on actin/crotical cytoskeleton just after electrofusion.	MASER Sounding Rocket (6 min)	Jongkind et al. (1996) [29]
A431 epidermoid carcinoma cells	Cell morphology changes. Remodeling of actin microfilaments may affect signal transduction. F-actin content is increased.	MASER Sounding Rockets (6 min)	Booonstra (1999) (rev.) [27]
T lymphocytes	Bundling of vimentin and tubulin filaments. Con A binds to cells within 30 s; cell motility and cell–cell contacts occur in microgravity. Changes not due to acceleration at $13–15g$ during launch. Results indicate a direct effect of microgravity on the cytoskeleton; little or no changes noted at cell membrane.	MASER Sounding Rockets (6 min)	Cogoli-Greuter et al. (1997) [15,28] Sciola et al. (1999) [16]
T lymphocyte leukemic cell line (Jurkat)	Significantly higher number of cells with altered vimentin (appeared as thick bundles). Thirty seconds after launch vimentin filaments discontinuous and points of stained protein (probably depolymerized vimentin) were visible in cytoplasm. No changes in structure of actin or co-localization of actin on inner side of cell membrane after Con A binding. Binding of con-A to cell membrane not affected in microgravity; patching is retarded and capping is probably slowed.	MAXUS Sounding Rocket (12.5 min)	Sciola et al. (1999) [16] Cogoli-Greuter et al. (1997) [15,28]
In vitro cell-free solution purified tubulin	Absence of self-assembly of purified tubulin into microtubules. Results show that self-organization of microtubules does not occur if gravity is absent during the first 13 min.	MAXUS sounding rocket (13 min)	Papaseit et al. (2000) [20]

Table 7

Cytoskeletal changes: cells flown on the space shuttle and orbiting spacecraft

Cell type	Effect	Space-flight	Ref.
Xenopus embryonic muscle cells	Marked changes in distribution and organization of actin filaments; increased cabling, decreased filament linearity, reduced acetylcholine receptor aggregates at contacts with polystyrene beads. Conclude possible involvement of cytoskeleton in gravi-sensing.	STS-52, 56	Gruener et al. (1993) [34]
Mouse MC3T3-E1 osteoblasts	Abnormal morphology of actin cytoskeleton, reduced number of stress fibers, nuclei 30% smaller in flown cells, oblong shape and fewer punctate areas (see Fig. 3)	STS-56	Hughes-Fulford and Lewis (1996) [32]
Mouse MC3T3-E1 osteoblasts	Elongated actin cytoskeleton, reduced actin cytoskeleton surface area compared to in-flight 1 g and ground controls, smaller nuclei with oblong shape, fewer pucntate areas than controls.	STS-76, 81, 84	Hughes-Fulford and Gilbertson (1999) [101]
HL60 human myeloid leukemic cells	Lack of microtubule polymerization. Microtubules appeared as globular dots in the cytoplasm. Differences in drug distribution in flown compared to ground control cells.	STS-69	Piepmeier et al. (1997) [25]
T lymphocyte leukemic cell line (Jurkat)	Altered microtubule cytoskeleton in cells fixed 4 h after activation in microgravity; coalesced microtubules not extending to cell membrane, MTOC disorganization. Microtubule cytoskeleton appeared to re-established within 24 h.	STS-76	Lewis et al. (1998) [12]
Sea urchin embryo cells	4% of all cells undergoing division in microgravity showed abnormalities in centrosome–centriole complex.	STS-77	Schatten et al. (1999) [35]
Human breast cancer cells (MCF-7)	Microtubules (MT) altered in many cells 1.5 h after launch; may be free tubulin subunits or very short microtubules. At 48 h, MT remained abnormal. Perinuclear cytokeratin network loosely woven. Proliferation reduced, prolonged mitosis possibly from altered MT self-organization. All three cytosleketal elements altered, cytoskeleton did not reorganize. Actin stress fibers less abundant than 1 g in-flight controls. Actin concentrated in small spots in cytoplasm not at cell periphery.	Foton capsule (orbiting space- craft)	Vassy et al. (2001) [36]
ROS 17/2.8 osteoblasts	Disorganized cytoskeleton associated with disassembly of vinculin and phosphorylated proteins in focal contacts. Cortical location of F-actin network. Conclusion that microgravity substantially alters integrin-mediated cell adhesion.	Foton capsule (orbiting spacecraft)	Guignandon et al. (2001) [37]
Human fetal osteoblastic cells (hFOB)	No changes in cytoskeletal morphology of cells grown on cytodex beads.	STS-80 Bion 10	Harris et al. (2000) [38]

Cogoli-Greuter [15,28] and Sciola [16] and their colleagues reported the presence of large bundles of vimentin in significantly higher numbers of T lymphocytes and Jurkat cells flown on MAXUS and MASER sounding rockets. Vimentin filaments and the microtubule network in Jurkat cells were altered after 30 s in microgravity. The vimentin filaments appeared as discontinuous points of stained protein (probably depolymerized vimentin) in the cytoplasm. No changes were noted in the structure of actin or the inner side of the cell membrane after binding of Con A and co-localization of actin with Con A receptors. Cytoskeletal changes were not due to acceleration at 13–15 g during launch of the rocket. Cogoli–Greuter concluded that the cytoskeletal anomalies were a direct effect of microgravity on the cytoskeleton. The results of the MASER flights also demonstrated that Con-A binding to the cell membrane is unaffected in microgravity but that patching and capping appeared to be slower in flown compared to ground controls. Additionally, these flights confirmed that cell motility and cell–cell contacts occur in microgravity. Membrane changes were not noted in Jurkat cells. This finding was supported by a cell fusion experiment flown on a MASER sounding rocket showing that launch and microgravity did not affect cell fusion or the cortical actin cytoskeleton, within the limits of the assay, after electrofusion of primary human skin fibroblasts and six minutes of microgravity [29].

Although conducted in a cell-free system, the sounding rocket experiments of Papaseit [20] and his co-workers have essential relevance to this chapter. In unit gravity, microtubules self-organize in solutions containing purified bovine brain tubulin and GTP when the temperature is increased from 7 to 36°C. Compared to in-flight 1 g centrifuge and ground controls, the microtubules in microgravity exhibited almost no self-organization and those that formed were disoriented and disordered relative to each other. Papaseit describes microtubule self-organization as the outcome of a gravity-triggered branch or bifurcation that fits the model of a reaction–diffusion process [30]. This is a complex process in which macroscopic self-organized forms appear progressively from initially homogeneous solutions. It is at the bifurcation point, about six minutes after initiation of the self-organization process, that organization of microtubules appears to be sensitive to gravity. Microtubule self-organization does not occur if gravity is absent during the first 13 minutes of the reaction. These results suggest that reaction–diffusion processes may be an underlying mechanism in the dependence of cellular functions on gravity [20]. (A description of these processes is presented by Dr. James Tabony in an earlier chapter.)

Studies on the space shuttle and orbiting spacecraft

Of the cells flown on the shuttle for which cytoskeletal changes are reported (Table 7), only HL-60 and Jurkat cells represent lymphocyte-type cells. Spaceflight caused significant changes in both of these malignant cell lines. In human myeloid HL-60 cells flown on the shuttle [25], the microtubules failed to polymerize and appeared as amorphic globular stained bodies of depolymerized tubulin or very short polymers of

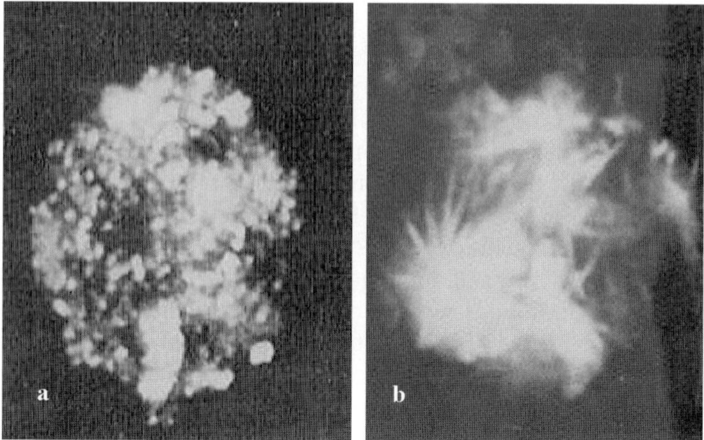

Fig. 1. The microtubule cytoskeleton in taxol treated HL-60 cells flown on the space shuttle. HL-60 suspension cultures were treated with taxol in microgravity and fixed with formalin-glutaraldehyde four days later. Microtubules were visualized by fluorescence microscopy post-flight after reacting cells with anti-tubulin antibody and immunostaining with FITC. The microtubules in flown cells (a) failed to polymerize and appeared as highly stained, short, globular material. In the ground control cells (b), microtubules formed filaments that radiated from microtubule organizing centers (MTOC) (magnification 1000×). (Courtesy of Dr. E.H. Piepmeier, Jr.)

tubulin (Fig. 1). This is in contrast to ground controls, maintained under the same conditions of time, temperature, and hardware, in which polymerized microtubules were apparent. Taken together, the dependence on gravity for tubulin polymerization *in vitro* [20] and absence of microtubule polymerization in HL-60 cells [25] provide strong evidence to support the concept that anomalous cytoskeletal organization occurs in microgravity.

The microtubule cytoskeleton was extensively disorganized in Jurkat cells fixed 20 h after launch and 4 h after increasing serum to 10% and temperature to 20°C on STS-76 (Fig. 2A) [12,31]. The microtubules in many of the flown cells were coalesced, prematurely terminated, and the MTOCs were disrupted. In ground controls, intact microtubules radiated from well-formed MTOCs toward the cell membrane (Fig. 2B). With time in microgravity, the microtubule cytoskeleton appeared to re-organize (Fig. 2C). This, and the presence of mitotic figures in a few cells at 48 h in microgravity, leads to the conclusion that some other cytoskeletal property associated with cell division is altered during spaceflight.

Mouse osteoblasts flown on STS-56 [32] showed clearly that actin cytoskeletal morphology was significantly altered when compared to ground controls in the same hardware and held at the same temperature as the flight experiments (Fig. 3). The flight cells had compressed actin stress fibers, perhaps due to decreased integrin attachment, and cytoskeletal morphology was altered compared to ground controls. Additionally, cell growth was retarded. Continuing this research on STS-76, STS-81 and STS-86, Hughes-Fulford et al. [33] noted elongation of the actin cytoskeleton

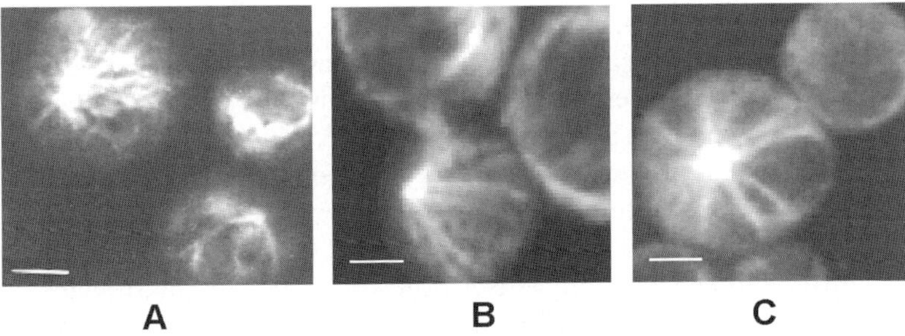

Fig. 2. The microtubule cytoskeleton in Jurkat cells flown on the STS-76 Biorack Payload. Jurkat cells were activated in microgravity 20 h after launch by increasing serum from 2 to 10% in the medium and raising the temperature from 20 to 37°C. After 4 and 48 h, cells in microgravity and the concurrent ground controls were filtered from medium and fixed with formalin. The microtubule cytoskeleton was visualized after flight by immunofluorescence staining and confocal microscopy. Ground control cells had well formed MTOCs with microtubules radiating toward the cell periphery (Fig. 2B). The microtubules in cells, fixed 4 h after activation of the experiment in microgravity (Fig. 2A), appeared as randomly oriented, thickened cables of prematurely terminated filaments and MTOCs were poorly organized. In cells fixed at 48 h in microgravity, microtubules appeared to have re-established normal morphology (Fig. 2C). Bars, 5 μm. (Fig. 2A reproduced from The FASEB Journal (1998) 12, 1007–1018 by copyright permission).

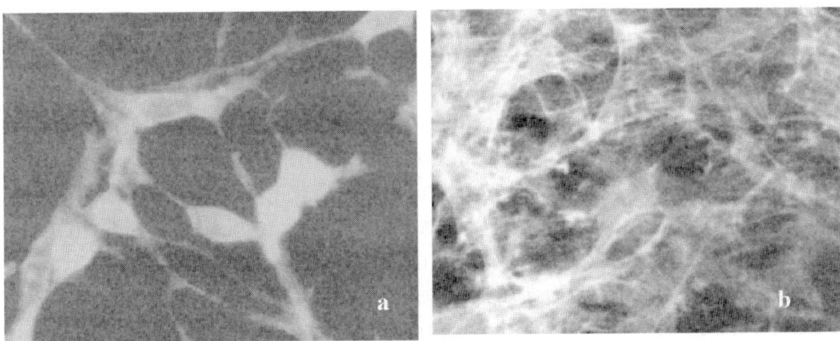

Fig. 3. The actin cytoskeleton in space-flown MC3T3 E-1 osteoblasts (a) and ground controls (b) stained with rhodamine-phalloidin and visualized by fluorescence microscopy. Cells attached to cover slips, were flown on STS-56 in automated sliding block hardware [32]. Growth was stimulated on orbit by increasing serum from 0.5 to 10% and temperature from 20 to 37°C. After four days in microgravity, cells were fixed by addition of formalin (3% final concentration). In flown cells (a), the number of F-actin fibers was reduced and filaments appeared to be coalesced while in the ground controls (b), the cytoskeleton showed multiple F-actin stress fibers characteristic of the MC3T3 cells.

and reduced actin cytoskeleton surface area compared to in-flight 1 *g* and ground controls. In addition to cytoskeletal anomalies, nuclei were smaller, oblong, and had fewer punctuate areas than controls.

Gruener et al. [34] reported marked changes in distribution of actin filaments in *Xenopus* embryonic muscle cells flown on STS-52 and STS-56. Actin filaments were thickened and filament linearity was decreased. This was associated with reduced acetylcholine receptor aggregation at focal contacts with polystyrene beads.

Microtubule organizing centers (MTOCs) are also disrupted during spaceflight. In an experiment flown on STS-77, Schatten et al. [35] found that 4% of the dividing cells in space-flown sea urchin embryos had abnormalities in the centriole-centrosome region. Lewis et al. [9] found increased gene expression for a centriole-related protein (C-NAP1) in Jurkat cells flown on the shuttle and cells subjected to simulated shuttle launch vibration. This is noteworthy because anomalies in MTOC structure and genes regulating MTOCs in microgravity may affect cell division.

Vassy et al. [36] reported apparent high sensitivity of microtubules to microgravity in human breast cancer MCF-7 cells flown in a Russian Photon orbiting space capsule. Cells were sampled at 1.5, 22, and 48 hours after launch for comparison to the $1\,g$ in-flight and ground controls. Only a few microtubule filaments were apparent in cells at 1.5 hours while the $1\,g$ controls had numerous well-polymerized filaments (Fig. 4). By 48 h, the $1\,g$ in-flight controls re-established the microtubule cytoskeleton whereas cells in microgravity did not. Many of the flown cells had microtubule abnormalities at 48 h and the perinuclear cytokeratin network appeared loosely woven. The labeling pattern in flown cells suggested numerous but very short microtubules or labeled free tubulin subunits. In flown cells, mitosis was prolonged (possibly explained by microgravity effect on microtubule self-organization), cell proliferation was reduced, and the cells were partially blocked in the G_2M cell cycle phase. Phosphotyrosine labeling at the ends of stress fibers at the cell periphery was reduced. In addition, though not quantitatively analyzed, microfilament organization into stress fibers and signal transduction from focal contacts appeared to be decreased compared to $1\,g$ in-flight and ground controls.

Guignandon et al. [37] found cytoskeletal disorganization in ROS 17/2.8 osteoblasts flown on the Russian Photon capsule in addition to sensitivity to gravity change noted during parabolic flight [26]. Vinculin disorganization was associated with disassembly of vinculin and phosphorylated proteins in focal contacts [37]. F-actin networks were located in the cortical area of cells. These investigators concluded that microgravity significantly affects integrin-mediated cell adhesion in these osteoblastic cells.

Ground-based simulated shuttle launch vibration

Knowledge of the effect of launch vibration on the cytoskeleton is key to distinguishing spaceflight versus microgravity effects. To assess the effect of vibration on space-flown cells, an accelerometer or other device for measuring vibrations in the actual vicinity of the experiment is recommended though this is not usually available. The level of vibration varies depending on the amount of insulation and location of the payload in the spacecraft. In each of the ground-based evaluations described in the following section, the vibration intensity and duration were set to mimic a typical shuttle launch.

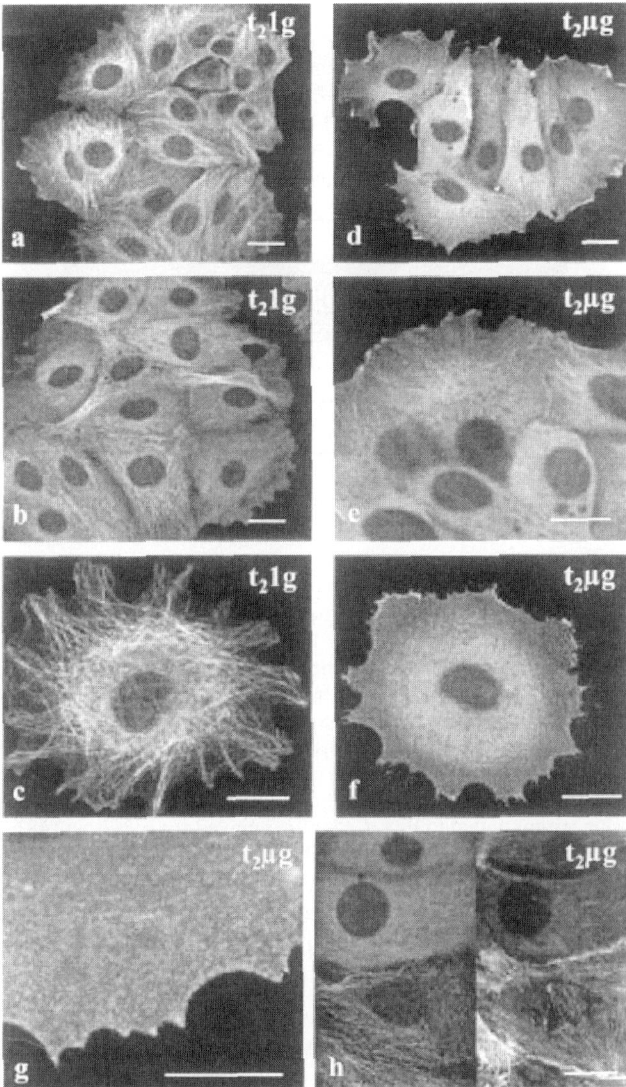

Fig. 4. Alpha-tubulin cytoskeleton in MCF-7 cells. Cells of the human breast cancer line, MCF-7 were flown in a Photon capsule and fixed after 48 h (t_2) in space. Confocal microscopy shows α-tubulin in the in-flight 1 g controls (a–c), and microgravity (d–h). Panels a–g were stained for α-tubulin alone, or double labeled (h) for α-tubulin (left, FITC) and microfilaments (right, Texas Red-phalloidin). Forty-eight hours after launch, the 1 g in-flight controls (a–c) re-established a normal microtubule cytoskeleton with long, strongly labeled microtubules that radiated throughout the cytoplasm. In contrast, the cells in microgravity (d–h) did not re-establish the cytoskeleton and only a few filaments were present in a gray background of either labeled free tubulin subunits or very short microtubules. The short microtubules had no preferential orientation and appeared to be perpendicular to the substratum in contrast to 1 g controls that oriented toward the cell periphery parallel to the coverslip surface. (Reproduced from Vassy et al., The FASEB Journal, 10.1096/fj.00-0527fje, by copyright permission).

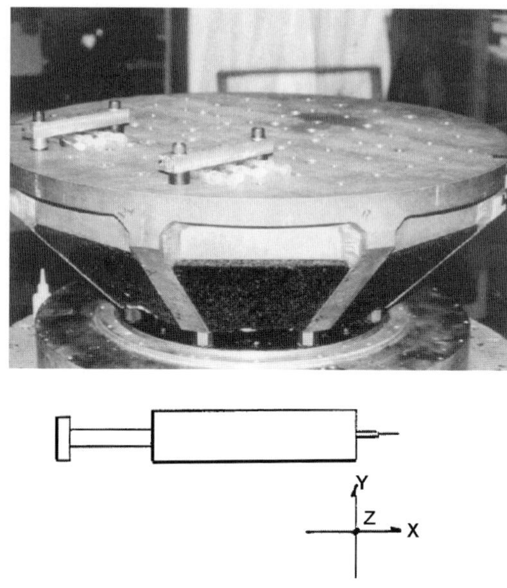

Fig. 5. A vibration table and schematic representation of vibrational axes. Jurkat cells, suspended at 1.5 million/ml in 3 ml of culture medium containing 2% fetal bovine serum (FBS), were loaded into duplicate sets of test and control syringes (10 ml) and maintained at 20°C for 24 h to mimic the pre-launch conditions of experiments flown on STS-80 and STS-95. The syringes were transported from the laboratory by car approximately eight miles to the NASA Marshal Space Flight Center Vibration Facility in Huntsville, Alabama. Syringes were bolted onto vibration tables and vibrated sequentially in the X, Y, and Z planes. Vibration profiles are shown in Fig. 6. Vibrated and non-vibrated control samples were taken immediately for RNA extraction and all other syringes were transported back to the laboratory where they were placed at 37°C after increasing serum content to 10%. Cells were sampled at 4 h, 24 h, and 48 h after vibration and RNA was extracted for gene expression analyses (Table 14).

To determine whether launch vibration could bring about cytoskeletal anomalies similar to those described in space-flown cells [9,12,31], Jurkat cells were subjected to simulated shuttle launch vibration at the NASA Marshal Space Flight Center vibration facility in Huntsville, Alabama. A vibration table with syringes (containing cells) bolted to the surface (Fig. 5) and vibration profiles in the X, Y, and Z planes (Fig. 6) are shown. Results clearly demonstrate that vibration disrupts MTOCs and microtubules (Fig. 7A and B). The disruption was similar to that observed in space-flown Jurkat cells [12,31] (Fig. 2A) and reported for other cell types [16,25, 35,36]. The flown (Fig. 2C) and vibrated (Fig. 7C and D) Jurkat cells appeared to reorganize the cytoskeleton within 24 h (vibrated) [9,31] and 48 h (time at which cells were sampled in microgravity) [12,31]. However, in contrast to the space-flown cells,

---

Opposite: Fig. 6. Representative vibration profiles in the X, Y, and Z planes for Jurkat cell vibration experiments. Parameters were set to mimic the STS-95 shuttle launch. (Profiles provided by the NASA Marshall Space Flight Center Vibration Laboratory, Huntsville, AL.)

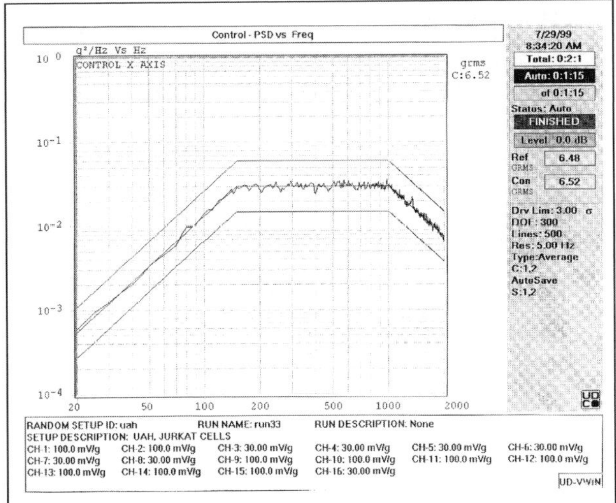

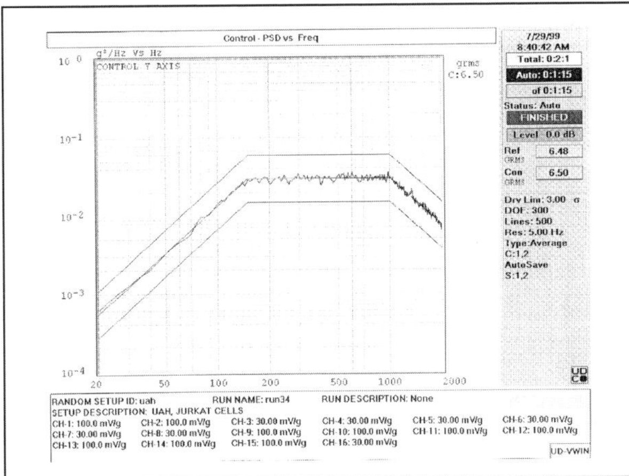

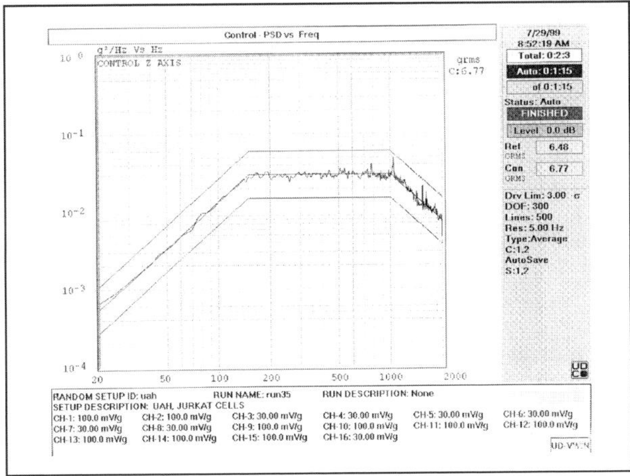

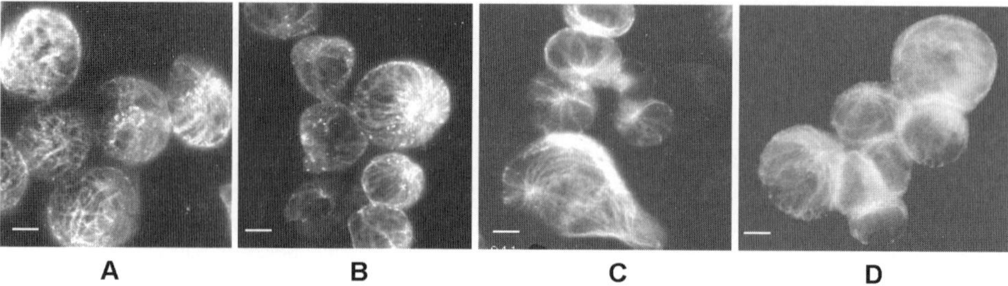

Fig. 7. Effect of simulated shuttle launch vibration on the microtubule cytoskeleton in Jurkat cells. Jurkat cells were subjected to vibration at the NASA Marshall Space Flight Center as described for Figs. 5 and 6. Duplicate samples of vibrated and non-vibrated cells were fixed with formalin at 10 min (A), and 4 h (B), 24 h (C) and 48 h (D) after vibration. Microtubules were visualized by immunofluorescence microscopy. In cells fixed at 10 min and 4 h after vibration the microtubules appeared to be coalesced, randomly distributed, and prematurely terminated. The strongly stained dots may represent depolymerized tubulin or disorganized MTOCs. By 24 h, the microtubule cytoskeleton appeared to be re-established. The lack of cytoskeletal staining at the focal contacts between cells, not evident in earlier samples, appeared to be restored at 48 h. Bar, 5 μm. (Reproduced by copyright permission of The FASEB Journal (June 18, 2001) 10.1096/fje.00-0820fje) and ESA-SP-1222, 1999, 55-68.)

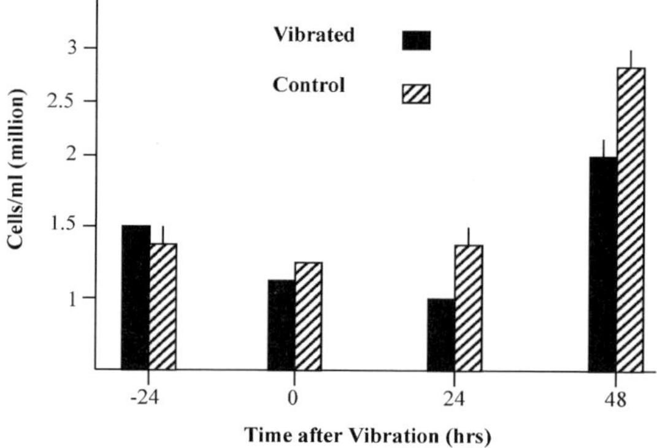

Fig. 8. Effect of vibration on cell growth. Cells were suspended in RPMI 1640 medium containing 2% fetal bovine serum (–24 h) and maintained at 20°C for 24 h before vibrating them to mimic the STS-95 shuttle flight experiment. Cells were counted 30 min after vibration at the time that serum was increased to 10% and temperature was raised to 37°C (0 h), and at 24 and 48 h. Bars represent the mean and standard error of the mean (SEM) of cells counted in five hemacytometer squares for duplicate cell samples at each time point. The vibrated cells proliferated actively although the number of cells at 24 and 48 h was approximately 12% less than non-vibrated controls reflecting the initial loss of cells during vibration. (Redrawn from Ref. [10]).

which remained growth arrested [12,31], the vibrated cells resumed active growth [10] (Fig. 8). Based on this information, the investigators concluded that lymphocyte growth arrest in the microgravity environment is not a direct result of cytoskeletal disruption caused by vibration during shuttle launch. However, even though cells appeared to reorganize the cytoskeleton with time in microgravity, it is not clear whether the microtubules reorganize properly or carry out normal cytoskeletal functions.

Based on ground-based tests simulating shuttle launch vibration, the observed disruption of the MTOCs, bundling of microtubule filaments, loss of filament alignment, lack of extension of filaments to the cell membrane, and depolymerization of microtubules may be a result of launch vibration [9,31].

Discussion and conclusions: the cytoskeleton

Based on the recent evidence of Papaseit et al. [20] that gravity is required for proper self-assembly of microtubules *in vitro* and that once committed, microtubules organize in a particular way according to the bifurcation paradigm [30], it is highly possible that in microgravity, the self-organization process may commit to propagate anomalous microtubules. The research of Lewis et al. [9,12] as well as that of Vassy [36] and others [3,28] is consistent with the reaction–diffusion process and bifurcation theory. If abnormal assembly is initiated in microgravity, then the cytoskeleton could continue to re-assemble abnormally. This would be extremely significant for the fate of the cytoskeleton disrupted by launch vibration. In microgravity, if assembly is miscued early in the re-organization process, then signal transduction, and all other cytoskeleton-related processes would be affected. This is attractive as an underlying mechanism to explain why, even after putative re-assembly of the microtubule cytoskeleton, lymphocytes (Jurkat) do not grow in microgravity. The cells subjected to simulated shuttle launch vibration were disorganized but resumed normal growth after re-establishing the cytoskeleton. Very small changes in cytoskeletal morphology may not be detectible with techniques like fluorescence microscopy yet these changes may be sufficient to block the transport of critical signal transduction molecules and other cytoskeletal processes.

Not all space-flown cells exhibit cytoskeletal anomalies. The SV40 stable transformed hFOB osteoblasts grown on cytodex beads on STS-80 did not show changes in cytoskeletal morphology [38]. However, the majority of cell types examined have exhibited significant cytoskeletal abnormalities. The microtubules in human breast cancer MCF-7 cells did not re-establish normal morphology in microgravity [36]. The presence of free tubulin subunits or short microtubules appears to be a common cytoskeletal anomaly in space-flown cells. Piepmeier et al. [25] reported depolymerized or very short polymers of microtubule protein in HL-60 cells (Fig. 1). Other descriptions of depolymerized cytoskeletal elements include discontinuous "points" of vimentin in Jurkat cells flown on sounding rockets [16,28], dots of microtubule protein in vibrated Jurkat cells [9,31] and shortened microtubules in space-flown cells [12].

Concurrent with cytoskeletal anomalies, proliferation was reduced in microgravity in MCD-7 cells [36] and mouse osteoblasts [32], and Jurkat cells were growth arrested [12]. These findings strengthen the concept that cytoskeletal modifications affect cell proliferation in microgravity. The lack of microtubule polymerization in

96

space-flown HL-60 cells [25], the failure of Jurkat cells to grow after apparent cytoskeletal re-organization in microgravity [12], reduced proliferation of MCF-7 cells, [36] and reduced response of lymphocytes to mitogen activation in microgravity [3–6] as well as the requirement for gravity in the self-organization process of microtubules *in vitro* demonstrated by Papaseit et al. [20] all support the concept of morphological bifurcation in the self-organization process of cytoskeletal elements. This principle, which predicts anomalous microtubule self-assembly in microgravity, is described by Tabony [30] and further discussed in this volume (Dr. J. Tabony).

Cytoskeleton-membrane associations

Under normal terrestrial conditions, microtubules interact with the cell membrane to maintain uniform cell surface tension and cells maintain shape through memb-rane-cytoskeletal interactions based on cytoskeletal polymerization dynamics [39]. Even very subtle cell volume changes resulting from hydrostatic pressure effects in low-g [40] could affect membrane–cytoskeleton association. Interruption of normal membrane–cytoskeleton association causes membrane blebbing, loss of cell surface receptors [41], loss of cellular integrity [42], and disruption of receptor-mediated signal transduction.

The cytoskeleton and signal transduction

The cytoskeleton functions in signal transduction and transition of cells from G_0 to G_1 as signaling molecules relay second messengers by signal amplification [43]. Protein kinase C (PKC) is a key molecule in signal transduction in lymphocytes. PKC is closely associated with the cytoskeleton [44,45,46]. In the case of cardiac myocytes, an isozyme of PKC is translocated along actin stress fibers [45] thus demonstrating the "highway" function of the cytoskeleton in transporting signaling molecules. Molecules required for binding and activation of PKC are also associated with these cytoskeletal elements [47]. The association of the microfilament cytoskeleton and EGF receptor-mediated signal transduction, including the role of actin microfila-ments in signal transduction through the G-proteins (Rac, Rho, and Cdc42 sub-families) is discussed by Boonstra [27]. Ingber points out that many of the enzymes and substrates responsible for DNA and protein synthesis, RNA processing, and glycolysis are bound and immobilized on cytoplasmic and nuclear cytoskeletal elem-ents [48]. If cytoskeletal components, disorganized by spaceflight, do not re-organize or organize improperly in microgravity, any downstream function in the signal transduction cascade, including growth and differentiation, would be compromised.

The cytoskeleton-associated proteins are also involved in signal transduction. Profilin and gelsolin form an important link between cytoskeletal dynamics and signal transduction [49]. Both gelsolin and profilin, regulate actin polymerization and filament formation. The function of these proteins is modulated by inositol di-phosphate (PIP) hydrolysis at the cell inner membrane. PIP binds gelsolin [49,50], profilin, and profilactin [51] and interacts specifically with profilin to dissociate the profilin–actin complex and thus mediate actin polymerization. Lewis [9] and colleagues found down-regulation, compared to ground controls, of message for gelsolin precursor in Jurkat cells flown on STS-95. Whether altered gelsolin gene expression is relevant in space-flown cells is yet to be determined.

Evidence that an intact cytoskeleton is required for proper signal transduction is provided, in part, from studies showing that signal transduction can be blocked by drugs that alter the cytoskeleton. Colchicine inhibits phosphatidylinositol turnover in Con A stimulated human peripheral lymphocytes by inhibiting inositol incorporation and directly interacting with tubulin or tubulin-like proteins in the membrane [52]. The EGF receptor signal transduction cascade is associated with actin filaments and when cytoskeletal filament polymerization is blocked with cytochalasin D, *c-fos* gene expression is affected [27,53].

Cytoskeletal disruption and cellular tensegrity

Cells respond to environmental change (e.g. altered gravity) by detecting physical or mechanical stimuli and converting these into biochemical signals (cytokines, integrins) to effect structural and molecular changes (cytoskeletal rearrangement, PKC translocation, mobilization of signal transduction molecules) to mediate physiological responses (growth, differentiation, apoptosis). The tensegrity paradigm, as applied to cellular response in altered gravity by Ingber [54], relates transduction of mechanical stimuli (gravity change) to biochemical signals within the cell. Tensegrity depends on the cytoskeleton and is patterned after geodesic architecture that maintains structural integrity by balancing interrelated and dynamic internal stresses against opposing forces or external compression from the environment (i.e. gravity). In this paradigm, the microtubules and cross-linked microfilaments, in association with contractile microfilaments and intermediate filaments, form a tension-dependent cellular infrastructure. This network is the conduit for mRNA, signaling molecules and other cellular components. Scaffolds of intermediate filament proteins serve as structures on which enzyme systems and organelles are arranged spatially. Mechanical stability of the cellular tensegrity system as a whole depends on connection of cytoskeletal elements to sites in the nucleus and to transmembrane receptors on the cell surface. If cellular tensegrity is interrupted by launch vibration or anomalous microtubule organization in microgravity, a breakdown in nuclear structure–function interrelationships may lead to altered gene expression [55] through decreased transduction of mechanical signaling by the elements of the cytoskeleton and cytokeratin network [56].

Apoptosis

Apoptosis is a normal process by which the body maintains a balance in specific cell populations assuring that supply does not outstrip demand and that specific cell types are available when needed. Though cell death has been described in the literature for more than a hundred years, Ellis and Horvitz in 1986 [57] instigated current understanding of the topic and in the five years between the mid- and late 1990s, the number of publications on apoptosis grew to over 20,000. Apoptosis plays a significant role during embryonic development and differentiation of hematopoietic cells.

An outcome of an intracellular cascade of genetically determined steps, apoptotic mechanisms are complex and may follow several different pathways involving activation of endogenous proteases [58]. The caspase-induced cascade of DNA fragmentation involves activation of at least thirteen cysteine proteases, or caspases (C: cysteine, ASP: aspartate, ASE: enzyme activity). The cellular process begins with nuclear shrinking followed by fragmentation of DNA into increasingly smaller bodies surrounded by small amounts of membrane-bound cytoplasm [59]. Apoptosis is further characterized by loss of mitochondrial membrane potential [60], disruption of the cytoskeleton, cell shrinkage, membrane blebbing, and the breakdown of the cell into apoptotic bodies [58,61,62]. The number of apoptotic cells in a population may be determined morphologically by staining cells with a fluorescent DNA dye, such as Hoechst, and viewing with a fluorescence microscope, or by terminal deoxynucleotidyl transferase-mediated deoxyuridine triphosphate-biotin nick end labeling (TUNEL) staining, flow cytometry, annexin V staining, DNA fragmentation gel analysis, or assays for caspases.

Apoptosis differs significantly from necrosis in that necrotic cells swell, rupture, and release cellular contents into extracellular spaces. Necrosis is generally induced by external factors including acute injury, pressure, desiccation, viral infection, or toxic chemicals. While apoptosis may be caused by environmental factors [63,64] it is also initiated by internal or genetically controlled factors [65,66]. Some inducers of apoptosis that may be encountered by space-flown cells include serum and nutrient depletion, binding of factors such as Fas/APO-1, TNF and other ligands to membrane receptors, DNA damage from ultraviolet and X-irradiation, heat shock, and altered regulation of oncogenes and tumor suppressor genes.

In human lymphocytes, apoptosis is commonly mediated by the cell death factor, Fas/APO-1, which produces a death signal when it binds to Fas ligand (Fas-L) [67,68] on the surface of various cell types. Described by Oehm et al. [69], Fas/APO-1 is also designated as CD95. Activated lymphocytes constitutively express Fas and intermittently express Fas-L on their surface. A member of the tumor necrosis factor (TNF) receptor family, human Fas includes a membrane-spanning region in the middle of the molecule and a 70 amino acid death domain extending into the cytoplasm is required for transduction of the apoptotic signal [67]. A variety of soluble Fas (sFas) proteins lacking the transmembrane region appear to be a product of alternative splicing [70,71] and may bind Fas-L prior to binding Fas [72] to protect against apoptosis [73,74].

Apoptosis in space-flown cells

There are only a few reports describing apoptosis in cells under altered gravity conditions and results are varied (Table 8). Lewis et al. found increased apoptosis and release of soluble Fas/APO-1 protein in medium of Jurkat cells flown on the

Table 8

Apoptosis in cells exposed to altered gravity

Cell type	Effect	Gravity condition	Ref.
T lymphocyte leukemic cell line (Jurkat)	Increased number of apoptotic cells 4 h after initiation of experiment in microgravity compared to ground. Centrifugation (1 g in space and 1.4 g ground control) increased number of apoptotic cells compared to non-centrifuged cultures. Time- and microgravity-related increase in release of soluble Fas/APO-1 protein into culture medium.	STS-76	Lewis et al. (1998) [12]
T lymphocyte leukemic cell line (Jurkat)	Repeatable time- and microgravity-dependent increase in release of soluble Fas/APO-1 from space-flown cells. Increase in number of cells expressing membrane-bound Fas; increase in number of apoptotic cells in microgravity. Maintaining cells at 20°C in medium with 2% FBS for ~30 h pre-launch did not increase soluble Fas/APO-1 release.	STS-80 STS-95	Cubano and Lewis (2000) [13]
Human breast cancer MCF-7 cells	No DNA condensation characteristic of apoptosis detected under conditions of the experiment. Conclude that apoptosis may not have occurred during the times studied.	Photon capsule	Vassy et al. (2000) [36]
Human colorectal carcinoma cell line MIP-101 (Cells in complete medium; experiment duration: 6 days)	Rotation in microgravity significantly decreased apoptosis and increased CEA expression compared to rotation in 1 g. Rotation appears to increase apoptosis and decrease proliferation compared to 3D static suspension and monolayer cultures.	Space shuttle; microgravity vs. ground; suspension cultures: (a) rotated 3D*; (b) static 3D; and monolayer.	Jessup et al. (2000) [78]
Chick mesenchymal/ cartilage cells	No significant change in percentage of apoptosis in flown compared to ground cells.	STS-95	Doty et al. (1998) [80]
Osteoblast-like ROS 17/2.8 cells	Clinorotation at 50 rpm for 6–24 h. At 24 h, 35% cells detached and became apoptotic; death associated with perinuclear distribution of cell-surface integrin beta-1 and disorganized actin cytoskeleton. Suggests vector-averaged gravity causes apoptosis by altering organization of the cytoskeleton.	Clinostat	Sarkar et al. (2000) [23]

continued

100

Table 8 (continuation)

Cell type	Effect	Gravity condition	Ref.
Rauscher murine erythroleukemia cells	No difference in apoptosis found in rotated versus static cultures. Doubling time in RWV and differentiation in presence of erythropoietin half that of static culture.	NASA rotating wall bioreactor (RWV)	Sytkowski and Davis (2001) [76]
Bone marrow BM CD34$^+$ cells	Cultures in RWV did not increase in number; exit G_0/G_1 phase of cell cycle slower. No difference in apoptosis compared to static control.	NASA rotating wall bioreactor (RWV)	Plett et al. (2001) [77]
Peripheral blood mononuclear cells (not activated)	Inhibition of radiation- and activation-induced apoptosis in rotating bioreactor.	NASA rotating wall bioreactor (RWV)	Risin and Pellis (2001) [79]

*3D = three-dimensional suspension culture.

shuttle [12,13,31]. At present, data indicate that cell type, cell density, nutrient availability, and other experiment-specific conditions may act alone or in concert with altered gravity to affect apoptosis [12,31]. Additional spaceflight experiments are needed before the relationship between programmed cell death and altered gravity can be established for T lymphocytes as well as for other cell types.

Vector-averaged studies in clinostats and rotating bioreactors

A review of the characteristics of clinostats and rotating wall vessel (RWV) bioreactors developed at NASA is presented by Klaus [75]. Clinostats and the RWV are functionally similar and neither actually removes the influence of gravity or produces microgravity. Both devices facilitate ground-based investigations in which the *g*-vector is averaged over time [75]. It is important to note that while the end results obtained with these devices may be similar to those in the true microgravity of space, the mechanisms leading to the observed effects may not be identical. The use of clinostats and centrifuges for biological research has been described previously (ASGSB Bulletin Vol. 5, No. 2, October 1992).

Apoptosis in clinorotated cells

In clinostat-rotated cultures of osteoblast-like ROS 17/2.8 cells, Sarkar et al. [23] found that 35% of the cells detached and underwent apoptotic cell death after 24 h of rotation at 50 rpm. Additionally, the actin cytoskeleton was disorganized leading to the conclusion that vector-averaged gravity causes apoptosis by altering the organization of the cytoskeleton.

Apoptosis in rotating wall vessel (RWV) bioreactors

The incidence of apoptosis in Rauscher murine erythroleukemia cells [76] and bone marrow (BM CD34$^+$) cells [77] was not increased in the RWV compared to static cultures. The CD34+ cells, cultured four to six days in the RWV, did not increase in number compared to static 1 g controls and were slow to exit the G_0/G_1 cell cycle phase. Cells taken from the RWV and cultured long-term, had a low incidence of apoptosis that was comparable to cultures taken from the static 1 g controls. Another experiment compared human colorectal carcinoma MIP-101 cells cultured for six days in microgravity and ground in three-dimensional (3D) rotated, 3D static, and monolayer cultures [78]. In general, rotation appeared to increase apoptosis but rotated 3D microgravity cultures had significantly fewer apoptotic cells than rotated 1 g cultures. Apoptosis was inhibited in radiation and activation treated populations of peripheral blood lymphocytes maintained in the RWV [79].

Apoptosis in space-flown cells

Very few experiments have addressed the incidence of apoptosis during spaceflight. Vassy [36] and co-workers did not detect apoptosis in the human breast cancer MCF-7 cells flown in a Photon capsule. They conclude that apoptosis may not have occurred under the conditions of their experiment and during the times evaluated. Using chick mesenchymal/cartilage cells, Doty [80] found no significant change in the number of apoptotic cells flown on STS-95 compared to ground controls. Human colorectal carcinoma MIP-101 cells, grown for six days in microgravity, had a reduced incidence of apoptosis in 3D cultures in rotated vessels compared to rotated ground controls [78].

In contrast our research with Jurkat cells consistently showed an increase in the number of apoptotic cells flown on three shuttle missions (STS-76 [12,31], STS-80 [13], and STS-95 [13]) (Fig. 9) and a microgravity- and time-dependent released of the cell death factor Fas/APO-1 into the culture medium (Fig. 10). The results of the three experiments were similar even though percentage of apoptosis was higher in cells flown on STS-76 than for STS-80 and STS-96. This was probably due to the longer time before activation of the experiment on orbit for STS-76 (20 h) compared to STS-80 and STS-95 (4 h) and to hardware-related culture differences. For all three shuttle flight experiments, the cells were suspended at approximately 1.5 million per ml and maintained for 24–30 h at 20°C in medium containing low serum (2%) to retard growth pre-launch. Characteristic appearance of apoptotic bodies in a population of space-flown Jurkat cells is shown in Fig. 11.

Apoptosis: discussion and conclusions

The fact that apoptosis was not increased in cells cultured in the RWV or in non-lymphocyte cells flown in space is not surprising. Martinez-Lopez et al. [81] report that Jurkat cells, stimulated with PHA, release cytotoxic molecules within 15

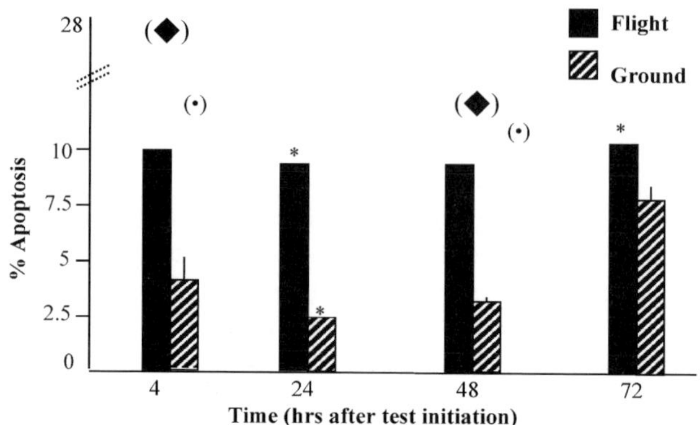

Fig. 9. Apoptosis in Jurkat cells flown on STS-76, STS-80, and STS–95. For all three flights, cells were suspended at approximately 1.5 million/ml at 20°C in medium containing 2% FBS for 20–30 h before launch. Experiments were initiated at launch plus either 20 h (STS-76) or 4.5 h (STS-80 and STS-95) by increasing temperature from 20 to 37°C and serum from 2 to 10%. Samples were fixed with formalin in microgravity at the times shown and cells were stained post-flight with Hoechst 33253. Ground controls were handled in exactly the same manner. A minimum of 300 cells in each of two samples were counted for each time and gravity condition and the percentage of apoptotic cells was determined [13]. Bars represent the average and range of two flight experiments except where only one was conducted (*). The higher percentages of apoptotic cells at 4 and 48 h for STS-76 [flight (◆) and ground control (•)] compared to STS-80 and 95 (bars) are presumably a result of the longer period before experiment initiation in microgravity, the different types of hardware used and related culture characteristics. Jurkat cells on STS-76 were flown in ESA Biorack Cytokines hardware [135]. The total volume of cell suspension in each of six replicate wells for each time point was approximately 0.5 ml. Hardware flown on both STS-80 and 95 consisted of manually operated, interconnected syringes (Bioprocessing Modules) [10] assembled at the University of Alabama in Huntsville. Each syringe contained 5 ml of cell suspension. Regardless of hardware and experiment specific conditions, apoptosis was consistently higher in space-flown cells.

min and maximal cytotoxic activity occurs within 1–7 h after growth stimulation. Thus, if samples are not taken within the first few hours after activation, apoptotic cells in microgravity could be missed. In accord with this, Lewis et al. found a higher incidence of apoptosis in cells sampled soon (4 h) after increasing serum to 10% and temperature to 37°C in microgravity than at later times [12,13]. Although flown populations exhibited more apoptosis than ground controls throughout the time in microgravity, centrifuged cells (1 g in-flight and 1.4 g ground controls) had a higher incidence than static cultures [12]. The time and microgravity-dependent increase in Fas/APO-1 in the culture medium suggests that apoptosis during spaceflight may be mediated by a Fas mechanism.

Some cell types may be more susceptible to programmed cell death in microgravity than others (Table 8). Normal populations of actively growing T lymphocytes are highly regulated by apoptosis and thus unstimulated cells would be less susceptible than mitogen-stimulated lymphocytes or actively growing Jurkat cells. Whereas we found higher incidence of apoptosis in space-flown Jurkat cells, Risin and Pellis [79] report that apoptosis in radiation and activation treated peripheral T lymphocytes was reduced in cells cultured in optimized medium in the RWV. This is not

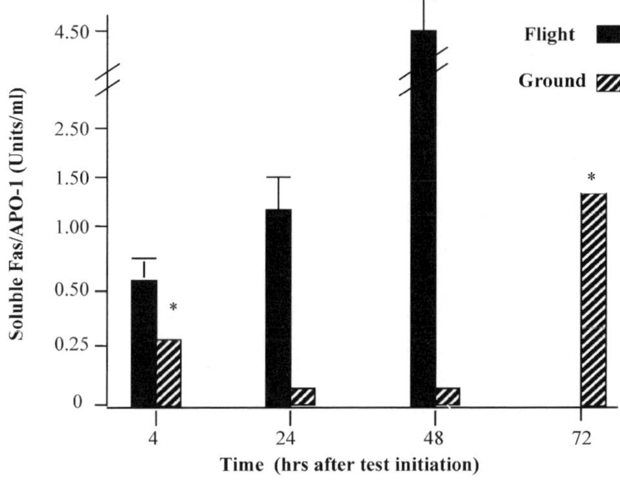

Fig. 10. Effect of spaceflight on soluble Fas/APO-1 protein. Culture medium samples from cells flown on STS-76, STS-80, and STS-95 were evaluated for soluble Fas/APO-1 protein by an enzyme linked immunosorbent assay (ELISA) [13]. Bars represent the average and range of the values from all three flight experiments except where samples were available for only one time point for a particular flight experiment (*). Soluble Fas/APO-1 increased in a time and microgravity dependent relationship between 4 and 48 h while ground control values remained near zero at 24 and 48 h. The increase in the 72 h ground control reflects overcrowding and nutrient depletion which were found in independent ground control experiments to increase soluble Fas/AP0-1 levels and apoptosis in Jurkat cells [13]. Absence of error bars indicates no significant difference among the three samples.

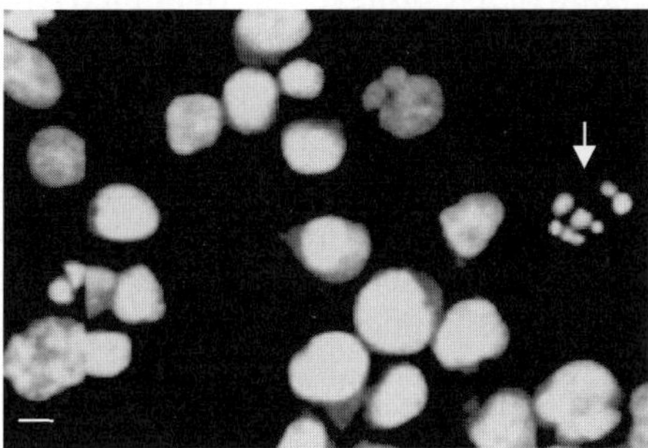

Fig. 11. A population of space-flown Jurkat cells. Cells flown on STS-76 were stained with Hoechst 33253 and visualized by fluorescence microscopy. Bodies characteristic of an apoptotic cell are shown by the arrow. (Bar, 5 μm).

surprising considering the difference in cell type and that responses of cells cultured in a RWV may not occur by the same mechanisms as in true microgravity [75]. The combination of microgravity with other factors including launch vibration, electro-magnetic fields, and radiation, as well as microgravity *per se* must be considered in evaluating apoptosis in space-flown cells.

Spaceflight-related factors such as pre-launch maintenance at low temperature in serum-reduced medium may also influence apoptosis. In ground-based tests, Cubano and Lewis [13] found an increase in number of apoptotic cells as well as increased sFas in culture medium when the cells were allowed to grow to high population density (approximately four million per ml). However, high population density did not appear to cause apoptosis in space-flown Jurkat cells since the cells were evaluated at 48 h when the density was approximately 2.5 million per ml. These investigators also tested the possibility that pre-flight conditions or low-serum medium at low temperature (20°C) might predispose cells to apoptosis. Apoptosis in the ground control cells returned to baseline values, less than 6%, 24 and 48 h after increasing temperature to 37°C and serum to 10%, however, apoptosis in cells flown on STS-80 and STS-95 remained at approximately 10% (Fig. 9). This indicates that holding cells at reduced temperature in medium with low serum content before launch did not of itself, cause the increase in apoptosis and sFas in microgravity. However, this does not rule out a combined effect of microgravity in conjunction with these factors. In ground-based tests, low serum concentration and sub-optimal temperature did not affect the level of sFas into the culture medium [13].

Tumor cells may produce sFas as a means of protecting against apoptosis [74]. Based on results of STS-80 and STS-95, it does not appear that sFas was protective since apoptosis remained at approximately 10% throughout the 48 h tests. During the same time, sFas levels in the medium increased significantly [13] (Fig. 10). The time-dependent increase in Fas/APO-1 appears to be a response to the microgravity environment. Based on previous observations we do not believe that medium stagnation or nutrient depletion were factors [12]. It is not clear whether this is directly microgravity-related or a residual effect of launch stresses, radiation or other factors alone or in concert with microgravity. In ground-based simulated shuttle launch vibration studies, the incidence of morphologically detectible apoptosis was increased in cells 4 h after vibration and returned to baseline values of less than 6% by 24 h. The sFas values remained near zero during the 48 h test (unpublished data).

There is a direct relationship between cytoskeletal integrity and apoptosis. Inhibition of actin filament assembly [82] and disruption of microtubules [83,84] increase apoptosis while stabilization of the microtubule cytoskeleton inhibits the apoptotic process [85]. Sarkar et al. [23] found that ROS 17/2.8 cells detached from surfaces during clinorotation and became apoptotic. In addition, the actin cytoskeleton was disorganized. From these investigations, he concluded that vector-averaged gravity induces apoptosis by altering the organization of the cytoskeleton. In the case of the Jurkat cells, launch vibration may exacerbate apoptosis during spaceflight by destroying cytoskeletal integrity as previously reported [12]. Further information on the role of the cytoskeleton in apoptosis is presented in a review by Atencia et al [86].

Conclusions:

1. Very few spaceflight experiments have evaluated apoptosis.
2. Some space-flown cell types (lymphocytes) appear to be more susceptible to apoptosis than others.
3. Low-shear environments of clinostats and rotating wall bioreactors may not pre-

dict apoptosis in microgravity since other factors (vibration, pre-launch maintenance, methods of activation) may act alone or in concert with microgravity to
induce apoptosis.
3. Space-flown Jurkat cells consistently showed an increase in apoptosis and the cell
death factor, Fas/APO-1, in three shuttle flight experiments suggesting a
Fas-related apoptotic mechanism in microgravity.
4. Apoptosis increased soon after optimizing culture conditions for Jurkat cells in
microgravity indicating that sampling soon after growth stimulation is necessary
to detect apoptosis in space-flown cells.
5. Current information is insufficient to establish whether the microgravity environment and spaceflight affect apoptosis.

Gene expression

The gene expression database for space-flown cells is increasing rapidly yet there is a
paucity of information for lymphocytes. Tables 9–11 summarize expression of a
variety of genes in cells subjected to centrifugation, vector-averaged gravity in
clinostats and rotating wall bioreactors, and orbiting spacecraft. This chapter
primarily describes cytoskeletal gene expression in lymphocytes. A comprehensive
description of gene regulation in space-flown osteoblasts is provided by Dr. Millie
Hughes-Fulford in this volume.

Techniques that have been used to evaluate gene expression in cells under altered
gravity include the reverse transcriptase polymerase chain reaction (RT-PCR) and
competitive RT-PCR, RNase protection assay, and cDNA microarray analyses. Use
of cDNA microarray makes it possible to evaluate expression of thousands of genes
in very small samples of RNA. A typical microarray membrane showing distribution
of ^{33}P-bound RNA hybridized to known cDNA spots is shown in Fig. 12 for RNA
extracted from Jurkat cells vibrated to simulate the STS-95 launch profile [9,31].

Studies in clinostats, rotating wall bioreactors, and centrifuges

In addition to concanavalin-activated lymphocytes and Jurkat cells, studies in
clinostats and rotating wall vessel bioreactors (RWV), and hypergravity (centrifuges)
have evaluated gene expression in bone cells (osteoblast-like MC3T3-E1), renal
cells, HeLa, HepG2 hepatoblastoma, A431 epidermoid carcinoma cells, and other
human tumor cells (Table 9).

Regulation of heat shock protein gene expression in Jurkat cells centrifuged at 3
g to simulate shuttle maximum launch acceleration and rotated in a RWV bioreactor
was evaluated using primers designed to detect inducible and constitutive forms of
HSP-70 [87]. Neither 3 g centrifugation nor rotation in the RWV up-regulated
HSP-70 gene expression. Fitzgerald and Hughes-Fulford [88] found that *c-fos*
expression increased 30 min after centrifugation at 3 g and returned to baseline
within 3 h in MC3T3-E1a cells. Three hours after centrifugation, osteocalcin message was decreased to 44% of control. There was no difference between control and

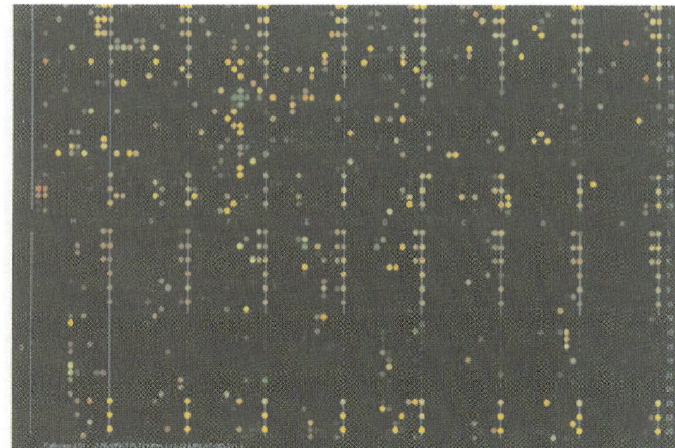

Microarray (above) and histogram (below) summarizing differences in hybridization of RNA extracted from spaceflown Jurkat cells compared to ground controls at 48 hours. (Red = genes up-regulated, green = genes down-regulated, and yellow = no differences of two-fold or greater between genes expressed in flight and ground control cells.

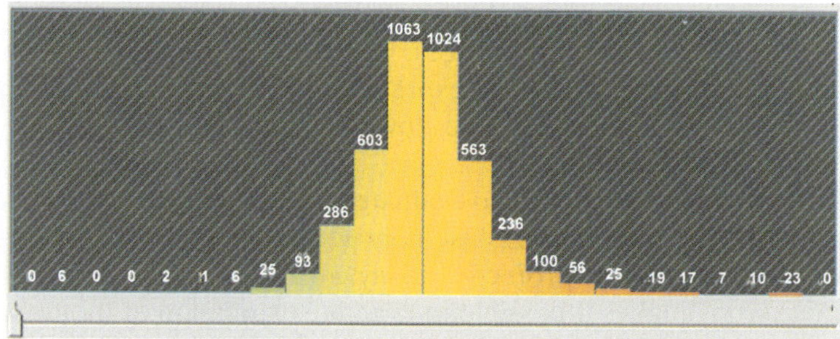

Fig. 12. Representative cDNA microarray membrane and histogram showing genes up- or down-regulated in Jurkat cells flown on STS-95 compared to ground controls. For the STS-95 spaceflight experiment, the Jurkat cells in microgravity were filtered from culture medium at 24 and 48 h and a guanidinium isothiocyanate (GITC) extraction solution was added. Cells were immediately frozen at $-80°C$ until return to the laboratory post-flight where RNA extraction was completed and isolated total RNA was labeled with [33]P. The labeled RNA from 24 h microgravity and ground control samples was hybridized to a membrane (GeneFilter™, Research Genetics, Huntsville, AL, USA) containing approximately 4300 cDNA spots representing known human genes or expressed sequence tags (EST). The 48 h samples were hybridized to each of five membranes (~20,000 genes). Labeling intensity was determined by a phosphor imaging system and Pathways™ software (Research Genetics) to determine messages, expressed as intensity units, in the cellular RNA. From this, the ratio of values for flown versus ground controls was determined. Intensity units were adjusted for background based on control spots located on each membrane. There was no difference (two-fold or greater) between flown and ground control cells in expression of approximately 98% of the 20,000 genes interrogated. Of the approximately two percent regulated differently, 11 were cytoskeleton-related genes (Table 13) [9] (Photograph by Dr. Phillip Bowman).

centrifuged cells in expression of message for *c-myc*, TGF beta 1 and beta 2, cyclophilin and actin. These investigators conclude that up-regulation of *c-fos* is a response to mechanical stimulation and that even small changes in mechanical loading can alter gene expression. De Groot et al. [89] and Rijken et al. [22] also

Table 9

Gene expression in cells in clinostats, rotating bioreactors and centrifuges

Cell type	Effect	Altered gravity source	Ref.
T lymphocyte leukemic cell line (Jurkat)	Hypergravity (centrifugation) simulating shuttle launch and culture in the rotating wall vessel bioreactor did not up-regulate mRNA for heat shock protein (HSP) 70. Culture at 42°C significantly up-regulated message for HSP-70. (RT-PCR)	Centrifugation and RWV bioreactor	Lewis and Hughes-Fulford (2000) [87]
Primary renal cell culture	10,000 genes evaluated by cDNA microarray in 6-day cultures. Cells handled same as for STS-90 shuttle flight experiment. Only 15 genes were expressed differently from static controls (compared to 1632 genes regulated differently in space-flown cells).	Centrifuge $3\,g$ at NASA Ames Research Center	Hammond et al. (2000) [95]
Mouse osteoblast MC3T3-E1a cells	Centrifuged at maximum of $3\,g$ to simulate space shuttle launch; gene expression assayed by RT-PCR. Significant (89%) increase in c-fos at 30 min of centrifugation return to control level 3 h after centrifugation. Osteocalcin message decreased (44% of control); no difference between control and centrifuged cells in expression of c-myc, TGF beta 1 and beta 2, cyclophilin and actin.	Centrifugation	Fitzgerald and Hughes-Fulford (1996) [88]
HeLa cells	Centrifugation at 18, 35, and 70 g at 37°C for up to 4 days. Proliferation increased, G1 phase duration decreased, enhanced expression of *c-myc* gene especially at $35\,g$ after 2 h.	Centrifugation	Kumei et al. (1989) [90]
Osteoblast-like MC3T3-E1	Clinostatic rotation and short-term microgravity on a sounding rocket inhibited EGF-induced *c-fos* gene expression in MC3T3–E1 cells. Phosphorylation of mitogen-activated protein kinase (MAPK) not affected. Conclude that microgravity action site in signal transduction may be downstream of MAPK.	Clinostat and short-term microgravity in a sounding rocket	Sato et al. (1999) [94]
Concanavalin-activated lymphocytes	IL-2 and IL-2 Receptor alpha gene expression significantly inhibited. Assayed by RT-PCR.	Fast rotating clinostat and the Random Positioning Machine	Walther et al. (1999) [91]

continued

Table 9 (continuation)

Cell type	Effect	Altered gravity source	Ref.
A431 epidermoid carcinoma cells	Microgravity decreased EGF-induced proto-oncogene *c-fos* expression by 20% in fast rotating clinostat and hypergravity (centrifuge at 10 *g*) increased expression of *c-fos* by about 18% compared to 1 *g* static controls.	Fast rotating clinostat and centrifuge	Rijken et al. (1989) [22]; de Groot et al. (1991) [89]
Primary renal cell culture	Expression of genes changed for giant glycoprotein scavenger receptors (cubulin and megalin); structural microvillar protein (villin) and classic shear stress response genes (ICAM, VCAM and MnSOD).	Rotating wall vessel (RWV) bioreactor	Kaysen et al. (1999) [96]
Primary renal cell culture	10,000 genes evaluated by cDNA microarray in 6-day cultures. Gene expression changes in the RWV were less marked and more transient than in space-flown cells.	Rotating wall vessel (RWV) bioreactor	Hammond et al. (2000) [95]
HepG2 cells (hepatoblastoma)	Changes in expression of 95 genes (upregulation of 85 and down-regulation of 10). Conclude that low-shear, high-mass transfer 3D enabling environment of RWV (characteristic of quiescent environment of microgravity) modifies expression of a number of genes. (cDNA microarray)	Rotating wall vessel (RWV) bioreactor	Khaoustov et al. (2001) [97]

reported up-regulation of *c-fos* in A431 cells in hypergravity. EGF-induced proto-oncogene *c-fos* expression increased at 10 *g* (centrifugation) by about 18% compared to 1 *g* static controls and decreased by 20% in a fast rotating clinostat. Kumei et al. found increased HeLa cell proliferation, reduced residence in the G1 phase of the cell cycle and an increase in *c-myc* gene expression [90]. In cells centrifuged at 18, 35, and 70 rpm, maximum *c-myc* expression occurred at 35 *g* after 120 minutes of centrifugation at 37°C.

Using a fast rotating clinostat and a Random Positioning Machine to simulate conditions of microgravity, Walther et al. [91] showed that IL-2 and the IL-2 receptor alpha genes were significantly inhibited in mitogen activated T lymphocytes. These data support the previous reports that IL-2 secretion and function are impaired in microgravity [4,92,93]. Sato et al. [94] compared the effect of clinorotation and sounding rocket flight on gene expression in MC3T3-E1 cells and found that EGF-induced *c-fos* expression was inhibited. Phosphorylation of mitogen-activated protein kinase (MAPK) was not affected. They concluded that the site of micro-gravity effect in signal transduction is downstream of MAPK. As a 3 *g* launch acceleration control for their STS-90 experiment, Hammond et al. [95] evaluated expression of 10,000 genes by cDNA microarray in 6-day cultures of primary renal cells centrifuged at 3 g at the NASA Ames Research Center. Only 15 genes were expressed differently compared to the static controls (compared to 1,632 genes regulated differently in space-flown cells).

Primary renal cells cultured in RWV bioreactors showed change, compared to static controls, in expression of genes for giant glycoprotein scavenger receptors (cubulin and megalin), a structural microvillar protein (villin), and shear stress response genes (ICAM, VCAM, and MnSOD) [96]. Of the 10,000 genes evaluated by cDNA microarray in 6-day primary renal cells cultured in the RWV, Hammond et al. [95] found that gene expression changes in the RWV were less marked and more transient than in the cells flown on STS-90. Hepatoblastoma HepG2 cells cultured in the RWV and evaluated by cDNA microarray, up-regulated 95 and down-regulated ten genes compared to static cultures suggesting that the low-shear, high-mass transfer three dimensional-enabling environment of RWV (characteristic of quiescent environment of microgravity) modifies expression of a number of genes [97]

Sounding rocket experiments

Table 10 summarizes gene expression in cells flown on sounding rockets. First reported for clinorotated cells by Rijken et al. [22] and deGroot et al. [89], reduction of *c-fos* oncogene expression in A431 epidermoid carcinoma cells was confirmed in cells exposed to 6 min of microgravity on MASER sounding rockets [99]. These results demonstrated that gravity moderates signal transduction pathways. The observed gravity change resulted from specific modulations of signal transduction in cells treated with EGF and 12-O-tetradecanoyl-phorbol-13-acetate (TPA). In the presence of EGF and TPA in microgravity the expression of *c-fos* and *c-jun* proto-

Table 10

Gene expression in cells flown on sounding rockets

Cell type	Effect	Flight	Ref.
A431 epidermoid carcinoma cells	Decreased EGF-induced proto-oncogene *c-fos* expression in microgravity. EGF and 12-O-tetradecanoyl-phorbol-13- acetate (TPA)-induced expression of *c-fos* and *c-jun* protooncogenes decreased; no effect of gravity changes in A23187- and forskolin-induced expression of these genes. Conclusion that gravity differentially modulates distinctive signal transduction pathways. (RNase protection assay)	Sounding Rocket MASER 3 and 4	Rijken et al. (1994) [99]; De Groot et al. (1991) [98]
Osteoblast-like MC3T3-E1	Clinostatic rotation and short-term microgravity on a sounding rocket inhibited EGF-induced *c-fos* gene expression. Phosphorylation of mitogen-activated protein kinase (MAPK) not affected. Conclude that microgravity action site in signal transduction may be downstream of MAPK.	Sounding Rocket and Clinostat	Sato et al. (1999) [94]

oncogenes in A-431 cells was decreased. Cells treated with A23187 and forskolin, which do not engage the same pathways, had no effect on *c-fos* or *c-jun* message. Expression of the beta-2-microglobulin gene, which is not involved in signal transduction, was not affected by gravity changes.

Sato et al. [94] provide additional evidence that expression of the *c-fos* gene is inhibited in microgravity. Their experiments with osteoblast-like MC3T3–E1 cells showed that both clinorotation and short-term microgravity on a sounding rocket caused inhibition of EGF-induced *c-fos* expression. Additionally, the point at which microgravity exerts its influence appears to be downstream of the phosphorylation of mitogen-activated protein kinase.

Orbital spaceflight

Non-lymphocyte cell types

A number of shuttle flight experiments have evaluated expression of a variety of genes in human, rat, and mouse cells [38,100–106]. These are summarized in Table 11. Results of studies with osteoblasts and osteoblast-like cultures are described in detail in this volume by Dr. Millie Hughes-Fulford.

Table 11

Gene expression in cells flown on the shuttle and orbiting spacecraft

Cell type	Effect	Spaceflight	Ref.
Human T lymphocyte leukemic cell line (Jurkat)	cDNA microarray analysis of >20,000 genes in space-flown compared to ground controls. Changes in mRNA levels for cytoskeleton, growth, transcription factors, tumor suppressors, cell cycle control. Up-regulation of cytoskeleton-related messages in microgravity: plectin, tropomodulin, C-NAP-1, calponin, myosins, ankyrin, dynactin, keratin 8, actin-like EST. Down-regulation of gelsolin precursor. Heat shock protein HSP-27 up-regulated and heat shock factor protein, expressed 3.3 times more than ground controls at 24 h, down regulated at 48 h.	STS-95	Lewis et al. (2001) [9]
Human T lymphocyte leukemic cell line (Jurkat)	Gene expression for heat shock protein (HSP) 70 inducible and constitutive were not up-regulated in space compared to ground controls. Message for HSP 27 was up-regulated in microgravity. (RT-PCR)	STS-95	Cubano and Lewis (2001) [10]
Primary human renal cortical cells	10,000 genes evaluated by cDNA microarray in 6-day cultures. 1632 genes regulated differently compared to ground. Those changed most in microgravity included many cytoskeletal genes and certain transcription factors; Wilms' tumor zink finger protein and vitamin D receptor were down-regulated in microgravity. Plectin was up-regulated.	STS-90	Hammond et al. (2000) [95]; (1999) [107]

continued

Table 11 (continuation)

Cell type	Effect	Spaceflight	Ref.
Primary Chick Calvaria osteoblasts	Reduced osteocalcin and collagen message in microgravity. Suggest that microgravity slows differentiation.	STS-59	Landis et al. (2000) [106]
MC3T3-E1 mouse osteoblast cells	Cells in microgravity increased *c-fos* mRNA; immediate early growth gene, cox-2 decreased; cox-1 was not detected in any samples; no significant difference in expression of actin between flight and 1 *g*. Data indicate that quiescent osteoblasts are slower to enter cell cycle in microgravity. (RT-PCR)	STS-76	Hughes-Fulford et al. (1998) [100]
MC3T3-E1 mouse osteoblast cells.	Fibronectin mRNA levels reduced soon after activation of cells by increasing serum medium content in microgravity compared to ground and the in-flight 1 *g* control at 2.5 h after activation. No difference between flown and 1 *g* controls at 27.5 h. Conclude, after 43 h spaceflight, fibronectin transcription, translation or altered matrix assembly are not the cause for observed cell shape alteration or matrix formation of osteoblasts. (RT-PCR)	STS-76	Hughes-Fulford and Gilbertson (1999) [101]
Normal rat osteoblast cultured from bone marrow	Increase in expression for prostaglandin G/H synthase-2 (PGHS-2), interleukin-6 mRNA and insulin-like growth factor binding protein (IGF BP-3). Decrease in IGF BP-4 and IGF-BP-4 mRNA in $1,25(OH)_2D_3$-treated cells. (RT-PCR)	STS-65	Kumei et al. (1996) [102]
Rat osteoblasts	mRNA for PDGF-beta receptor, growth factor receptor adaptor protein (*Shc*), and *c-fos* decreased compared to ground. EGF receptor mRNA levels not altered in microgravity. Signal transduction via growth factor receptors appears impaired (quantitative RT-PCR)	STS-65	Akiyama et al. (1999) [103]
Human fetal osteoblastic (hFOB) cells	No change in mRNA levels: osteonectin, alkaline phosphatase and osteocalcin. After landing, reduction in mRNA level for cytokines and skeletal growth factors. IL-6 and IL-1α decreased 8 h post flight. Modest reduction in message for TGF-β1 and 2. (RNase protection assay)	STS-80 17 day mission	Harris et al. (2000) [38]
Human osteosarcoma cell line MG-63	Reduced gene expression in differentiation-stimulated cells for collagen Iα 1 (51% of 1 *g* control), osteocalcin (19% of 1 *g* control), and alkaline phosphatase increased by a factor of 5 compared to tenfold increase at 1 *g*. Conclusions, microgravity reduces differentiation of MG-63 cells in response to systemic hormones and growth factors. (competitive RT-PCR)	Unmanned Foton 10 9 days	Carmeliet et al. (1997) [105]; (1998) [104]

112

Using cDNA microarray, Hammond et al. [95,107] evaluated genes expressed in primary human renal cortical cells flown on the shuttle. Ground controls were evaluated in a RWV, static culture, and on a short arm centrifuge (NASA Ames Research Center in Mountain View, CA) simulating the 3 g acceleration characteristics of shuttle launch. Of 10,000 genes evaluated, those for which expression differed most included genes encoding adhesion molecules, apoptosis, cytoskeletal proteins, differentiation mediators, drug metabolizing proteins, heat shock proteins, intracellular signaling proteins, receptors, transcription factors and electron transport chain related genes. Of 1,632 genes in which expression differed by three-fold or more in microgravity compared to static cultures on the ground, expression changed for many cytoskeletal genes and certain transcription factors. Many of the same genes that were expressed differently in microgravity compared to static ground controls were also expressed differently in RWV culture; however, changes were either less, or more transient, than genes expressed in microgravity. In centrifuged cells, only 15 genes changed more than two-fold compared to static controls.

Lymphocytes

Only two investigations of gene expression in T lymphocytes have been conducted in space so far and both evaluated regulation of genes in Jurkat cells flown on STS-95 [9,10] (Table 11). Cubano and Lewis [10] evaluated the RNA messages for the heat shock proteins HSP-27 and the constitutive and inducible forms of HSP-70 by RT-PCR to determine if spaceflight up-regulates these stress-related genes. Neither spaceflight nor simulated shuttle launch vibration up-regulated the HSP-70 proteins, however, analysis by both cDNA microarray and RT-PCR showed two-fold increase in message for HSP-27 in space-flown Jurkat cells compared to ground controls at 24 h after activation in space. In addition, cDNA microarray screening revealed that heat shock factor protein 1 was expressed 3.3 times more than controls at 24 h and 4.1 times less than controls at 48 h [9]. This cytoplasmic protein binds to heat shock gene promoters to activate other heat-shock elements such as HSP-70A [108]. Proteins of the heat shock family protect mammalian cells exposed to heat or oxidative stress against necrotic and apoptotic cell death. HSP-27 is suggested as an inhibitor of Fas/AP0-1-induced apoptosis and is also reported to confer resistance to apoptotic cell death induced by the protein kinase C (PKC) inhibitor, staurosporine [109]. Since anomalies in PKC function are well documented in T lymphocytes in microgravity, the up-regulation of HSP-27 at 24 h may be a protective response against apoptosis.

Using cDNA microarray to evaluate RNA message for 4,324 genes at 24 h and more than 20,000 genes and expressed sequence tags (EST) at 48 h in Jurkat cells flown on STS-95, Lewis et al. [9] found 11 cytoskeletal genes regulated differently in space-flown compared to ground control cells. These are shown in Table 12. Eight of these genes were evaluated qualitatively by RT-PCR to confirm cDNA results (Fig. 13). Except for two genes for which specific primer sequences were used in the RT-PCR, the cDNA microarray results were similar.

Table 12

Cytoskeletal gene expression in space-flown Jurkat cells

Genes up- or down-regulated	Ratio flt:ground	Sample time (h after act.)
Myosin light polypeptide (MYL5)	11.3	24
Plectin (PLEC)	4.7	48
EST similar to Ankyrin	4.5	48
Nek2 assoc protein 1 C-NAP1)	4.0	48
EST Actin-like protein	3.4	48
Calponin	3.0	48
Tropomodulin	2.8	48
Myosin (MLC-2)	2.5	48
Dynactin	2.0	48
Keratin 8	2.0	48
Gelsolin precursor	−2.0	48

Ninety-eight percent of genes expressed in Jurkat cells flown on STS-95 were similarly expressed in the ground controls. Differences of two-fold or greater between flight and ground were found for genes that regulate cytoskeletal organization (discussed in the following sections) metabolism, signal transduction, adhesion, transcription, apoptosis, and tumor suppression [9]. A tumor suppressor gene, *Tsc2*, encoding tuberin was expressed 11.7 times more in space at 24 h than in ground controls. Tuberin is a large membrane-associated GTPase activating protein that is involved in the disease state, tuberous sclerosis [110]. Cyclin-dependent kinase 6 (CDK6) was up-regulated at 24 h by a factor of five, and message for the G1/S cell cycle inhibitor, prohibitin, was increased at 48 h. A complete listing of the genes expressed in space-flown and static and vibrated control cells is available as a PDF attachment to the full text online publication (Lewis et al., FASEB J (June 18, 2001) 10.1096/fj.00-0820fje) [9]. The identity of the cDNA spots incorporated into the

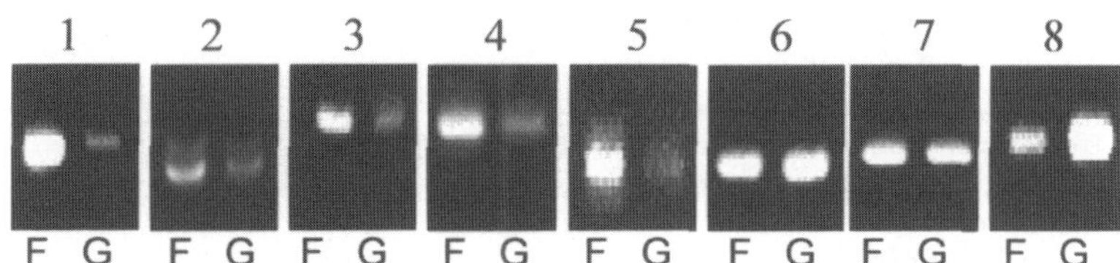

Fig. 13. RT-PCR gels of cytoskeletal genes expressed in space-flown Jurkat cells. The relative expression of messages in flown and ground control cells was determined for eight cytoskeleton-related genes (numbers in parentheses represent base pair size of primers used in the RT-PCR): 1. Tropomodulin (499); 2. Plectin (498); 3. C-NAP1(280); 4. Calponin (334); 5. MLC-2 (360); 6. Ankyrin (266); 7. Dynactin (332); 8. Gelsolin precursor (326). Based on semi-quantitative density scans, five messages were up-regulated by a factor of two or more in flown compared to ground control cells and gelsolin precursor was down-regulated. The cDNA results were qualitatively similar to RT-PCR with the exception that ankyrin and dynactin were up-regulated when tested by cDNA but not RT-PCR in which specific primers for ankyrin and dynactin were used. (Redrawn from Ref. [9]).

microarray GeneFilters™ (Research Genetics) is available at http://www.resgen.com.

Ground-based simulated shuttle launch vibration studies

To better understand whether observed gene expression changes in cells flown on the shuttle originate from vibration during launch or from characteristics of the micro-gravity environment, several types of cells (Table 13) have been subjected to simulated shuttle launch vibration in ground-based experiments either at NASA Ames Research Center, Moffett Field, CA or NASA Marshall Space Flight Center in Huntsville, Alabama.

In ground-based simulated shuttle launch vibration studies, Tjandrawinata et al. [111] found that mRNA for *c-fos* and *c-myc* in osteoblastic MC3T3 cells, vibrated at the NASA Ames facility, was up-regulated 30 min after vibration. Expression levels returned to baseline after three hours. Messages for osteocalcin and transforming growth factor beta 1 were decreased an hour after vibration and there were no

Table 13

Gene expression: ground-based simulated space shuttle launch vibration

Cell type	Effect	Vibration	Ref.
Human T lymphocyte leukemic cells (Jurkat)	Heat shock protein (HSP) 70 inducible and constitutive and HSP 27 were not up-regulated in vibrated compared to non-vibrated controls (RT-PCR). HSP transcription factor 4 was up- and HSP A2 was down-regulated.	Simulated shuttle launch vibration	Cubano and Lewis (2001) [10]
Human T lymphocyte leukemic cells (Jurkat)	cDNA microarray analysis of more than 4000 genes in cells subjected to simulated shuttle launch vibration was compared to non-vibrated controls. Plectin and C-NAP1 up- regulated in vibrated and flown Jurkat cells suggests role in vibration damage repair. Seven cytoskeleton-related genes were up-regulated and six were down-regulated four hours after vibration; two up-regulated at 48 hours; one up- and two down-regulated at 24 hours (Table 14). Cells resumed growth after vibration unlike space-flown cells which did not grow in microgravity.	Simulated shuttle launch vibration	Lewis et al. (2001) [9]
MC3T3E1 Mouse osteoblast cells	Up-regulation of *c-fos* and *c-myc* within 30 min after vibration; osteocalcin and transforming growth factor-beta 1 decreased within 3 h after vibration; no changes in beta-actin, histone H4, or cytoplasmic phospholipase A2; no basal levels of cyclooxygenase 1 detected. Conclusion: these changes in gene expression in vibrated cells are due to direct mechanical effect. (RT-PCR).	Simulated shuttle launch vibration	Tjandrawinata et al. (1997) [111]

changes in beta-actin, histone H4, or cytoplasmic phospholipase A2. No change in cyclooxygenase 1 expression was detected. These data clearly show that vibration induces changes in gene expression and this must be considered when evaluating the effect of spaceflight on cells.

Vibration studies at the NASA Marshall Space Flight Center Vibration Facility, show that vibration does not up-regulate heat shock protein HSP-70 constitutive or inducible genes or HSP-27 in Jurkat cells [10] However, message for human mRNA for heat shock transcription factor 4 in vibrated cells was double that of ground control and the HSP A2 gene was down-regulated in cells sampled 48 h after vibration.

Lewis et al. [9] screened expression of approximately 4,300 genes by cDNA microarray in Jurkat cells sampled 4, 24, and 48 h after vibration simulating the STS-95 launch profile. A list of genes expressed in vibrated versus non-vibrated cells is accessible as a PDF attachment to the online publication (Lewis et al., FASEB J (June 18, 2001) 10.1096/fj.00-0820fje) [9]. This chapter summarizes only the expression of cytoskeletal genes. The cytoskeleton-related genes expressed in vibrated compared to non-vibrated controls are shown in Table 14. When these are compared to cytoskeletal genes expressed in Jurkat cells flown on STS-95 (Table 12), messages for only two genes, plectin and C-NAP1, were up-regulated in both flown and vibrated cells.

The vibrated cells up-regulated more genes soon after vibration (4 h) whereas most of the genes up-regulated in space-flown cells were expressed at 48 h after launch. Because cell handling procedures and hardware were exactly the same in the

Table 14

Cytoskeletal gene expression in Jurkat cells subjected to simulated space shuttle launch vibration

Genes expressed	Ratio vibe:non-vibe	Sampled after vibration (h)
Plectin(PLEC)	6.2	4
Plectin(PLEC)	−2.2	24
Plectin(PLEC)	2.3	48
NeK-2 assoc protein 1(C-NAP1)	3.8	4
Human cytoplasmic beta actin	3.8	4
Keratin Type 1 (cytoskeletal 14)	3.6	4
Keratin Type 1 (cytoskeletal 14)	2.3	48
Keratin Type 1 (cytoskeletal 20)	2.1	24
Human alpha tubulin mRNA	3.4	4
Actin-related protein	2.8	4
Keratin 5 (KRT-5)	2.4	4
Human alpha-tubulin isotype H2	−2.7	4
Actin depolymerization factor	−2.4	4
Human mRNA ankyrin motif	−2.3	4
Myosin polypeptide 3	−2.1	4
Myosin light polypeptide 2 regulatory	−3.3	4
14-3-3 protein tau	−2.6	24

116

flight and vibration experiments, it appears that microgravity, or some factor in the microgravity environment, affects cytoskeletal gene expression and slows the transcription process.

Discussion and conclusions: cytoskeletal gene expression

Possible significance of plectin message up-regulation

The cytoskeleton is comprised of interconnecting structural proteins that link the three filaments to each other and to other cellular components. Plectin, in addition to regulating actin filament dynamics [112] and functioning as a cytoskeleton-membrane anchoring protein, [113] is a ubiquitous and multi-functional cytoskeletal linker protein that connects intermediate filaments, microtubules, and actin to each other and to the cell membrane [114,115]. In this way, plectin contributes to mechanical strength, structure, and tensegrity of cells. Figure 14 shows a schematic diagram [115] of the interconnections of plectin and other cytoskeletal elements. Plectin appears to play a role in reorganization of the cytoskeleton disrupted by shuttle launch vibration since it was up-regulated in both flown and vibrated cells [9]. The increased expression of plectin at 48 h suggests that although the cytoskeleton

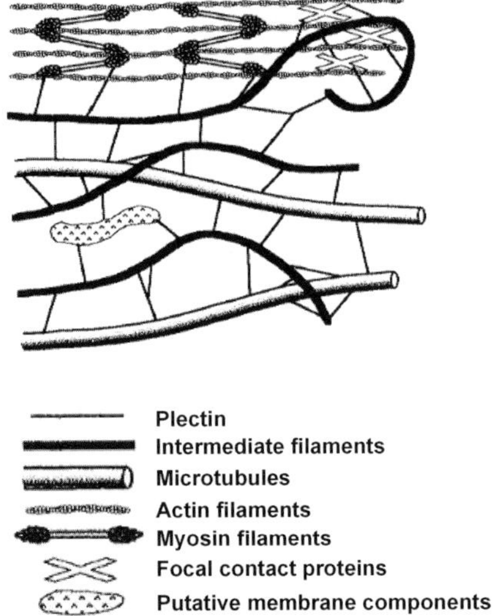

Fig. 14. Schematic diagram of the role of plectin in maintaining cytoskeletal integrity. Plectin links intermediate filaments to microtubules, myosin II minifilaments in stress fibers, and probably links these elements to the cell membrane. In several types of space-flown cells, the microtubules appeared to be coalesced or collapsed together. This and the finding that gene expression for plectin was up-regulated in space-flown and vibrated Jurkat cells implies that plectin is involved in re-organization of the cytoskeleton and suggests that plectin may have a critical role in support of cytoskeletal structure and function in microgravity. (Reproduced from Svitkina et al., The Journal of Cell Biology, 1996, Vol. 135, 991–1007, by copyright permission of The Rockefeller University Press and author).

appears to reassemble in microgravity [12,31], it may not reorganize properly in cells. The work of Papaseit [20] and his colleagues showing inhibition of *in vitro* microtubule self-assembly in microgravity establishes that the cytoskeleton does not reorganize properly in the absence of gravity. Though this evidence strongly supports the occurrence of aberrant cytoskeletal assembly in space-flown cells, future experiments must determine whether cytoskeletal re-organization in living cells is impaired in microgravity.

Signal transduction

Plectin also associates with PKC, which in turn, associates with intermediate filaments and stress fibers [116]. Since PKC is implicated in reduced lymphocyte growth response during spaceflight [117–119], up-regulated plectin gene expression may be important to signal transduction. If the cytoskeleton is not properly organized in microgravity, the molecules involved in PKC signal transduction may not be in the proper spatial proximity required for transduction of mechanical stress into biochemical mediators [54]. The outcome could prevent PKC function and consequently, inhibit cell growth. Flusberg et al. [120] found that simultaneous disruption of microfilaments and microtubules enhanced apoptosis to effectively control cell survival in capillary endothelial cells. Because plectin is involved in interconnecting these elements, anomalies in plectin function may also contribute to apoptosis.

In space-flown Jurkat cells, messages for myosin and dynactin, which are important in organelle positioning and transport, were up-regulated compared to ground controls. Dynactin and the microtubule motor protein, dynein, function in positioning the Golgi complex [121] and myosin, dynein, and dynactin transport molecules and organelles to specific locations in the cell. Many of the molecules involved in signal transduction are immobilized on the insoluble cytoskeletal scaffold [54]. When signaling molecules are not in proper spatial relationship to each other, cells can become growth arrested or apoptotic.

Filament elongation and polymerization

Gelsolin, down-regulated at 24 h in space-flown cells [9], normally binds actin and nucleates actin polymerization [122]. Plectin also facilitates actin filament dynamics [114]. Tropomodulin is a tropomyosin-binding protein that regulates tropomyosin–actin interaction [123]. The differential expression of these genes that regulate filament elongation and polymerization suggests potential for inadequate actin filament polymerization in microgravity.

Cytoskeletal–membrane interactions

Several genes that regulate cytoskeletal-membrane association are expressed differently during spaceflight. Message for calponin and an EST similar to ankyrin were increased at 48 h [9]. Calponin links actin and vinculin to anchor F-actin to the cell

118

membrane, [124] functions in transduction of signals from membrane receptors to contractile proteins, and binds actin, tropomyosin, tubulin [125] and a human myosin light chain protein (MLC-2) [126], which was also up-regulated during spaceflight. Calponin may also regulate actin-myosin interactions. Ankyrin, a peripheral membrane protein that binds to the spectrin-based membrane cytoskeleton, directly links cytoplasmic structural proteins to a membrane protein [127].

Comparison of gene regulation in space-flown and vibrated cells

Up-regulation of messages for plectin and C-NAP1 in both vibrated and flown cells [9] implies their function in reorganization of the cytoskeleton and MTOCs after disruption by vibration. Schatten et al. also found disorganization of the centriole region in cells of space-flown sea urchin embryos [35]. Investigation of C-NAP1 protein message regulation in other cell types will determine whether there is a relationship between this gene and recovery of MTOCs in microgravity. Except for plectin and C-NAP1, other cytoskeletal genes were regulated differently in flown compared to ground control cells [9]. Expression of 11 times more message for the myosin regulatory protein (MYL5) in flight and absence of expression in vibrated cells, suggests that some characteristic of the microgravity environment such as microgravity *per se*, space radiation, electromagnetic fields, or other factors may affect myosin gene expression. Other cytoskeleton-related cDNAs incorporated in microarray filters including actins, tubulins, dynein, vinculin, filamin, spectrin, profilin and kinesin were not detected in flown or vibrated cells [9]. The transient nature of gene expression may explain why expression of some genes may not have been detected at the times sampled.

Comparison of genes expressed in space-flown lymphocytes and kidney cells

It is not possible to directly compare gene expression in Jurkat and kidney cells since the cell types, sampling times (six days for kidney cells and 24 and 48 h for Jurkat), and the type of cDNA microarrays used were different. However, both kidney [107] and Jurkat cells [9] up-regulated plectin message and expressed messages for keratins and myosin light chain proteins differently in flown compared to ground controls. Thus, some of the same types of genes may be involved in gravity response in cells of totally different origin. Messages for an EST similar to ankyrin, C-NAP1, calponin, tropomodulin, and dynactin were up-regulated in Jurkat but not in the kidney cells and a number of genes expressed in the kidney cells were not expressed in Jurkat cells.

Future experiments are needed to determine if these observed differences in cytoskeletal gene expression in microgravity specifically contribute to growth arrest in lymphocytes. Hughes-Fulford found that *cox-2* message was reduced in space-flown osteoblasts and suggests that cells may not be entering into the growth phase of the cell cycle since *cox-2* is depressed [33]. Differential regulation of other genes such as tumor suppressor and cell cycle genes may also affect lymphocyte cell growth response in altered gravity environments.

Conclusions and summary

Research conducted during the past 15 years conclusively confirms that the cytoskeleton is affected by spaceflight and altered gravity. All three cytoskeletal elements (microtubules, actin filaments, and intermediate filaments) in a variety of cell types are changed when the gravity environment is modified. Anchorage dependent *Xenopus* embryonic myocytes, osteoblasts (MC3T3 E-1, ROS 17/2.8, hFOB), A431 epidermoid carcinoma cells, primary skin fibroblasts, and human breast cancer cells (MCF-7), as well as suspension cultures of T lymphocytes (primary peripheral lymphocytes), Jurkat leukemic T cells, and HL60 myeloid leukemic cells manifest cytoskeletal anomalies under gravity-altered conditions.

Cytoskeletal abnormalities in altered gravity include microtubule disorganization, coalescing, cabling or bundling of filaments, de-polymerization, absence of polymerization, non-extension of filaments to the cell membrane, lack of filament orientation, irregular arrangement, and disorganization of the microtubule organizing centers. The actin cytoskeleton is disorganized, cellular F-actin content is changed, actin filament distribution and organization are altered, cabling is increased, filament linearity is decreased, the number of stress fibers is reduced, and the actin surface area in cells is decreased. Microgravity also caused bundling and probable depolymerization of vimentin filaments, loosely woven cytokeratin network, and disassembly of vinculin and phosphorylated proteins in focal contacts.

Apparent cellular manifestations of cytoskeletal changes in altered gravity include cell shape changes (rounding), apoptosis, altered focal contacts and focal reorganization, altered integrin-mediated adhesion, growth reduction or arrest, changes in cell cycle phases, and increased cell death factor, Fas/APO-1. Based on results of numerous experiments, investigators propose that implications to cells would include an impact on signal transduction, tensegrity, apoptosis, metabolism, growth, and gene expression.

The recent discovery that gravity is required for the *in vitro* self-assembly of microtubules, the differences in microtubule assembly in low-*g* and 2 *g* phases of parabolic flight, failure of cells to polymerize microtubules, lack of microtubule reorganization in cells in microgravity and consideration of the bifurcation theory of microtubule assembly strongly support the conclusion that abnormal cytoskeletal organization occurs in microgravity. Some cell types (e.g. human breast cancer MCF-7) did not re-establish the microtubule cytoskeleton and microtubules failed to polymerize in HL60 cells. Conversely, lymphocytes (Jurkat) appeared to reorganize microtubule structure and MTOCs in microgravity. However, the absence of cell growth supports the conclusion that the cytoskeleton, when reorganized in microgravity, may not function normally. Though current evidence strongly supports this view, it is yet to be proven whether cellular responses in the absence of gravity are directly linked to changes in cytoskeletal dynamics, structure, and function. And it remains to be determined whether cytoskeletal self-organization in living cells is impaired in microgravity though evidence strongly supports this. More sophisticated analytical methodology is needed to discern whether the cytoskeleton, reorganized in

120

microgravity, is structurally and functionally normal.

Simulated shuttle launch vibration can significantly disrupt the cytoskeleton. Vibrated cells reorganize the microtubule cytoskeleton and proliferate. Because space-flown Jurkat cells remain in growth arrest in microgravity, the conclusion may be drawn that absence of growth is not a direct result of cytoskeletal disruption caused by vibration during shuttle launch but rather to some other cytoskeletal abnormality deriving from microgravity or the microgravity environment.

Apoptosis was increased in space-flown Jurkat cells and in clinostat cultures of ROS 27/2.8 osteoblast-like cells suggesting that altered gravity may cause apoptosis by altering organization of the cytoskeleton. This is supported by the finding that simultaneous disruption of actin filaments and microtubules leads to apoptosis [120]. Apoptosis in chick cartilage cells and human colorectal carcinoma cells flown on the shuttle did not increase. And the colorectal carcinoma cells as well as Rauscher murine erythroleukemia, bone marrow CD34$^+$ cells, and peripheral blood mono-nuclear cells failed to show an increase in apoptosis when cultured in gravity vector-averaged rotating bioreactors (RWV). Based on the summarized data available, it can be concluded that apoptosis in microgravity is a function of cell type, time sampled after growth stimulation in serum reduced cultures, experiment specifics and culture vessel characteristics. Although RWV bioreactors provide a low-shear environment for optimized culture and achieve some aspects of microgravity, they may not predict the responses of cells that may be more sensitive to perturbations of spaceflight (vibration, radiation, microgravity *per se*). There is insufficient evidence to date to determine conclusively whether or not apoptosis is increased in micro-gravity though results with lymphocytes seem to indicate that apoptosis may be a factor in population growth response in these cells during spaceflight.

The cytoskeletal genes expressed in space-flown and vibrated Jurkat cells indi-cate that genes encoding the ubiquitous and multi-functional cytoskeletal linker protein, plectin, and a centriole-associated protein are involved in reorganization of the cytoskeleton after vibrational damage. The elevation of plectin message at 48 h in microgravity suggests that the cytoskeleton may not be properly organized and that a feedback mechanism may not terminate plectin message transcription. This is yet to be determined. Differences in genes up-regulated in cells in microgravity compared to those expressed in vibrated cells, identify the genes most likely to be affected by altered gravity and implicate candidate genes and pathways for future investigations. The differential expression of cytoskeletal genes in space-flown compared to 1 *g* controls supports the results of studies reviewed in this chapter describing alterations in membrane-cytoskeletal, centriole, and cytoskeletal filament associations in microgravity.

Acknowledgements

The author expresses sincere appreciation to Dr. Luis A. Cubano for his dedicated work in the areas of apoptosis and gene expression. Appreciation is also expressed to collaborators: Drs. B. de Sales Lawless, Edward Piepmeier, Jr., Phillip Bowman, and

Baiteng Zhao, Hong-Khanh Dinh, and Johnathan Pabalan. The author acknowledges the crews of STS-76 and STS-95, especially Senator John Glenn and Scott Parazinski for conducting the experiments in space. Appreciation is expressed to Julie Reynolds, Kathy Summer, and C.A. Yancey for technical assistance and to Karen Murphy for technical and editorial contributions. This work was supported in part by NASA Grants NAG2-985 and NAGW-812.

References

1. Konstantinova, I., Antropova, Y.N., Legen'kov, V.I. and Zazhirey, V.D. (1973) Study of reactivity of blood lymphoid cells in crew members of the Soyuz 6, 7, and 8 space ships before and after flight. Space Biol. Aerospace Med. 7, 45–55.

2. Taylor, G.R., Konstantinova, I., Sonnenfeld, G. and Jennings, R. (1998) Changes in the immune system during and after spaceflight. Advances in Space Biology and Medicine, Vol. 6, pp. 1–32.

3. Cogoli, A. and Gmunder, F.K. (1991) Gravity effects on single cells. In: Advances in Space Biology and Medicine, Vol. 1 (S.L. Bonting, ed.) pp. 183–248.

4. Cogoli, A. and Cogoli-Greuter, M. (1997) Activation and proliferation of lymphocytes and other mammalian cells in microgravity. In: Advances in Space Biology and Medicine, Vol. 6 (S.L. Bonting, Ed.) 6, 33–79.

5. Cogoli, A. (1996) Gravitational physiology of human immune cells: A review of *in vivo*, *ex vivo* and *in vitro* studies. J. Gravitational Physiol. 3, 1–9.

6. Cogoli, A., Tschopp, A. and Fuchs-Bislin, P. (1984) Cell sensitivity to gravity. Science 225, 228–230.

7. Schmitt, D.A, Hatton, J.P., Emond, C., Chaput, D., Paris, H., Levade, T., Cazenave, J.P. and Schaffar, L. (1996) The distribution of protein kinase C in human leukocytes is altered in microgravity. FASEB J. 10, 1627–1634.

8. Hatton, J.P., Gaubert, F., Lewis, M.L., Darsel, Y., Ohlmann, P., Cazenave, J-P. and Schmitt, D. (1999) The kinetics of translocation and cellular quantity of protein kinase C in human leukocytes are modified during spaceflight. FASEB J. 13, S23–S33.

9. Lewis, M.L., Cubano, L.A., Zhao, B., Dinh, H-K., Pabalan, J.G., Piepmeier, E.H. and Bowman, P.D. (2001) cDNA microarrray reveals altered cytoskeletal gene expression in space-flown leukemic T lymphocytes (Jurkat). FASEB J. (June 18, 2001) 10.1096/fj.00–0820fje.

10. Cubano, L.A. and Lewis, M.L. (2001) Effect of vibrational stress and spaceflight on regulation of heat shock proteins hsp70 and hsp 27 in human lymphocytes (Jurkat). J. Leukoc. Biol. 69, 755–761.

11. Grove, D.S., Pishak, S.A. and Mastro, A.M. (1995) The effect of a 10-day space flight on the function, phenotype, and adhesion molecule expression of splenocytes and lymph node lymphocytes. Exp. Cell Res. 219,102–109.

12. Lewis, M.L., Reynolds, J.L., Cubano, L.A., Hatton, J.P., Lawless, B.D. and Piepmeier, E.H. (1998) Spaceflight alters microtubules and increases apoptosis in human lymphocytes (Jurkat). FASEB J. 12, 1007–1018.

13. Cubano, L.A. and Lewis, M.L. (2000) Fas/APO-1 protein is increased in spaceflown lymphocytes (Jurkat). Exp. Gerontol. 35, 389–400.

14. Limouse, M., Manie, S., Konstantinova, I., Ferrua, B. and Schaffar, L. (1991) Inhibition of phorbol ester-induced cell activation in microgravity. Exp. Cell Res. 197, 82–86.

15. Cogoli-Greuter, M., Cogoli, A., Spano, A., Sciola, L. and Pippia, P. (1997) Influence of microgravity on mitogen binding and cytoskeleton in Jurkat cells: experiment on MAXUS-2. Proceedings 13th ESA Symposium on European Rocket and Balloon Programmes and related Research. ESA SP-397, 73–76.

16. Sciola, L., Cogoli-Greuter, M., Spano, A. and Pippia, P. (1999) Influence of microgravity on mitogen binding and cytoskeleton in Jurkat cells. Adv. Space Res. 24, 801–805.

17. Hymer, W.C., Grindeland, R., Farrington, M., Fast, T., Hayes, C., Motter, K. and Paul, I. (1985) Microgravity associated changes in pituitary growth hormone (GH) cells prepared from rats flown on spacelab. Physiologist 28, 377.

18. Grindeland, R., Vale, W., Hymer, W., Sawchenko, P., Vasques, M., Krasnov, M.I., Kaplansky, A. and Povova, I. (1990) Growth hormone regulation, synthesis and secretion in microgravity. In: Final Reports of the U.S. Experiments Flown on the Cosmos Biosatellite 1887 (J.P. Connolly, R. Grindeland and R. Ballard, eds.). NASA Technical Memorandum 102254.

19. Cogoli, A., Bechler, B., Cogoli-Greuter, M., Criswell, S.B., Joller, H., Joller, P., Hunzinger, E. and Muller, O. (1993) Mitogenic signal transduction in T lymphocytes in microgravity. J. Leukoc. Biol. 53, 569–575.

20. Papaseit, C., Pochon, N. and Tabony, J. (2000) Microtubule self-organization is gravity-dependent. Proc. Natl. Acad. Sci. USA 97, 8364–8368.

21. Hoeger, G. and Gruener, R. (1990) Cytoskeletal properties are sensitive to vector-free gravity. ASGSB Bulletin 4, 42, Abstr # E-l.

22. Rijken, P.J., de Groot, R.P, Briegleb, W., Verkleij, A.J., de Laat, S.W., Kruijer, W. and Boonstra, J. (1989) Influence of simulated gravity changes on epidermal growth factor induced cell rounding, epidermal growth factor–receptor clustering and fos transcription. ELGRA News 11, 3.

23. Sarkar, D., Nagaya, T., Koga, K., Nomura, Y., Gruener, R. and Seo, H. (2000) Culture in vector-averaged gravity under clinostat rotation results in apoptosis of osteoblastic ROS 17/2.8 cells. J. Bone Mineral Res. 15, 489–498.

24. Moos, P.J., Graff, K., Edwards, M., Stodieck, L.S., Einhorn, R. and Luttges, M.W. (1988) Gravity-induced changes in microtubule formation. ASGSB Bulletin 2, 55.

25. Piepmeier, E.H., Kalns, J.E., McIntyre, K.M. and Lewis, M.L. (1997) Prolonged weightlessness affects promyelocytic multidrug resistance. Exp. Cell Res. 237, 410–418.

26. Guignandon, A,. Usson, Y., Laroche, N., Lafage-Proust, M.H., Sabido, O., Alexandre, C. and Vico, L. (1997) Effects of intermittent or continuous gravitational stress on cell–matrix adhesion: quantitative analysis of focal contacts in osteoblastic ROS 17/2.8 cells. Exp. Cell Res. 236, 66–75.

27. Boonstra, J. (1999) Growth factor-induced signal transduction in adherent mammalian cells is sensitive to gravity. FASEB J. 13, S35–S42.

28. Cogoli-Greuter, M., Bechler, B., Lorenzi, Cogoli, A, Sciola, L., Pippia, P. and Sechi, G. (1997) Mitogen binding, cytoskeleton patterns and motility of T lymphocytes in microgravity. ESA SP 1206. http://esapub.esrin.esa.it/sp/sp1206/cogol/htm

29. Jongkind, J.F., Visser, P. and Verkerk, A. (1996) Cell fusion in space: plasma membrane fusion in human fibroblasts during short term microgravity. Adv. Space Res. 17, 21–25.

30. Tabony, J. (1994) Morphological bifurcations involving reaction-diffusion processes

during microtubule formation. Science 264, 245–248.

31. Lewis, M.L. (1999) Mechanisms of gravity sensing and response in haematopoietic cells: effect of microgravity on the microtubule cytoskeleton, apoptosis, and population dynamics in human lymphocytes (Jurkat). European Space Agency SP-1222 Biorack on Spacehab, 55–68.

32. Hughes-Fulford, M. and Lewis, M.L. (1996) Effects of microgravity on osteoblast growth activation. Exp. Cell Res. 224, 103–109.

33. Hughes-Fulford, M. (2001) Changes in Gene expression and signal transduction in microgravity. J. Gravitational Physiol. 8(1), 1–4.

34. Gruener, R., Roberts, R. and Reitstetter, R. (1993) Exposure to microgravity alters properties of cultured muscle cells. ASGSB Bulletin 7, 65.

35. Schatten, H., Chakrabarti, A., Taylor, M., Sommer, L., Levine, H., Anderson, A., Runco, M. and Kemp, R. (1999) Effects of spaceflight conditions on fertilization and embryogenesis in the sea urchin Lytechinus pictus. Cell Biol. Int. 23, 407–415.

36. Vassy, J., Portet, S., Beil, M., Millot, G., Fauvel-Lafève, F., Karniguian, A., Gasset, G., Irinopoulou, T., Calvo, F., Rigaut, J.P and Schoevaert, D. (2001) The effect of weightlessness on cytoskeleton architecture and proliferation of human breast cancer cell line MCF-7. FASEB J. 15, 1104–1106.

37. Guignandon, A., Lafage-Proust, M-H., Usson, Y., Laroche, N., Caillot-Augusseau, A., Alexandre, C. and Vico, L. (2001) Cell cycling determines integrin-mediated adhesion in osteoblastic ROS 17/2.8 cells exposed to space-related conditions. FASEB J. (July 24, 2001) 10.1096/fj.00–0837fje.

38. Harris, S.A., Zhang, M., Kidder, L.S., Evans, G.L., Spelsberg, T.C. and Turner, R.T. (2000) Effects of orbital spaceflight on human osteoblastic cell physiology and gene expression. Bone 26, 325–331.

39. Osborn, M. and Weber, K. (1976) Cytoplasmic microtubules in tissue culture cells appear to grow from an organizing structure towards the plasma membrane. Proc. Natl. Acad. Sci. 73, 867–871.

40. Todd, P. (1989) Gravity-dependent phenomena at the scale of the single cell. ASGSB Bulletin 2, 95–113.

41. Kaplan, J. and Keough, E.A. (1982) Temperature shifts induce the selective loss of alveolar-macrophage plasma membrane components. J. Cell Biol. 94, 12.

42. Ingber, D.E. (1997) Tensegrity: the architectural basis of cellular mechanotransduction. Ann. Rev. Physiol. 59, 575–99.

43. Carraway, K.L. and Carraway, C.A.C. (1989) Membrane–cytoskeleton interaction in animal cells. Biochim. Biophys. Acta 988, 147–71.

44. Kiley, S.C. and Parker, P. J. (1995) Differential localization of protein kinase C isozymes in U937 cells: evidence of distinct isozyme functions during monocyte differentiation. J. Cell Sci. 108, 1003–1016.

45. Mochly-Rosen, D.C., Henrich, J., Cheever, K., Khaner, H. and Simpson, P.C. (1990) A protein kinase C isozyme is translocated to cytoskeletal elements on activation. Cell Reg. 1, 693–706.

46. Murti, K.G., Kaur, K. and Goorha, R.M. (1992) Protein kinase C associates with intermediate filaments and stress fibers. Exp. Cell Res. 202, 36–44.

47. Vemuri, B. and Singh, S.S. (2001) Protein kinase C isozyme-specific phosphorylation of profilin. Cell Signal 13, 433–439.

48. Ingber, D.E. (1993) The riddle of morphogenesis: a question of solution chemistry or molecular cell engineering? Cell 75, 1249–1252.

49. Lind, S.E., Janmey, P.A., Chaponnier, C., Herbert, T.J. and Stossel, T.P. (1987) Reversible binding of actin to gelsolin and profilin in human platelet extracts. J. Cell Biol. 105, 833–842.

50. Yin, H. L., Iida, K. and Janmey, P.A. (1988) Identification of a phosphoinositide-modulated domain in gelsolin which binds to the sides of actin filaments. J. Cell Biol. 106, 805–812.

51. Lassing, I. and Lindberg, U. (1985) Specific interaction between phosphatidylinositol 4,5-biphosphate and profilactin. Nature 314, 472–474.

52. Schellenberg, R.R. and Gillespie, E. (1977) Colchicine inhibits phosphatidylinositol turnover induced in lymphocytes by concanavalin A. Nature 265, 741–742.

53. Zambetti, G., Ramsey-Ewing, A., Bortel, R and Stein, G. (1991) Disruption of the cytoskeleton with cytochalasin D induces c-fos gene expression. Exp. Cell Res. 192, 93–101.

54. Ingber, D. (1999) How cells (might) sense microgravity. FASEB J. 13, S2–S22.

55. Stein, G.S., Van Wijnen, A.J., Stein, J.L., Lian, J.B., Pockwinse, S.H. and McNeil, S. (1999) Implications for interrelationships between nuclear architecture and control of gene expression under microgravity conditions. FASEB J. 13, S157–S166.

56. Wang, N. and Stamenovic, D. (2000) Contribution of intermediate filaments to cell stiffness, stiffening and growth. Am. J. Physiol. Cell Physiol. 279, C188–C194.

57. Ellis, H.M. and Horvitz, H.R. (1986) Genetic control of programmed cell death in the nematode C. elegans. Cell 44, 817–829.

58. Thompson, C.B. (1995) Apoptosis in the pathogenesis and treatment of disease. Science 267, 1456–1462.

59. Wyllie, A. (1980) Glucocorticoid-induced thymocyte apoptosis is associated with endogenous endonuclease activation. Nature 284, 555–556.

60. Susin, S., Zamzami, N. and Kroemer, G. (1998) Mitochondria as regulators of apoptosis, doubt no more. Biochim. Biophys. Acta 1366, 151–165.

61. Cohen, J. (1993) Apoptosis. Immunology Today 14, 126–130.

62. Squier, M.K., Sehnert, A.J. and Cohen, J.J. (1995) Apoptosis in leukocytes. J. Leukoc. Biol. 57, 2–10.

63. Guchelaar, H., Vermes, A., Vermes, I. and Haanen, C. (1997) Apoptosis: molecular mechanisms and implications for cancer chemotherapy. Pharm. World Sci. 19, 119–125.

64. Schulze-Osthoff, K., Ferrari, F., Los, M., Wesselborg, S. and Peter, M. (1998) Apoptosis signaling by death receptors. Eur. J. Biochem. 254, 439–459.

65. Evan, G. and Littlewood, T. (1998) Apoptosis. Science 281, 1317–1321.

66. McKenna, S., McGowan, A. and Cotter, T. (1998) Molecular mechanisms of programmed cell death. Adv. Biochem. Eng. Biotechnol. 62, 1–31.

67. Nagata, S. and Goldstein, P. (1995) The Fas Death Factor. Science 267, 1449–1456.

68. Boise, L.H. and Thompson, C.B. (1996) Hierarchical control of lymphocyte survival. Science 274, 67–68.

69. Oehm, A., Behrmann, I., Falk, W., Pawlita, M., Maier, G., Klas, C., Li-Eever, M., Richards, S., Dhein, J. and Trauth, B. (1992) Purification and molecular cloning of the APO-1 cell surface antigen, a member of the tumor necrosis factor/nerve growth factor receptor superfamily. Sequence identity with the Fas antigen. J. Biol. Chem. 267,

10709–10715.

70. Cascino, I., Fiuccci, G., Papoff, F. and Ruberti, G. (1995) Three functional soluble forms of the human apoptosis-inducing Fas molecule are produced by alternative splicing. J. Immunol. 154, 2706–2713.

71. Cascino, I., Papoff, F., Eramo, A. and Ruberti, G. (1996) Soluble Fas/Apo-1 splicing variants and apoptosis. Front Biosci. 1, d12-8.

72. Lee, S., Kim, S., Lee, J., Shin, M., Dong, A., Na, W., Park, Q., Kim, K., Kim, C., Kim, S. and Yoo, N. (1998) Detection of soluble Fas mRNA using *in situ* reverse transcription–polymerase chain reaction. Lab Invest. 78, 453–459.

73. Cheng, J., Zhou, T., Liu, C., Shapiro, J., Vrauer, M., Kiefer, M., Varr, P. and Mounts, J. (1994) Protection from Fas-mediated apoptosis by a soluble form of the Fas molecule. Science 263, 1759–1762.

74. Owcn-Schaub, L., Angelo, L., Radinsky, R., Ware, C., Gerner, T. and Bartos, D. (1995) Soluble Fas/APO-1 in tumor cells: a potential regulator of apoptosis? Cancer Lett. 94, 1–8.

75. Klaus, D.M. (2001) Clinostats and Bioreactors. Gravitational and Space Biology Bulletin 14, 55–64.

76. Sytkowski, A.J. and Davis, K.L. (2001) Erythroid cell growth and differentiation in vitro in the simulated microgravity environment of the NASA rotating wall vessel bioreactor. In Vitro Cell Dev. Biol. Anim. 37, 79–83.

77. Plett, P.A., Frankovitz, S.M., Abonour, R. and Orschell-Traycoff, C.M. (2001) Proliferation of human hematopoietic bone marrow cells in simulated microgravity. In Vitro Cell Dev Biol Anim, 37, 73–78.

78. Jessup, J. M., Frantz, M., Sonmez-Alpan, E., Locker, J., Skena, K., Waller, H., Battle, P., Nachman, A., Webber, M.E., Thomas, D.A., Curbeam Jr, R.L., Baker, T.L. and Goodwin, T.J. (2000) Microgravity culture reduces apoptosis and increases the differentiation of a human colorectal carcinoma cell line. In Vitro Cell Dev. Biol. Anim. 36, 367–73.

79. Risin, D. and Pellis, N.R. (2001) Modeled microgravity inhibits apoptosis in peripheral blood lymphocytes. In Vitro Cell Dev. Biol. Anim. 37,66–72.

80. Doty, S.B., Stiner, D. and Telford, W.G. (1999) The effect of spaceflight on cartilage cell cycle and differentiation. J. Grav. Physiol. 6, P89–P90.

81. Martinez-Lorenzo, M.J., Alava, M.A., Anel, A., Pineiro, A. and Naval, J. (1996) Release of preformed Fas ligand in soluble form is the major factor for activation-induced death of Jurkat T cells. Immunology 89, 511–517.

82. Kolber, M.A., Broschar, K.O. and Landa-Gonzalez, B. (1990) Cytoclasin B induces cellular DNA fragmentation. FASEB J. 4, 3021–3027.

83. Martin, S.J. and Cotter, G.G. (1990) Disruption of microtubules induces an endogenous suicide pathway in human leukemia HL-60 cells. Cell Tissue Kinet. 23, 545–559.

84. Gajate, C., Barasoain, I., Andreu, J.M. and Mollinedo, F. (2000) Induction of apoptosis in leukemic cells by the reversible microtubule-disrupting agent 2-methoxy-5-(2′,3′,4′-trimethoxyphenyl)-2,4,6-cycloheptatrien-1-one: protection by Bcl-2 and Bcl-X(L) and cell cycle arrest. Cancer Res. 60, 2651–2659.

85. Matylevich, N.P., Korol, B.A., Nelipovich, P.A., Afanasev, V.N. and Jmansky, S.R. (1991) D_2O inhibition of interphase thymocyte death. Radiobiologia 31, 27–32.

86. Atencia, R., Asumendi, A. and Garcia-Sanz, M. (2000) Role of cytoskeleton in apoptosis. Vitam. Horm. 58, 267–297.

87. Lewis, M.L. and Hughes-Fulford, M. (2000) Regulation of heat shock protein message in Jurkat cells cultured under serum starved and gravity altered conditions. J. Cell Biochem. 77, 127–134.

88. Fitzgerald, J. and Hughes-Fulford, M. (1996) Gravitational loading of a simulated launch altered mRNA expression in osteoblasts. Exp. Cell Res. 228, 168–171.

89. de Groot, R.P., Rijken, P.J., Boonstra, J., Verkleij, A.J., de Laat, S.W. and Kruijer, W. (1991) Epidermal growth factor-induced expression of c-fos is influenced by altered gravity conditions. Aviat. Space Environ. Med. 62, 37–40.

90. Kumei, Y., Nakajima, T., Sato, A., Kamata, N. and Enomoto, S. (1989) Reduction of G1 phase duration and enhancement of c-myc gene expression in HeLa cells at hypergravity. J. Cell Sci. 93, 221–226.

91. Walther, I., Pippia, P., Meloni, M.A., Turrini, F., Mannu, F. and Cogoli, A. (1998) Simulated microgravity inhibits the genetic expression of interlukin-2 and its receptor in mitogen-activated T lymphocytes. FEBS Lett. 436, 115–118.

92. Licato, L.L. and Grimm, E.A. (1999) Multiple interlukin-2 signaling pathways differentially regulated by microgravity. Immunopharmacology 44, 273–279.

93. Crucian, B.E., Cubbage, M.L. and Sams, C.F. (2000) Altered cytokine production by specific human peripheral blood cell subsets immediately following space flight. J. Interferon Cytokine Res. 20, 547–556.

94. Sato, A., Hamazaki, T., Oomura, T., Osada, H., Kakeya, M., Watanabe, M., Nakamura, T., Nakamura, Y., Koshikawa, N., Yoshizaki, I., Aizawa, S., Yoda, S., Ogiso, A., Takaoki, M., Kohno, Y.K. and Tanaka, H. (1999) Effects of microgravity on c-fos gene expression in osteoblast-like MC3T3-E-1 cells. Adv. Space Res. 24, 807–813.

95. Hammond, T.G., Benes, E., O'Reilly, K.C., Wolf, D.A., Linnehan, R. M., Taher, A., Kaysen, J. H., Allen, P.L. and Goodwin, T.J. (2000) Mechanical culture conditions effect gene expression: gravity-induced changes on the space shuttle. Physiological Genomics 3, 163–173.

96. Kaysen, J.H., Campbell, W.C., Majewski, R.R., Goda, F.O., Navar, G.L., Lewis, F.C., Goodwin, T.J. and Hammond, T.G. (1999) Select de novo gene and protein expression during renal epithelial cell culture in rotating wall vessels is shear stress dependent. J. Membr. Biol. 168, 77–89.

97. Khaoustov, V.I., Risin, D., Pellis, N.R. and Yoffe, B. (2001) Microarray analysis of genes differentially expressed in HepG2 cells cultured in simulated microgravity: preliminary report. In Vitro Cell Dev. Biol. Anim. 37, 84–88.

98. de Groot, R.P., Rijken, P.J., Den Hertog, J., Boonstra, J., Verkleij, A.J., De Laat, S.W and Kruijer, W. (1991) Nuclear responses to protein kinase C signal transduction are sensitive to gravity changes. Exp. Cell Res. 197, 87–90.

99. Rijken, P.J., Boonstra, J., Verkleij, A.J., de Laat, S.W. (1994) Effects of gravity on the cellular response to epidermal growth factor. Advances in Space Biolology and Medicine, Vol. 4, pp. 159–188.

100. Hughes-Fulford, M., Tjandrawinata, R., Fitzgerald, J., Gasuad, K. and Gilbertson, V. (1998) Effects of microgravity on osteoblast growth. Gravitat. Space Biol. Bull. 11, 51–60.

101. Hughes-Fulford, M. and Gilbertson, V. (1999) Osteoblast fibronectin mRNA, protein synthesis, and matrix are unchanged after exposure to microgravity. FASEB J. 13, S121–S127.

102. Kumei, Y., Shimokawa, H., Katano, H., Hara, E., Akiyama, H., Hirano, M., Mukai, C.,

Nagaoka, S., Whitson, P.A. and Sams, C.F. (1996) Microgravity induces prostaglandin E2 and interleukin-6 production in normal rat osteoblasts: role in bone demineralization. J. Biotechnol. 47, 313–324.

103. Akiyama, H., Kanai, S., Hirano, M., Shimokawa, H., Katano, H., Mukai, C., Nagaoka, S., Moriata, S. and Kumei, Y. (1999) Expression of PDGF-beta receptor, EGF receptor, and receptor adaptor protein Shc in rat osteoblasts during spaceflight. Mol. Cell Biochem. 202, 63–71.

104. Carmeliet, G., Nys, G., Stockmans, I. and Bouillon, R. (1998) Gene expression related to the differentiation of osteoblastic cells is altered by microgravity. Bone 22, S139– S143.

105. Carmeliet, G., Nys, G. and Bouillon, R. (1997) Microgravity reduces the differentiation of human osteoblastic MG-63 cells. J. Bone Miner. Res. 12, 786–794.

106. Landis, W.J., Hodgens, K.J., Block, D., Toma, C.D. and Gerstenfeld, L.C. (2000) Spaceflight effects on cultured embryonic chick bone cells. J. Bone Miner. Res. 15, 1099–1112.

107. Hammond, T.G., Lewis, F.C., Goodwin, T.J., Linnehan, R.M., Wolf, D.A., Hire, K.P., Campbell, W.C., Benes, E., O'Reilly, K.C., Globus, R.K. and Kaysen, J.H. (1999) Letter: Gene expression in space. Nature Med. 5, 359. (http://www.tmc.tulane.edu/ astrobiology/ microarray/default.html.).

108. Soncin, F., Prevelige, R. and Calderwood, S.K. (1997) Expression and purification of human heat-shock transcription factor 1. Protein Expr. Purif. 9, 27–32.

109. Mehlen, P., Schulze-Osthoff, K. and Arrigo, A-P. (1996) Small stress proteins as novel regulators of apoptosis. J. Biological Chem. 271, 16510–16514.

110. Fang, J., Wienecke, R., Xiao, G-H., Maize Jr., J.C., DeClue, J.E and Yeung, R.S. (1996) Suppression of tumorigenicity by the wild-type tuberous sclerosis 3 (Tsc2) gene and its C-terminal region. Proc. Nat. Acad. Sci. USA 93, 9154–9159.

111. Tjandrawinata, R.R., Vincent, V.L. and Hughes-Fulford, M. (1997) Vibrational force alters mRNA expression in osteoblasts. FASEB J. 11, 493–497.

112. Andra, K., Nikolic, B., Stocher, M., Drenckhahn, D. and Wiche, G. (1998) Not just scaffolding: plectin regulates actin dynamics in cultured cells. Genes Dev. 12, 3442–3451.

113. Smith, F.J. Eady, R.A., Leigh, I.M., McMillan, J.R., Rugg, E.L., Kelsell, D.P., Bryant, S.P., Spurr, N.K., Geddes, J.F., Kirtschig, G., Milana, G., de Bono, A.G., Owaribe, I.K., Wiche, G., Pulkkinen, L., Uitto, J., McLean, W.H. and Lane, E.B. (1996) Plectin deficiency results in muscular dystrophy with epidermolysis bullosa. Nat. Genet. 13, 450–457.

114. Wiche, G. (1998) Role of plectin in cytoskeleton organization and dynamics. J. Cell Sci. 111, 2477–2486.

115. Svitkina, T. M., Verkhovsky, A.B. and Borisy, G.G. (1996) Plectin sidearms mediate interaction of intermediate filaments with microtubules and other components of the cytoskeleton. J. Cell Biol. 135, 991–1007.

116. Murti, K.G., Kaur, K. and Goorha, R.M. (1992) Protein kinase C associates with intermediate filaments and stress fibers. Exp. Cell Res. 202, 36–44.

117. Limouse, M., Manie, S., Konstantinova, I., Ferrua, B. and Schaffar, L. (1991) Inhibition of phorbol ester-induced cell activation in microgravity. Exp. Cell Res. 197, 82–86.

118. Schmitt, D.A., Hatton, J.P., Emond, C., Chaput, D., Paris, H., Levade, T., Cazenave, J.P. and Schaffar, L. (1996) The distribution of protein kinase C in human leukocytes is altered in microgravity. FASEB J. 10, 1627–1634.

119. Hatton, J.P., Gaubert, F., Lewis, M.L., Darsel, Y., Ohlmann, P., Cazenave, J.P. and

Schmitt, D. (1999) The kinetics of translocation and cellular quantity of protein kinase C in human leukocytes are modified during spaceflight. FASEB J. 13, S23–S33.

120. Flusberg, D.A., Numaguchi, Y. and Ingber, D.E. (2001) Cooperative control of Akt phosphorylation, bcl-2 expression, and apoptosis by cytoskeletal microfilaments and microtubules in capillary endothelial cells. Mol. Biol. Cell 12, 3087–3094.

121. Burkhardt, J.K. (1998) The role of microtubule-based motor proteins in maintaining the structure and function of the Golgi complex. Biochim. Biophys. Acta 1404, 113–126.

122. Kwiatkowski, D.J., Janmey, P.A., Mole, J.E. and Yin, H.L. (1985) Isolation and properties of two actin-binding domains in gelsolin. J. Biol. Chem. 260, 15232–15238.

123. Fowler, V.M. (1990) Tropomodulin: a cytoskeletal protein that binds to the end of erythrocyte tropomyosin and inhibits tropomyosin binding to actin. J. Cell Biol. 111, 471–481.

124. Masuda, H., Tanaka, K., Takagi, M., Ohgami, K., Sakamaki, T., Shibata, N. and Takahashi, K. (1996) Molecular cloning and characterization of human non-smooth muscle calponin. J. Biochem. (Tokyo) 120, 415–24.

125. Fujii, T., Hiromori, T., Hamamoto, M. and Suzuki, T. (1997) Interaction of chicken gizzard smooth muscle calponin with brain microtubules. J. Biochem. (Tokyo) 122, 344–351.

126. Szymanski, P.T. and Goyal, R.K. (1999) Calponin binds to the 20- kilodalton regulatory light chain of myosin. Biochemistry 38, 3778–3784.

127. Bennett, V. (1982) The molecular basis for membrane–cytoskeleton association in human erythrocytes. J. Cellular Biochem. 18, 49–65.

128. Allen, P.G. and Janmey, P.A. (1994) Gelsonin displaces phalloidin form actin filaments —a new fluorescence method shows that both Ca^{2+} and Mg^{2+} affect the rate at which gelsolin severs F-actin. J. Biol. Chem. 260, 32916–32923.

129. Rodriguez del Castillo, A., Vitale, M.L., Tchakarov, L. and Trifaro, J.M. (1992) Human platelets contain scinderin, a Ca^{2+}-dependent actin filament-severing protein. Thrombosis and Haemostasis 67, 248–251.

130. Onoda, K. and Yin, H.L. (1993) gCap39 is phosphorylated. Stimulation by okadaic acid and preferential association with nuclei. J. Biol. Chem. 268, 4106–4112.

131. Bockholt, S.M., Otey, C.A., Glenney, J.J. and Burridge, K. (1992) Localization of a 215-kDa tyrosine-phosphorylated protein that cross-reacts with tensin antibodies. Exp. Cell Res. 203, 39–46.

132. Bockholt, S.M. and Burridge, K. (1993) Cell spreading on extracellular matrix proteins induces tyrosine phosphorylation of tensin. J. Biol. Chem. 268, 14565–7.

133. Muller, S,C., Yeh, Y. and Chen, W.T. (1992) Tyrosine phosphorylation of membrane proteins mediates cellular invasion by transformed cells. J. Cell Biol. 119, 1309–1325.

134. Eger, A., Stockinger, A., Wiche, G. and Foisner, R. (1997) Polarisation-dependent association of plectin with desmoplakin and the lateral submembrane skeleton in MDCK cells. J. Cell Sci. 110, 1307–1316.

135. Hatton, J.P., Lewis, M.L., Roquefeuil, S.B., Chaput, D., Cazanave, J.P. and Schmitt, D.A. (1998) Use of an adaptable cell culture kit for performing lymphocyte and monocyte cell cultures in microgravity. J. Cell Biochem. 70, 252–267.

Cell Biology and Biotechnology in Space
A. Cogoli (editor)
Published by Elsevier Science B.V.

Physiological Effects of Microgravity on Osteoblast Morphology and Cell Biology

Millie Hughes-Fulford

Laboratory of Cell Growth, Department of Medicine, University of California San Francisco, Dept. of Veteran's Affairs Medical Center, San Francisco, California, USA

Introduction

When we think of modern space travel, we have images of the *USS Enterprise* outfitted with artificial gravity and all the comforts of home. But in the real world we do not flip a switch and turn on artificial gravity. The result is that during a long-term space flight like the Mars Mission, humans will experience several physiological changes including: space adaptation syndrome (motion sickness), loss of muscle, reduced immune function and loss of calcium and bone (up to 19% loss of bone after 12 months of flight). Some of these changes may be due to systemic or hormonal changes in the body; others may be a direct result of lack of mechanical stress in the microgravity environment. In this chapter we will discuss bone and osteoblast physiology and the effect of microgravity on the osteoblast.

For the first three decades of space flight scientists studied the systemic changes in the body, which included loss of bone, increases of calcium, and changes in the endocrine system such as the increase in glucocorticoids. From these animal and human studies investigators found that loss of weight bearing bone is as high as 1% per month and this loss is primarily due to lack of new osteoblast growth in space flight. It is likely that studies of the osteoblast in microgravity will give us new target molecules for development of pharmacological agents that will stimulate bone growth. A projected bone loss of 20–30% in astronauts is one of the major physiological 'show stoppers' in the proposed 30-month manned mission to Mars.

Advances in cell and molecular biology over the past ten years have brought a new understanding of osteoblast cell physiology. Although flight opportunities have been limited, the volume of data about microgravity effects on cell proliferation and differentiation is accumulating. Altered gene expression, reduction in transcription factors and a reduction in growth factor related proteins have been observed in microgravity. Specifically, it has been demonstrated that cell growth and differentiation of osteoblasts is inhibited or altered in the microgravity environment. Cytoskeletal changes in actin, intermediate filament and microtubule networks have been found in microgravity as well. These changes in the osteoblast physiology suggest that

130

the decrease in bone formation in astronauts may be partly due to microgravity-induced alterations at the molecular level in osteoblast biology, gene expression and signal transduction.

Background human data: the importance of studying osteoblasts

Skeletal changes and loss of total body calcium have been documented in both humans and animals exposed to microgravity from 7 to 237 days. During the Apollo and Skylab missions, photon absorptiometry was used to assess pre- and post-flight bone mineral mass. For the 12 crew members of Gemini 4, 5, and 7, and Apollo 7 and 8, the average post-flight loss from the os calcis (heel) was 3.2% over an average of 8.5 days [1–4]. Analysis of in-flight urine, fecal, and plasma samples from Skylab missions revealed changes in urinary output of hydroxy-proline indicating degradation of the collagenous matrix substance of weight bearing bones. Elevated concentrations of urinary calcium were noted in the early studies of Skylab astronauts starting during the first days of flight. In many of the astronauts urinary calcium concentrations remained at elevated levels throughout the mission. A direct effect of microgravity is the loss of mechanical stress on the skeletal system. Although in-flight exercise is used by astronauts as a countermeasure, the greatest bone mass losses in the US flight program (pre-Shuttle/Mir) occurred in the 84-day Skylab 4 mission even though exercise was regularly performed. Crew members using exercise as a countermeasure still lost an average of 4% of bone over the 84-day mission period [5,6]. Additional evidence of bone loss was found in a 19.5 day flight on Cosmos 782 where rats formed significantly less periosteal bone than controls [7,8].

In the 237-day Soviet Soyuz T-10 mission, the cosmonauts lost bone in spite of 2–4 hours of daily exercise. Both compact and trabecular bone was lost from the os calcis during this mission. Bone loss appears to increase in general proportion to mission length, from 4 to 19.8% over an 84 to 184 day period [5,6,9,10]. Collet et al. showed that there was a slight decrease seen after one month of space flight and a marked loss of trabecular and cortical bone in the tibia after six months of microgravity exposure. Overall, the lower weight-bearing bones appeared to be more sensitive than the upper ones in terms of space-flight-induced bone loss, moreover the recovery of the bone was not complete even six months after return to gravity [9]. In studies conducted on Euro Mir 95, after 180 days in space-flight bone formation and resorption biological markers in cosmonauts were measured. Parathyroid hormone (PTH) was found to decrease by almost half. Bone alkaline phosphatase (BAP), intact osteocalcin (iBGP) and type 1 procollagen propeptide (PICP) were decreased to approximately 30% of preflight ground control, showing for the first time that there is an uncoupling of bone remodeling between formation and resorption (11). Vico and colleagues measured bone mineral density (BMD) at the distal radius and tibia in 15 cosmonauts who spent one, two or six months on the Russian Mir space station. They found that neither the cancellous nor cortical bone of the radius was significantly changed at any of the time points. In contrast, at the weight-bearing tibial site, cancellous BMD loss was seen after two months of microgravity. After six months, loss of cancellous bone was more pronounced than cortical bone. In some individuals, the tibial deterioration was high; moreover, BMD loss did not seem to depend on previous exposure to microgravity. Moreover, tibial bone loss persisted after return to earth and was not recovered during the post flight study period, suggesting the recovery time is greater than the time spent in microgravity [12]. Since microgravity-induced bone loss is not fully recovered after return to gravity, this is a significant complication of long-term missions and must be addressed before a Mars Mission.

Exercise alone is not a total answer to the countermeasures, since paradoxically, excessive exercise aggravates bone loss. Stein et al. reported a negative energy balance in longer-term missions [13–17]. This strongly suggests the need for a programmatic and efficient exercise program that would not result in a catabolic state. The cause of bone loss in response to reduction of mechanical stress is not yet known, but several ground studies have demonstrated that prostaglandins may be involved. Forwood et al. found that prostaglandins are released with exercise and are key regulators in exercise-induced bone growth *in vivo* [18]. The specific cyclo-oxygenase-2 inhibitor, NS-389, completely blocked mechanical stress induced bone formation *in vivo*. Other studies have demonstrated that PGE_2 augments bone growth *in vivo* [19] and *in vitro* [20]. Mechanical stress also causes increases in cytosolic Ca^{++}, remarkably only *one cycle* of pressure can cause a 57% increase in proliferation demonstrating that very little force is needed to stimulate proliferation [21]. It has also been demonstrated that 97% of exercise stimulation occurred during as little as 2% of a 24-hour period [22]. The potential role of gravity and PGE_2 in microgravity-induced bone loss is strengthened by the findings of Stein et al. who discovered that PGE_2 synthesis is significantly decreased by 50% during a 12-day flight and remained significantly depressed by almost 50% until return to gravity. After landing, synthesis of PGE_2 rebounded to 140% of preflight values for the six days of post-flight testing [17], suggesting a direct correlation between gravity and PGE_2.

Phases of growth and stages of osteoblast differentiation

Bone development is regulated by sequential and stringently controlled gene expression of the osteoblast. Three principal stages of development are key elements of the temporal expression of genes that control cell growth and differentiation as illustrated in Fig. 1. In the proliferation phase, early immediate genes are up-regulated along with cyclins, cyclin inhibitors and kinases associated with phases of the cell cycle. In addition, the synthesis of histone-4 and fibronectin are associated with the osteoblast growth stage. During the growth stage, osteoblast proliferation is divided into four phases, G1, S-phase (DNA synthesis), G2 and M (Mitosis). During G1, growth factors (or mechanical stress) activate MAPK pathways within seconds; immediate early genes are induced and transcription factors are activated within minutes. The cells then transition from G1 to S-phase, where PGE_2 and histone-4 synthesis peaks allowing osteoblast DNA replication. After DNA content is doubled,

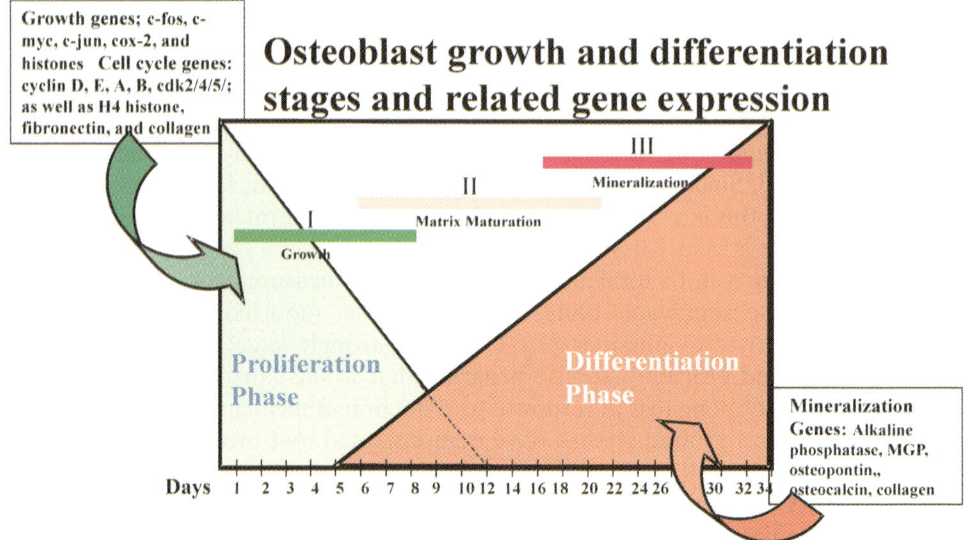

Fig. 1. Relationship of growth and differentiation gene expression. Cell growth is regulated early in the growth period, with growth genes *c-fos*, *c-myc*, *cox-2 egr-1* and transcription factors being actively made by the cell in the cell cycle during proliferation. Cell cycle is regulated partly by the cyclins, cyclin kinases (cdk), cyclin kinase inhibitors (cdki) and other cell cycle genes. During the very late proliferation stage and early portion of the differentiation stage, the collagen and mineralization genes are upregulated then alkaline phosphatase and matrix GLA proteins are upregulated in order for the extracellular matrix to mature. The final stages of mineralization occur during final differentiation where osteopontin and osteocalcin and collegians are upregulated. Adapted from [23].

the cells then go into G2 where proteins necessary for cell replication and cell division are made. The cell then passes into mitosis where the osteoblast divides into two daughter cells. Key regulators in growth are the cyclins (Fig. 2), which are induced in a temporal sequence of gene expression during the cell cycle. Once the osteoblasts have been replicated enough times to make a monolayer, the cells are ready for the second developmental stage of matrix maturation. During matrix maturation, genes for cell cycle and cell growth are downregulated and expression of genes needed for extracellular matrix maturation and organization begins. The third stage of bone development is the onset of extracellular matrix mineralization where osteoblasts mature into bone. These stages have been experimentally established [23] and defined as restriction points during osteoblast differentiation as depicted in Fig. 1. Throughout the sequential stages of osteoblast growth and differentiation there are specific requirements for proliferation; post-confluent proliferation accompanies osteoblast extracellular matrix biosynthesis. Post-confluent growth results in a monolayer of cells associated with type I collagen, which supports multilayering of cells in developing bone nodules. Osteopontin and osteocalcin exhibit maximal expression during the maturation and mineralization period when bone tissue-like organization is ongoing [23]. Establishing the basis of cellular

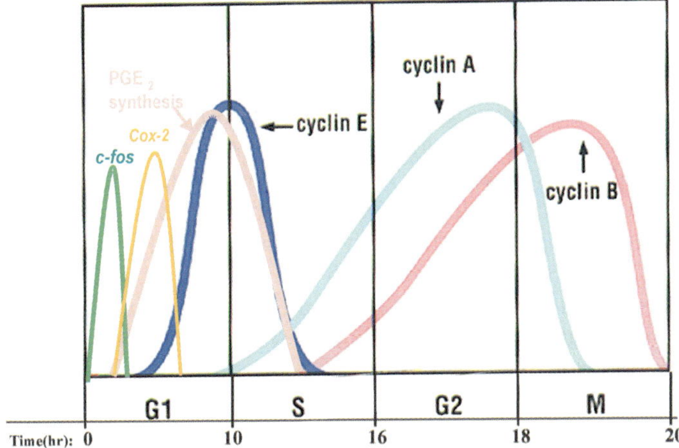

Fig. 2. Cell cycle regulation of immediate early genes and cyclins in the osteoblast. Cell growth is defined by the cell cycle, which is divided into four phases, G1, S, G2 and M. When a cell is not growing, it is resting in the G1 or G0 phase. Growth activation is regulated at the border of the G1 phase just before the start of DNA synthesis (S-phase). The G1/S-phase transition in osteoblasts can be triggered either by a chemical signal like addition of sera or growth factors [95] or by mechanical stress [18,21,63,64,96–98]. Once a cell enters into S-phase, it is committed to complete cell replication resulting in the production of two daughter cells at mitosis. Evidence from previous space flights suggest that the osteoblast is delayed in G1 presumably due to the lack of normal gravity force. Cells are first stimulated with growth factors (fetal calf sera) that stimulate synthesis of *c-fos, c-myc* and *cox-2*. Synthesis of PGE$_2$, is induced prior to entry into S-phase. S-phase is associated with an increase of cyclin E and spike of histone-4 synthesis and DNA synthesis. Once DNA replication is completed, cells enter G2 where proteins necessary for division are made before entry into mitosis.

competency for progression to the mature osteoblast necessitates the identification of the signaling pathways operative for each stage in order to understand the process. This is especially true of studies in microgravity, where investigators have used a range of models to study proliferation, matrix maturation and mineralization. For more details on osteoblast growth and bone development, see the review by Stein et al. [23].

Osteoblast growth and gene expression in microgravity

Many human and animal studies have demonstrated that loss of bone formation (osteoblast proliferation) may be the root cause of space-osteoporosis. The problem of decreased osteoblast proliferation in microgravity is a key issue in cell biology that may determine if we can explore space in the long term. Over the past ten years evidence of changes in cell biology have been noted in a variety of cells. Specifically, it has been demonstrated that cell growth of lymphocytes [24–27] (see review chapter by Lewis in this volume) and osteoblasts [28–40] is inhibited or altered in a microgravity environment.

134

The first microgravity studies on mammalian cells were completed in 1973 on Skylab-3 using human embryonic lung cells (WI-38). Cells grew more slowly in microgravity, with the flight cells taking two hours longer to complete cell cycle. No changes were seen in the karyotype; however, glucose utilization was lower in the flight samples [41]. A review of early cell biology can be found in publications by Hughes-Fulford [42] and Lewis and Hughes-Fulford [43].

In vitro osteoblast growth was first measured on STS-56 in December 1992. The MC3T3-E1 osteoblasts (a mouse line derived from calvarial cells) were launched in a quiescent state (1% FCS) at launch. Growth was initiated by adding fresh media with 10% fetal calf serum in microgravity. Ground controls were treated in a parallel manner in identical equipment. Samples were taken at 16 hours and fixed at four days. Microgravity grown cells grew slowly and cell number was reduced by 60% when compared to ground after four days of flight. Prostaglandin synthesis was significantly increased at 16 hours after growth activation, however it returned to control levels after four days. Glucose utilization was significantly less than ground controls due to a lower number of cells. The cell viability was normal with glucose utilization being comparable on a per cell basis for flight and ground experiments [28]. Confirming evidence was reported by Kumei et al. showing a 136-fold increase in PGE_2 synthesis in flight compared to ground controls. Moreover, they found that osteoblast growth was inhibited on day 4 by (9.2 ± 2 μg of DNA in flight vs. ground 17.4 ± 6 μg DNA) and on day 5 (19.9 ± 6 μg DNA in flight vs. ground 34.2 ± 11 μg DNA) [29]. Studies by Vassy using MCF-7 cells demonstrated slowing of cell growth with mitosis being delayed as well as a lowering of phosphotyrosine signal transduction in microgravity. Table 1 shows more experimental details on microgravity growth and gene expression studies discussed in this section. Taken together, the data suggests that osteoblast growth is significantly inhibited in microgravity.

Studies in microgravity on lymphocytes have also shown a marked growth inhibition in lymphocytes [26,44–46]. This inhibition of lymphocyte growth is associated with an enrichment of cells in G2/M after four hours of flight. After 48 hours there were no significant cell cycle differences between microgravity and flight 1-*g* centrifuged cells [27]. Recent studies by Vassy et al. have demonstrated that cell proliferation is reduced in breast cancer cells grown in microgravity as a consequence of a prolonged mitosis [47]. Changes in cell cycle accompanied by increased apoptosis have also been noted in osteoblasts ROS 17/ 2.8 cells grown in a vector averaged gravity clinostat [38] and in Jurkat cells grown in microgravity [27].

Changes in molecular biology osteoblast in microgravity

Importance of signal transduction in immediate early gene induction

DeLaat's group reported the first study of the early gene expression in short duration microgravity. Studying signal transduction and gene expression in A431 cells in microgravity using sounding rockets, they found a reduction of response to Epidermal Growth Factor (EGF) after approximately six minutes of microgravity

Table 1

Effects of microgravity on osteoblast growth and gene expression

Osteoblast cell type	Hardware, control and samples per data point	Flow	Media type	Collection and storage	Change in growth or mRNA/time in orbit	Ref.
Mouse metatarsal (STS-42)	Biorack; bags inside Type I container	No	1%αMEM	Collected after 4 days in microgravity and stored at −10°C	Glucose utilization and mineralization were decreased under microgravity. Mineral resorption was increased. No change in percent length increase or collagen synthesis in microgravity. Authors suggested that bone loss was result of impaired mineralization.	[85]
MC3T3-E1 (STS-56)	CMIX Ground control $N=4$	No	1%αMEM then activation with 10%αMEM	Collected in microgravity on days 1 and 4 and stored at 20°C	Glucose utilization decreased per well; however there was no change if calculated on a per cell basis. Reduced cell proliferation by day 4. Increased prostaglandin synthesis day 1 in microgravity; no difference day 4.	[28]
Primary femur marrow cells (STS-65)	Ground control	No	10% FCS αMEM with ascorbic acid, dexamethasone β-glycerol-phosphate	Collected in microgravity on day 5 and stored at −20°C	Reduced cell growth (μg DNA) in microgravity. Increases in PGE_2 synthesis in flight samples compared to ground. Increases in cox-2 enzyme message and IL-6 mRNA were increased up to 9.3-fold.	[29]
MG-63 (Foton-10)	Plunger box. Ground and flight 1-g control $n=4$	No	10% FCS DMEM	Collected in microgravity day 9 stored at +26°C	Response to 1,25(OH)$_2$ D$_3$ and TGFβ2 was significantly reduced. Expression of alkaline phosphatase and osteocalcin was decreased, while that of collagen 1a was not. Message expression for ALK, osteocalcin and Col I was significantly reduced.	[66]
Primary femur marrow cells (STS-65)	CCU. Ground control $n=2$	No	10% FCS αMEM with ascorbic acid, dexamethasone β-glycerol-phosphate	Collected in microgravity days 4 and 5 at −20°C	Response to 1,25(OH)$_2$ D$_3$ was altered in flight which caused increased levels of IGF-BP3 in media; mRNA levels were also increased. mRNA of IGFBP-4 and IGF-BP5 were decreased in flight.	[65]

continued

Osteoblast cell type	Hardware, control and samples per data point	Flow	Media type	Collection and storage	Change in growth or mRNA/time in orbit	Ref.
MC3T3-E1 (STS-76,81,84)	Plunger box. Ground control & flight 1-g controls $n=4$	No	1%αMEM; activation with 10% αMEM	Collected in μg at 3 h and 24 h stored at 20°C	*cox-2* message was decreased in the microgravity-activated cells compared to ground or 1-g flight controls. Cox-1 was not detected in any of the samples. There were no significant differences in the expression of actin mRNA between the 0-g and 1-g samples. Increased PGE$_2$ was present in media at 3 and 24 h.	[33]
MG-63 (Foton-10)	Plunger box. Ground and flight 1-g control $n=4$	No	10% FCS DMEM changed at 3 and 6 days	Collected Day 9 kept at 12°C until landing	Differentiation markers OC and collagen 1aI were decreased to 51% of 1-g control. Alkaline phosphatase mRNA was reduced compared to 1-g control. Osteocalcin mRNA was 19% of 1-g control.	[31]
MC3T3-E1 (sounding rocket)	CCEM	No	1% αMEM then activation with 10% αMEM	Collected at $t=0$ and $t=10$ min	Response to EGF in sounding rocket experiment showed a reduced *c-fos* response in microgravity. No changes were seen in MAPK phosphorylation during the short period of microgravity.	[58]
MC3T3-E1 (STS-76,81,84)	Plungerbox. Ground control & flight 1-g controls $n=4$	No	1% αMEM then activation with 10% αMEM	Collected in μg 3 h and 24 h and stored at –20°C	Synthesis of FN mRNA, protein and matrix were measured after activation in microgravity. FN mRNA synthesis was significantly reduced in microgravity (0-g) when compared to ground (GR) osteoblasts flown in a centrifuge simulating earth's gravity (1-g) field 2.5 h after activation. However 27.5 h after activation there were no significant differences in mRNA synthesis. A small but significant reduction of FN protein was found in the 0-g samples 2.5 h after activation. Total FN protein 27.5 h after activation showed no significant difference between any of the gravity conditions; however there was a fourfold increase in absolute amount of protein synthesized during the incubation.	[75]

continued

Osteoblast cell type	Hardware, control and samples per data point	Flow	Media type	Collection and storage	Change in growth or mRNA/time in orbit	Ref.
Primary rat femur marrow cells (STS-65)	CCU. Ground control $n=4$	No	10% FCS αMEM with ascorbic acid, dexamethasone β-glycerol-phosphate	Collected Day 4 & 5 at −20°C	mRNA levels for *PDGFβ* receptor, *EGFr*, *Shc* and *c-fos* were measured. *PDGFβ* receptor, *Shc* and *c-fos* were decreased by approximately 50% by Day 5. EGFr was not changed. The authors suggested that there is a change in signal transduction in microgravity.	[87]
hFOB (STS-80)	CCM on Cytodex™ beds in hollow-fiber cartridges. Ground control $n=2$	Yes	10% DMEM and Hams F12	Media collected on days 8 and 12 in-flight stored at RT. Days 18 and 19 were collected postflight.	*IL-1α* and *IL-6* were decreased 8 h after landing and returned to normal by 24 h post landing. TGFβ messages were modestly reduced 8 hours post flight but were not significant. No change in growth, PGE_2 or glucose utilization was noted.	[86]
Primary chicken calvaria osteoblasts (STS-59)	Cellmax capillary culture units. Osteoblasts grown on Cytodex™ beads $n=2$ or 4	Yes	αMEM10% FCS aMEM with ascorbic acid, β-glycerol-phosphate	Media samples removed during flight. Cartridges collected at landing and stored at −80°C.	No change in flight and ground control glucose utilization; however there was a reduction in flight total RNA compared to ground. Reduced osteocalcin and collagen mRNA and protein expression suggesting that microgravity slows progression of differentiation.	[32]
ROS 17/2.8 (clinostat)	Clinostat experiment-cells grown in 25 cm² falcon flasks	No	10% DMEM	Rotated on clinostat for 6, 12 and 24 h. Fixed or frozen at time of collection	Clinorotation induced apoptosis of osteoblasts suggested that cell death was associated with cytoskeleton disorganization.	[38]

138

[48–51]. Initiation and termination of the experiment was accomplished automatically, allowing the study of microgravity effects in the early phases of EGF induced signal transduction. Several events of the EGF signal cascade were characterized and quantified under normal and microgravity conditions. An excellent end-point of EGF signaling is the induction of immediate early genes. These genes are the first to be transcribed after stimulation of cells with growth factors or stress; examples of the genes are *c-fos, c-jun, erg-1, c-myc* and *cox-2*. De Laat used analysis of *c-fos* and *c-jun* because of their prominent role in cell proliferation and differentiation. Quantitative analysis demonstrated that *c-fos* and *c-jun* expression was reduced by approximately 50% when compared to the normal gravity control samples. During normal growth stimulation of cells by EGF, there is dramatic redistribution of EGFR within five minutes resulting in receptor clustering. When the EGF binding and clustering was measured in microgravity, there was no change compared to 1-*g*, suggesting a point down-stream for signal transduction [52,53]. In testing the effect of the different pathways on EGF signal transduction in the A431 cells with PMA (PKC pathway), A23187 (Ca^{++} pathway) and forskolin (PKA pathway), they found that only the PKC pathway was altered by microgravity [53]. It was concluded that the EGF signal transduction by microgravity occurred downstream of EGFR clustering, but upstream of *c-fos* transcription, and was most likely mediated by PKC-dependent signal transduction in the A431 cell, which is a human epidermoid carcinoma line.

Cell growth is regulated by both intracellular and extracellular signal transduction and it has recently been shown that MAPK is one of the key proteins involved in signal-transduction via growth factor mediated phosphorylation. This is illustrated in Fig. 3 where we see that many of the growth factors, G-proteins, and integrins act through a common MAPK pathway. Once phosphorylated, MAPK (ERK1/2) translocates to the nucleus in late G1 [54–57]. When mitogens are removed there is no MAPK phosphorylation or translocation and cell cycle progression is inhibited [56]. Taken together, these studies suggest that microgravity may inhibit growth by blocking signal transduction and inhibiting cell cycle in G1 or the G2/M phase.

The effect of microgravity on *c-fos* gene expression in the osteoblast (MC3T3-E1) was first studied by Sato et al. [58]. The cells were initially studied in simulated microgravity induced by clino-rotation. They found that EGF induced *c-fos* expression was reduced in both the MC3T3-E1 osteoblast and HeLa carcinoma cells, with the osteoblast being more sensitive to gravity than the HeLa cell. In a second set of experiments, the effect of EGF on osteoblasts was studied in short-term microgravity condition in the sounding rocket (~6 min). Cell lysates from flight and ground groups were analyzed for *c-fos* induction and MAPK phosphorylation. The induction of *c-fos* was reduced in microgravity, while no significant difference in MAPK phosphorylation was seen between the flight and ground samples. It is not known which pathway is associated with inhibition of *c-fos* induction.

Several reports have described gravity-specific changes in mRNA levels following exposure of a variety of cultured cells and whole animals to varying periods of microgravity in space flight, sounding rockets and clinostats. Studies by Hammond et al. using microarray have demonstrated that gene expression in human renal cells is

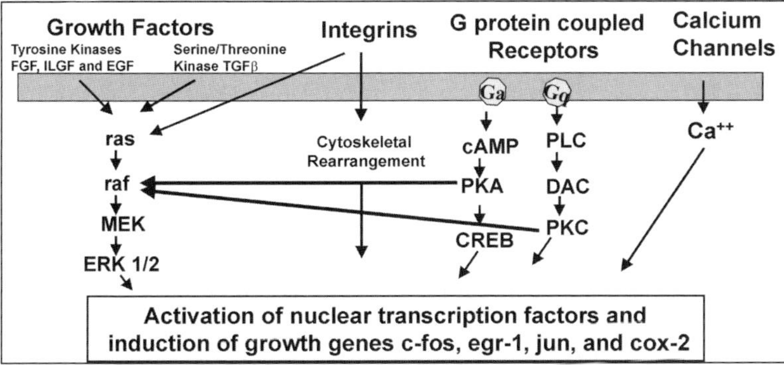

Fig. 3. Anabolic pathways for initiation of cell proliferation. Tyrosine kinases such as FGF, ILGFs and EGF, and Serine/threonine kinase growth factors like TGFβ stimulate ras, raf and finally extracellular signal regulated kinase (ERK) are phosphorylated and translocated to the nucleus and activate nuclear transcription factors and induces growth related genes. The integrins can also stimulate ras and the MAPK pathway. Finally, studies have demonstrated that both PKA and PKC stimulate the MAPK pathway via raf. All pathways lead to activation of nuclear transcription factors and induction of immediate early genes.

altered in microgravity [59,60] with the expression of transcription factors showing the largest changes. Studies by Lewis et al. also using micro arrays showed that the expression of cytoskeleton related genes in T-cells were changed in microgravity [61].

Here on Earth, serum activation of the osteoblast causes induction of the immediate early gene *c-fos* at 30 minutes and cyclooxygenase (*cox-2*) and cytosolic phospholipase A2 (*cpla2*) mRNA at three hours (Fig. 4) [62]. In microgravity experiments completed on STS 76, 81 and 84, early gene activation by sera growth factors failed to fully up regulate *cox-2* and *cpla2* (Hughes-Fulford, unpublished data). The importance of signal transduction in *cpla2* protein transcription has been demonstrated with data from the Boonstra laboratory showing that cpla2 activity is dependent upon MAPK phosphorylation in early G1 [57]. Our laboratory examined quiescent cells for gravity-dependent changes in mRNA levels for several genes involved in bone cell growth and maturation. We asked the question: Does gravity affect the gene expression of cox-2 in flight? This is an essential question because the cox-2 enzyme is directly responsible for the synthesis of PGE_2, which is directly linked to exercise induction of bone formation [18,63,64). In our experiments on STS-76, 81 and 84 we studied osteoblast gene expression in microgravity and have found a reduction of *cox-2* mRNA in microgravity; *cox-2* expression was recovered in 1-*g* centrifuge flight samples. Expression of EGFr, 18S and cyclophilin were constitutive and unchanged by microgravity. We found that *cox-2* mRNA is decreased in the 0-*g* samples when compared to 1-*g* controls. The copy number for *cox-1* mRNA was too low to detect. Finally, the lack of induction of *cox-2* in microgravity suggests that the cells subjected to the 0-g environment did not enter the growth phase of the cell cycle (G1-S transition). This agrees with previous studies showing reduced cell growth in microgravity.

140

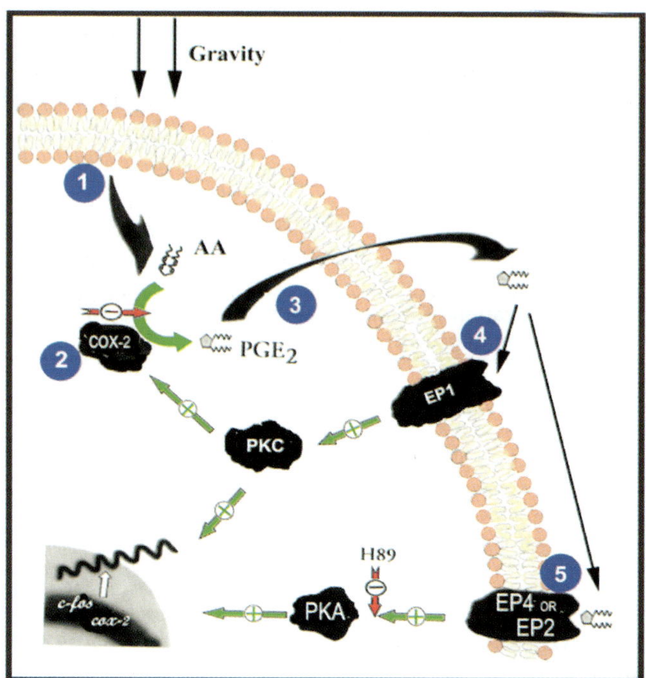

Fig. 4. Possible sites of regulation of G1-S transition. (1) Stimulation of cell membrane by gravity causing release of arachidonic acid (AA) from membrane, perhaps through activation of c-PLA$_2$. Mechanical stimulation also activates Ca^{++} channels and G-proteins. (2) AA conversion into PGE$_2$ by cyclooxygenase and or PGI$_2$ by PGI$_2$ synthase. (3) Export of the newly synthesized PGE$_2$/ PGI$_2$ out of the osteoblast. (4) Activation of seven-domain transmembrane receptors by PGE$_2$ and or by gravity interactions with the cytoskeleton. (5) Activation of the seven-domain transmembrane EP2/4 receptors by PGE$_2$ or by gravity interaction with the cytoskeleton.

Alteration in stem cell commitment to osteoblast phenotype in microgravity

While calvarial-derived osteoblasts support exploration of mechanisms associated with osteoblast growth and differentiation, marrow cell cultures permit evaluation of mechanisms regulating recruitment of stem cells to the osteoblast lineage [23]. Microgravity studies by Kumei et al. using primary femur marrow cells demonstrated that marrow cells had increased prostaglandin E$_2$ synthesis in flight compared to ground. These cells were grown in differentiating medium (αMEM with β-glycerol phosphate and ascorbic acid). Analysis demonstrated up to a nine-fold increase in cox-2 and IL-6 mRNA in differentiating femur marrow cells after five days microgravity [29]. This suggests increased stem cell recruitment to the osteoblast phenotype in microgravity and low levels of dexamethasone to facilitate recruitment to osteoblast phenotype.

In other studies, responses of primary rat marrow cells to 1,25(OH)$_2$D$_3$ were tested after four and five days of microgravity exposure on STS-65. These studies examined alterations of insulin-like growth factor expression in microgravity. After 20-h treatment with 1α, 25-dihydroxyvitamin D_3, conditioned media were harvested and cellular DNA and/or RNA were fixed on board. Insulin-like growth factor binding protein-3 (IGF BP-3) levels in the media were three- and ten-fold higher than ground controls on the fourth and fifth flight days, respectively, as quantitated by Western blotting and radioimmunoassay. The increased IGF BP-3 protein levels correlated with two to three-fold elevation of IGF BP-3 mRNA levels, obtained by reverse transcription-polymerase chain reaction (RTPCR). The IGF BP-5 mRNA levels in flight cultures were 33–69% lower than in ground controls. The IGF BP-4 mRNA levels in flight cultures were 75% lower than in ground controls on the fifth day but were not different on the fourth day. There was no change in the EGFr in flight, however PDGFr, and *c-fos* were significantly inhibited in microgravity. The glucocorticoid receptor mRNA levels in flight cultures were increased by three to eight-fold on the fourth and fifth days compared with levels in ground controls [65]. These studies suggest an altered response of marrow derived osteoblast cells in microgravity.

Differentiation studies by Carmeliet et al. examined the effect of microgravity using the osteocarcoma MG-63 cell line to analyze gene induction related to matrix formation and maturation both at the protein and mRNA levels. Cells that were untreated and hormone-treated with 1,25 dihydroxyvitamin D_3 [1,25(OH)$_2$D$_3$], at a 10^{-7} M concentration and transforming growth factor beta 2 (TGF-β2), 10 ng/ml) were cultured for nine days under microgravity conditions aboard the Foton-10 satellite. In-flight data was compared with ground and in-flight unit-gravity cultures. The expression of alkaline phosphatase (ALP) activity following treatment in microgravity increased only by a factor of 1.8 compared with the 3.8-fold increase at unit-gravity ($p < 0.01$), whereas no alteration was detected in the production of collagen type I between unit-gravity and microgravity. In addition, gene expression for collagen Iα I, ALP, and osteocalcin following treatment at microgravity was reduced to 51, 62, and 19%, respectively, of unit-gravity levels ($p < 0.02$). The lack of correlation between collagen type I gene and protein expression induced by microgravity is most likely related to the different kinetics of gene and protein expression observed at unit-gravity. Following treatment with 1,25(OH)$_2$D$_3$ and TGF-β2, collagen I-alpha1 mRNA increased gradually during 72 h, but collagen type I production was already maximal after treatment for 48 h [66]. Other studies by Carmeliet et al. again using the osteocarcoma line MG-63 examined the effect of gravity on osteoblast differentiation. Using competitive RTPCR in untreated and hormone treated cells, they found that collagen IαI was reduced to 51% of normal treated. In addition osteocalcin message was reduced to 19% of 1-*g* in-flight control [31].

Changes in gene expression in human fetal osteoblast (hFOB) SV-40 transformed cells were examined in microgravity. The hFOB cells were cultured on Cytodex 3 beads™ inside hollow fiber cartridges that were continually fed media with peristaltic pumps with 10% FCS media. Glucose utilization was essentially the same for flight and ground cells. Samples taken in-flight on days 8 and 12 showed very little PGE$_2$, most probably due to the fact that there was no refrigeration available for the

Cell type	Hardware, control and number of replicates per data point	Flow	Media type	Collection and storage	Change in growth or mRNA/Time in orbit	Ref.
ROS 17/2.8 (parabolic flight)	Plungerbox	No	10% FCS-DMEM	Cells collected at 36 h, fixed in 4% para-formaldehyde or 70% ethanol.	Global reductions seen in cell adhesion parameters during flight. Decreased cell area was associated with focal contact plaque re-organization.	[30]
MC3T3-E1 (STS-76,81 and 84)	Plungerbox. Ground control & flight 1-g controls $n=4$	No	1% αMEM then activation with 10% αMEM	Collected in microgravity 3 h and 24 h. Stored at –20°C	Elongated nuclei and stellar cytoskeleton present in microgravity samples, 1-g in-flight centrifugation reversed morphology changes to resemble ground controls.	[77]
Primary chicken calvaria osteoblasts (STS-59)	5-Cellco Cellmax capillary culture units. Osteoblasts grown on Cytodex™ beads	Yes	αMEM10% FCS aMEM with ascorbic acid, β-glycerol-phosphate	Media samples removed during flight. Cartridges collected at landing and stored at –80°C	No change in flight and ground control glucose utilization, reduction in flight total RNA compared to ground. Morphology showed fewer osteoblasts and matrix development around beads and hollow fibers in microgravity.	[32]
hFOB (STS-80)	CCM on Cytodex™ beads in hollow-fiber cartridges. Ground control $n=2$	Yes	10% DMEM/Hams F12 (growth)	Media collected on days 8 and 12 in-flight stored at RT. Days 18 and 19 were postflight on the ground and stored in freezer.	No obvious change in cytoskeleton or morphology of cells growing on Cytodex™ beads.	[86]
ROS 17/2.8 (clinostat)	Clinostat experiment-cells grown in 25 cm^2 falcon flasks	No	10% DMEM	Rotated on clinostat for 6, 12 and 24 h. Fixed or frozen at time of collection	Cell viability decreased significantly over 24 hours of vector-averaged gravity. Clinorotation associated with disorganized actin cytoskeleton, condensation of F-actin and microspikes. Localization of integrin β_1 was perinuclear.	[38]

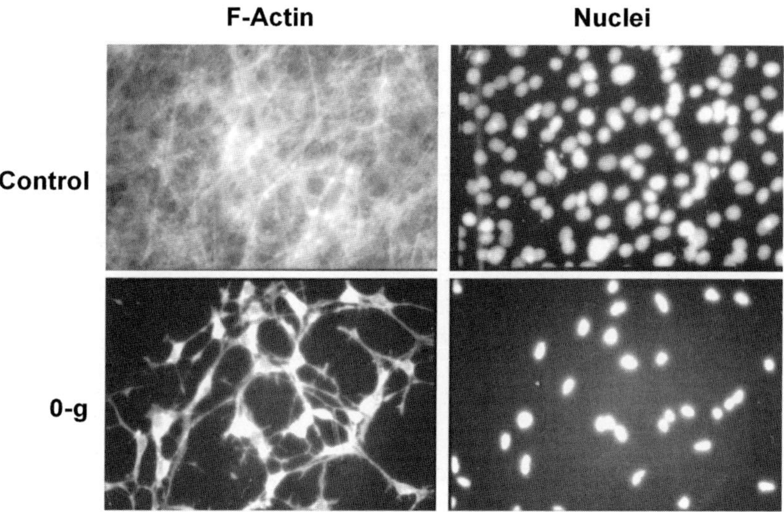

Fig. 5. Changes in MC3T3-E1 cytoskeleton and elongation of nuclei in microgravity. Structure of the F-actin cytoskeleton of 0-g and ground controls using a Zeiss Neofluar 100× oil objective using rhodamine phalloidin. Nuclei were stained with Hoechst 33258. Photos were from STS-56 samples.

over a four-day period (Fig. 5). We also found changes in nuclear shape after four days of space flight, while glucose metabolism per cell was unchanged [28]. As mentioned previously, the cox-2 enzyme was not induced in microgravity, but paradoxically, media PGE_2 content was significantly increased in flight in both the static and 1-*g* samples. This could be due to increased release of PGE_2 from the membrane during the 18 h in a microgravity environment before the start of the experiment or to increased half-life of the enzyme, or to residual *cox-2* activity present at activation. In-flight studies by Stein have demonstrated a reduction in PGE_2 in astronauts [78]. Normally, PGE_2 is cleared by the kidneys within seconds, and must be made continually to maintain high sera/urine levels. In isolated osteoblasts, PGE_2 is degraded by the enzyme 15-hydroxyprostaglandin dehydrogenase; it is possible that the activity of the degrading enzyme or alterations in the degradation process may be inhibited in microgravity.

These data point to an inhibition of anabolic stimuli in the absence of gravity. There are several anabolic signals that are regulated by mechanical stress and these signals seem to be downregulated in microgravity. Figure 3 shows the potential anabolic transducers tyrosine kinase and serine/threonine kinase growth factor receptors, seven domain transmembrane receptors (such as the prostaglandin E_2 receptors), integrins and calcium channels. As illustrated, the receptors and the integrins have a common stimulation of the MAPK (Mitogen Activated Protein Kinase) pathway. Mechanical stress can also activate the MAPK pathway, possibly through the G-proteins or the cytoskeleton.

Table 3

Sequential responses of osteoblasts to gravity-induced signaling and gene expression

Seconds	Minutes	1–8 hours	>18 hours
• G protein activation			
• Map Kinase Activation ⟶			
• PGE$_2$ release	• *c-fos c-myc, egr-1* and *cox-2* ⟶		
	• Induction actin cytoskeleton signaling	• Cytoskeletal changes ⟶	
	• Activation of nuclear transcription factors	• FAK formation	
		• Autocrine growth factor synthesis	• Cell doubling

Growth factors and stress forces cause changes in cell function in a sequenced manner over seconds, minutes and hours as illustrated in Table 3. If the cells are unable to recognize growth factors, or to have a normal 1-g vector, alterations in the signal transduction would occur. More recently we have presented preliminary data showing that gene expression, cytoskeleton and nuclear structure is changed when compared to on flight 1-g and ground controls [33,75,79]. Studies by Lewis et al. found that there were significant changes in the cytoskeleton gene expression and increased apoptosis in the Jurkat T cell line [27]. Researchers examining the effect of weightlessness on breast cancer cells found that microtubules were changed in altered gravity. In addition, they noted that the cytokeratin network was loosely woven in some of the cells modified by microgravity [47]. This laboratory recently found that osteoblast cyclin E mRNA was down regulated in microgravity (unpublished data). This finding correlates well with other reports of inhibition of cyclin E when actin cytoskeleton is disrupted by dihydrocytochalasin B accompanied by cell cycle arrested at late G1 [80]. We do not think the shape change is due to changes in fibronectin matrix since we found that fibronectin mRNA, protein synthesis and extracellular matrix organization (Fig. 7) is not changed in microgravity [75].

The cause of the elongation of nuclei and elongation of cytoskeleton as shown in Figs. 5 and 6 is not known, but it is possible that it is a result of a microgravity-induced inability of anabolic stimulation. The only example of a similar nuclear elongation was caused by limiting cell access to mitogens. Studies by Wang and Walsh demonstrated that when monocytes undergo growth factor withdrawal and start to differentiate, a portion of the cells form elongated nuclei after three days, a shape much like those seen in space flight after four days. In addition, there was spindle shaped elongation of the myocytes accompanied the elongation of the nucleus (Fig. 8). Other studies using anti-integrins demonstrated an elongation of the actin cytoskeleton, suggesting that a deficit of focal adhesions or integrins could be causing the alteration of cytoskeleton [81]. This is supported by the observation of Guignandon et al. showing that intermittent exposure to gravity causes flight induced shape changes that included focal contact plaque reorganization. When cells were blocked

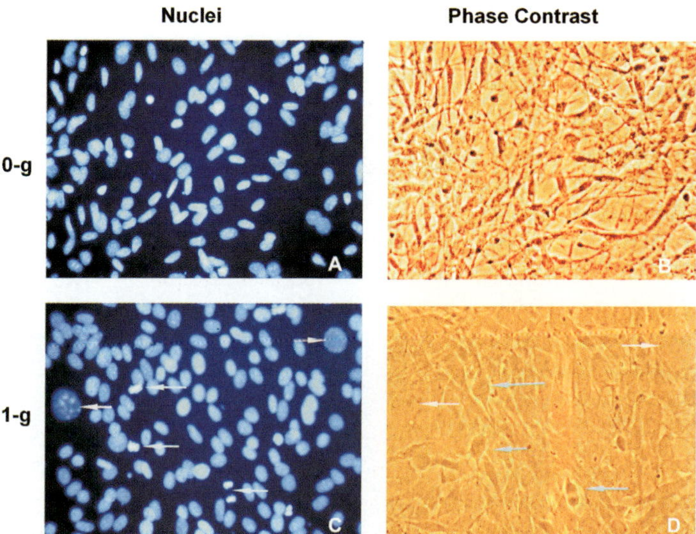

Fig. 6. Osteoblast grown in microgravity or in 1-*g* centrifuge in flight. Osteoblasts were flown on STS-76, photos taken from samples 24 hours after addition of serum to quiescent cells in microgravity. Nuclei of the 0-*g* are partly elongated and phase contrast shows elongation of the whole cell and few cells in S/G2 stage. Also there were few mitotic figures in the 0-*g* sample. One-*g* flight samples have several cells that are in G2 (pink arrows) or just entering or undergoing mitosis (light blue arrows). Photos were taken with 20× objective.

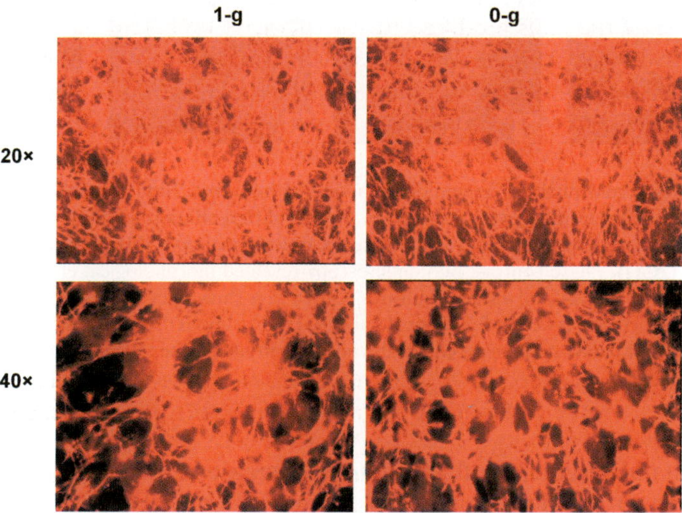

Fig. 7. Fibronectin network in osteoblasts grown in 1-*g* or 0-*g* environment. Osteoblasts were flown on STS-76, photos taken of samples 24 hours after addition of serum to quiescent cells in microgravity. Fibronectin was imaged with rabbit anti-fibronection with a secondary G-anti rabbit rhodamine antibody. No significant difference was seen in the fibronectin matrix in microgravity.

148

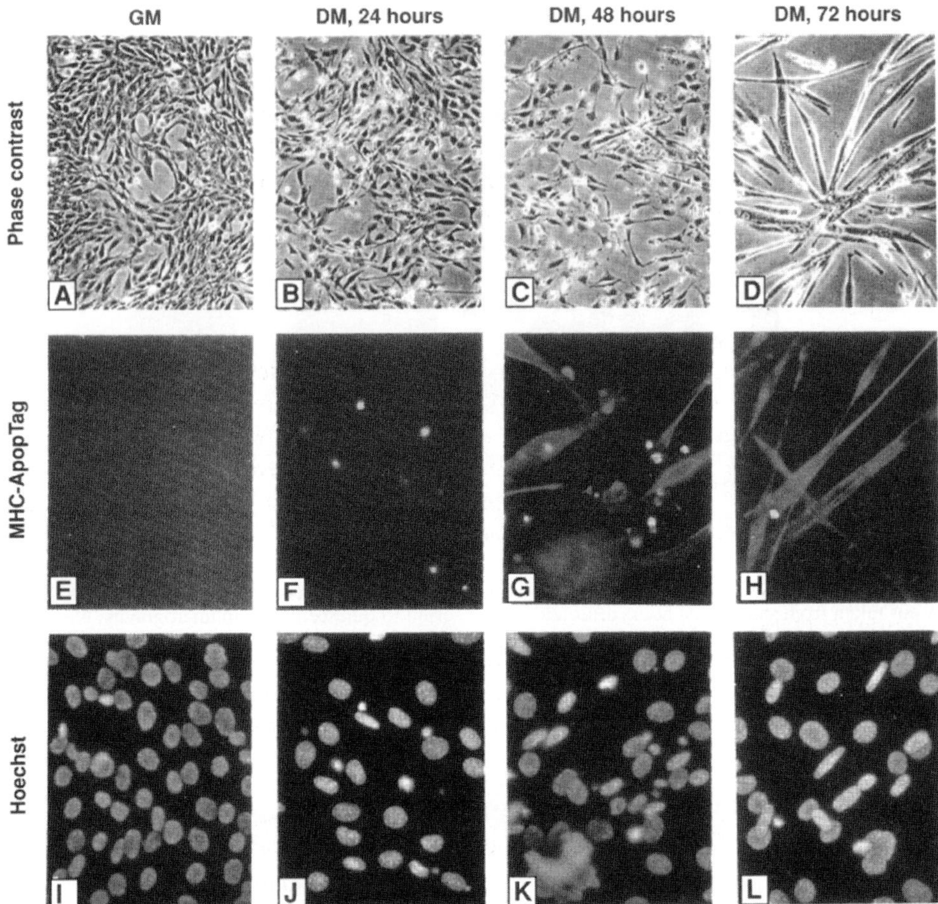

Fig. 8. Cytoskeleton and nucleus elongation with lack of mitogens. Induction of either apoptosis or terminal differentiation in monocytes cultured in the 2% differentiation medium (DM) which causes mitogen deprivation. Proliferating C2C12 monocytes in growth medium (GM) were shifted to DM for 24, 48 or 72 hours (A–D). Phase contrast photomicroscopy revealed morphological changes, with floating cells most evident in DM 24 and 48 hour cultures. Green tags show apoptotic cells, Red shows muscle differentiation (MCH red) as well as cell shape details (E–H). Nucleus is displayed in (I–L), apoptotic cells seen in F, G and H also have condensed chromatin and cell shrinkage. Reproduced with permission from Science [99].

in G2/M with nocodazole, the flight-induced decrease in adhesion was ameliorated [30] suggesting a role for the microtubule in gravity induced changes.

Other studies by Guignandon et al. demonstrated that microgravity induced changes in cell shape after four days of microgravity, with osteoblasts becoming rounder and covered with microvilli. At the end of the flight, cells had mixed morphological types, including stellar shapes. In addition, alkaline phosphatase

(ALK) was increased two-fold in microgravity when compared to controls [82]. Finally, work of Vassy working with breast cancer MCF-7 in culture in microgravity found that microtubules were altered in microgravity, this observation was also made in the test tube, when Papaseit et al. demonstrated that isolated tubulin auto-polymerization is sensitive to the absence of gravity [83] during parabolic flight. Taken together these studies suggest that cell cytoskeleton may be involved in the changes in gene expression and function seen in osteoblasts in reduced gravity.

Discussion

Cell growth and differentiation are fundamental in bone development, and as such they act as master regulators of bone development and remodeling. In space flight there are numerous effects on tissues and cells, some of which, such as bone and muscle loss, could be related to cell proliferation and development anomalies in microgravity. More specifically, preliminary data from our own space-flight gene expression experiments in osteoblasts [33,75,79] as well as others suggest that proliferation-related genes are downregulated in microgravity [48–51,54,58,59,61]. Development is also affected with many of the matrix genes downregulated in microgravity [30–32,66,72,74,82].

Bone proliferation and mineralization are two of the most tightly regulated processes in the cell and multiple checkpoints during the cell cycle and development ensure that cells have normal development. Among those checkpoints are growth inhibitors that cause DNA transcription arrest and programmed matrix organization. Studies have demonstrated that cell proliferation is not constitutive, it requires promoters, such as the extracellular matrix and serum growth factors that activate mitogenic kinase signaling pathways. In many cases, such as in skin fibroblasts, cells are normally quiescent until serum exposure due to wounding or another factor causes proliferation. In bone and muscle tissues, however, there is a continuous balance of cell proliferation and death to allow for dynamic remodeling. The dynamic nature of skeletal tissue allows it to quickly grow stronger with exercise and mechanical stimulation and to self-repair when bone breaks occur. Weight bearing and other mechanical stresses of earth gravity modulate the constant remodeling of these tissues. Therefore bone tissues will also dynamically decrease in mass if normal mechanical stimulation at 1-gravity is interrupted, such as during human bed rest, animal hind limb unloading [84], and space flight. Mechanical stimulation in skeletal tissues, therefore, appears to be accompanied by increased cell proliferation, and thus we hypothesize that force and proliferation are causally related. As a consequence, we expect to observe decreased proliferative rates in musculoskeletal tissues in microgravity. Conversely, increased gravity promotes cell proliferation. Separate ground studies that address this question using centrifugation are already underway (Hughes-Fulford, in press [100]).

Several investigators have reported reductions in glucose utilization in micro-gravity [28,41,85], however, this may be due to a reduced cell number caused by a

150

reduction in cell growth and slowing of the cell cycle [25,27–29,41,44,46,47]. The causes of reduced growth may relate directly to the reduction of induction of the immediate early growth genes in the cell cycle, namely reduction of *c-fos, c-myc,* and *cox-2* induction in microgravity [33,34,48,49,51,54]. Some authors reported no change in osteoblast cell growth, but most of these exceptions were examining mineralization or were using culture systems with continual media flow over the cells. The resulting sheer, even at low levels of flow, can stimulate cells [32,47,86].

Growth factor receptor EGFr was reported to be unchanged in several cell lines by several investigators [34,52,54,87]. There is a general consensus that the majority of changes are caused not due to loss of receptor number, rather to a change in signal transduction in microgravity. Several investigators have suggested that there is a microgravity induced alteration in signal transduction with altered gene expression being a major portion of the evidence. At the molecular level we have already discovered that mechanical stimulation by centrifugation of osteoblasts promotes the MAPK pathway [62]. The MAPK pathway is a key growth factor and integrin linked signaling pathway, well known for its ability to promote proliferation and cell survival. The fact that mechanical stimulation from increased G levels also activates these molecules is suggests that gravity promotes cell growth through the MAPK pathway.

The role of the cytoskeleton in growth and its relationship to cell cycle and the MAPK pathway is being actively investigated in many laboratories. Results from several investigators suggest that the actin cytoskeleton is a necessary component of the cell cycle, especially for the transition from G1-S phase and for anchorage dependent DNA synthesis [88,89]. Restrictions to the cytoskeleton formation inhibit growth and can induce differentiation [90–92]. The obvious changes in the cytoskeleton in microgravity may be a key component in regulation of growth and differentiation in microgravity as well as on Earth. The increase in PGE2 seen by several investigators may be a result of the altered cytoskeleton. Reports from Sawyer et al. showed a 6–11 fold increase of PGE_2 when the actin microfilament cytoskeleton is disrupted by cytochalasin D or latrunculin A in human-umbilical vein endothelial cells (HUVEC). Disruption of microtubules did not cause a change in prostaglandin release [93].

Continued studies of osteoblasts in microgravity will facilitate our understanding of mechanisms controlling anchorage-dependent anabolic signaling and regulation of bone growth in a gravity free environment. Since all terrestrial organisms evolved in a 1-*g* environment, understanding the effect of Earth's gravity on cytoskeleton, cell growth and regulation will give us insight into the fundamental biological laws underlying gravity-based life.

Acknowledgements

This work was supported by NASA grants NAG-2-1086 and NAG-2-1286 and a VA Merit Review awarded to MHF. I thank Vicki Gilbertson for her excellent technical skill. I thank Marian Lewis, Jason Hatton, Janice Voorsluys and George Fulford for their thoughtful comments during the writing of the manuscript. I also appreciate the hard work of ESA, NASA, NASDA their employees and our cosmonauts and astronauts for making our collective discoveries possible. I thank the Research Service of the Department of Veteran's Affairs Merit Review for my support while writing this chapter.

References

1. Mack, P. (1971) Bone density changes in a Macaca Nemestrina monkey during the Biosatellite II project. Aerospace Med., 42, 828–833.
2. Nicogossian, A., Huntoon, C. and Pool, S. (1989) Space Physiology and Medicine. Lea and Febiger, Philadephia.
3. Vogt, P.M.F. (1971) Roentgenographic bone density changes during representative Apollo space flight. Am. J. Roentenol. 113, 621–623.
4. Vose, G. (1975) Review of roentgenographic bone demineralization studies of Gemini space flights. Am. J. Roentgenol. 121.
5. Johnson, P.R.R. (1979) Prolonged weightlessness and calcium loss in man. Acta Astronautica 6, 1113–1122.
6. Cann, C. (1981) Determination of spine mineral density using computerized tomograms: a report: NASA CIF US/USSR Joint Working Group Meeting on Space Biology and Medicine, pp. 9–22.
7. Morey, E.R. and Baylink, D.J. (1978) Inhibition of bone formation during space flight. Science 201, 1138–1141.
8. Wronski, T.J. and Morey, E.R. (1983) Effect of spaceflight on periosteal bone formation in rats. Am. J. Physiol. 244, R305–309.
9. Collet, P., Uebelhart, D., Vico, L., Moro, L., Hartmann, D., Roth, M. and Alexandre, C. (1997) Effects of 1- and 6-month spaceflight on bone mass and biochemistry in two humans. Bone 20, 547–551.
10. Oganov, V.S. Voronin, L.I., Rakhmanov, A.S., Bakulin, A.V., Schneider, V.S., Leblank, A.D. (1972) Bone loss in microgravity. Aviakosm. Ekolog. Med. 6, 20–24.
11. Caillot-Augusseau, A., Lafage-Proust, M.H., Soler, C., Pernod, J., Dubois, F. and Alexandre, C. (1998) Bone formation and resorption biological markers in cosmonauts during and after a 180-day space flight (Euromir 95). Clin. Chem. 44, 578–585.
12. Vico, L., Collet, P., Guignandon, A., Lafage-Proust, M.H., Thomas, T., Rehaillia, M. and Alexandre, C. (2000) Effects of long-term microgravity exposure on cancellous and cortical weight-bearing bones of cosmonauts. Lancet 355, 1607–1611.
13. Stein, T.P., Leskiw, M.J., Schluter, M.D., Hoyt, R.W., Lane, H.W., Gretebeck, R.E. and LeBlanc, A.D. (1999) Energy expenditure and balance during spaceflight on the space shuttle. Am. J. Physiol. 276, R1739–1748.
14. Stein, T.P., Leskiw, M.J., Schluter, M.D., Donaldson, M.R. and Larina, I. (1999) Protein kinetics during and after long-duration spaceflight on MIR. Am. J. Physiol. 276, E1014–1021.
15. Stein, T.P., Schluter, M.D., Leskiw, M.J. and Boden, G. (1999) Attenuation of the protein wasting associated with bed rest by branched-chain amino acids. Nutrition 15, 656–660.

152

16. Stein, T.P. and Schluter, M.D. (1999) Plasma amino acids during human spaceflight. Aviat. Space Environ. Med. 70, 250–255.
17. Stein, T.P., Schluter, M.D. and Moldawer, L.L. (1999) Endocrine relationships during human spaceflight. Am. J. Physiol. 276, E155–162.
18. Forwood, M.R. (1996) Inducible cyclo-oxygenase (COX-2) mediates the induction of bone formation by mechanical loading in vivo. J. Bone Miner. Res. 11, 1688–1693.
19. Tang, L.Y., Cullen, D.M., Yee, J.A., Jee, W.S. and Kimmel, D.B. (1997) Prostaglandin E2 increases the skeletal response to mechanical loading. J. Bone Miner. Res. 12, 276–282.
20. Hughes-Fulford, M., Appel, R., Kumegawa, M. and Schmidt, J. (1992) Effect of dexamethasone on proliferating osteoblasts: inhibition of prostaglandin E2 synthesis, DNA synthesis, and alterations in actin cytoskeleton. Exp. Cell Res. 203, 150–156.
21. Brighton, C.T., Fisher Jr. J.R., Levine, S.E., Corsetti, J.R., Reilly, T., Landsman, A.S., Williams, J.L. and Thibault, L.E. (1996) The biochemical pathway mediating the proliferative response of bone cells to a mechanical stimulus. J. Bone Joint Surg. Am. 78, 1337–1347.
22. Konieczynski, D.D., Truty, M.J. and Biewener, A.A. (1998) Evaluation of a bone's *in vivo* 24-hour loading history for physical exercise compared with background loading. J. Orthop. Res. 16, 29–37.
23. Stein, G.S., Lian, J.B., Stein, J.L., Van Wijnen, A.J. and Montecino, M. (1996) Transcriptional control of osteoblast growth and differentiation. Physiol. Rev. 76, 593–629.
24. Cogoli, A., Tschopp, A. and Fuchs-Bislin, P. (1984) Cell sensitivity to gravity. Science 225, 228–230.
25. Cogoli, A. (1993) The effect of hypogravity and hypergravity on cells of the immune system. J. Leukoc. Biol. 54, 259–268.
26. Cogoli, A. and Cogoli-Greuter, M. (1997) Activation and proliferation of lymphocytes and other mammalian cells in microgravity. Adv. Space Biol. Med. 6, 33–79.
27. Lewis, M.L., Reynolds, J.L., Cubano, L.A., Hatton, J.P., Lawless, B.D. and Piepmeier, E.H. (1998) Spaceflight alters microtubules and increases apoptosis in human lymphocytes (Jurkat). FASEB J. 12, 1007–1018.
28. Hughes-Fulford, M. and Lewis, M.L. (1996) Effects of microgravity on osteoblast growth activation. Exp. Cell Res. 224, 103–109.
29. Kumei, Y., Shimokawa, H., Katano, H., Hara, E., Akiyama, H., Hirano, M., Mukai, C., Nagaoka, S., Whitson, P.A. and Sams, C.F. (1996) Microgravity induces prostaglandin E2 and interleukin-6 production in normal rat osteoblasts: role in bone demineralization. J. Biotechnol. 47, 313–324.
30. Guignandon, A., Usson, Y., Laroche, N., Lafage-Proust, M.H., Sabido, O., Alexandre, C. and Vico, L. (1997) Effects of intermittent or continuous gravitational stresses on cell–matrix adhesion: quantitative analysis of focal contacts in osteoblastic ROS 17/2.8 cells. Exp. Cell Res. 236, 66–75.
31. Carmeliet, G., Nys, G., Stockmans, I. and Bouillon, R. (1998) Gene expression related to the differentiation of osteoblastic cells is altered by microgravity. Bone 22, 139S–143S.
32. Landis, W.J., Hodgens, K.J., Block, D., Toma, C.D. and Gerstenfeld, L.C. (2000) Spaceflight effects on cultured embryonic chick bone cells. J. Bone Miner. Res. 15, 1099–1112.
33. Hughes-Fulford, M., Tjandrawinata, R., Fitzgerald, J., Gasuad, K. and Gilbertson, V. (1998) Effects of microgravity on osteoblast growth. Gravit. Space Biol. Bull. 11, 51–60.

34. Hughes-Fulford, M. and Gillbertson, V.G. (2001) Microgravity caused decreased gene induction of growth factors, oncogenes and cos-2 in osteoblasts. Abstracts from NASA cell biology meeting. Meeting Proceedings.

35. Hughes-Fulford, M. (1997) Prostaglandin regulation of gene expression and growth in normal and malignant tissues. Adv. Exp. Med. Biol., 269–278.

36. Marie, P.J., Jones, D., Vico, L., Zallone, A., Hinsenkamp, M. and Cancedda, R. (2000) Osteobiology, strain, and microgravity: part I. Studies at the cellular level. Calcif. Tissue Int. 67, 2–9.

37. Kunisada, T., Kawai, A., Inoue, H. and Namba, M. (1997) Effects of simulated microgravity on human osteoblast-like cells in culture. Acta Med. Okayama 51, 135–140.

38. Sarkar, D., Nagaya, T., Koga, K., Nomura, Y., Gruener, R. and Seo, H. (2000) Culture in vector-averaged gravity under clinostat rotation results in apoptosis of osteoblastic ROS 17/2.8 cells. J. Bone Miner. Res. 15, 489–498.

39. Zerath, E., Godet, D., Holy, X., Andre, C., Renault, S., Hott, M. and Marie, P.J. (1996) Effects of spaceflight and recovery on rat humeri and vertebrae: histological and cell culture studies. J. Appl. Physiol. 81, 164–171.

40. Zerath, E., Holy, X., Roberts, S.G., Andre, C., Renault, S., Hott, M. and Marie, P.J. (2000) Spaceflight inhibits bone formation independent of corticosteroid status in growing rats. J Bone Miner Res 15, 1310–1320.

41. Montgomery, P., Cox, J., Reynolds, R., Paul, J., Hayflick, L., Stock, D., Shulz, W. and Kimzey, S. (1977) The response of single human cells to zero gravity. In: Biomedical Results from Skylab (Johnson R.S. and Dietlein, L.F., eds). Superintendent of Documents US Government Printing Office, Washington, DC.

42. Hughes-Fulford, M. (1991) Altered cell function in microgravity. Exp. Gerontol. 26, 247–256.

43. Lewis, M. and Hughes-Fulford, M. (1997) Cellular Responses to Microgravity. Harvard University, Cambridge, MA.

44. Cogoli, A., Bechler, B., Cogoli-Greuter, M., Criswell, S.B., Joller, H., Joller, P., Hunzinger, E. and Muller, O. (1993) Mitogenic signal transduction in T lymphocytes in microgravity. J. Leukoc. Biol. 53, 569–575.

45. Cogoli, A. and Gmunder, F.K. (1991) Gravity effects on single cells: techniques, findings, and theory. Adv. Space Biol. Med. 1, 183–248.

46. Cogoli, A., Valluchi-Morf, M., Mueller, M. and Briegleb, W. (1980) Effect of hypogravity on human lymphocyte activation. Aviat. Space Environ. Med. 51, 29–34.

47. Vassy, J., Portet, S., Beil, M., Millot, G., Fauvel-Lafeve, F., Karniguian, A., Gasset, G., Irinopoulou, T., Calvo, F., Rigaut, J.P. and Schoevaert, D. (2001) The effect of weightlessness on cytoskeleton architecture and proliferation of human breast cancer cell line MCF-7. FASEB J. 15, 1104–1106.

48. de Laat, S.W., de Groot, R.P., den Hertog, J., Rijken, P.J., Verkleij, A.J., Kruijer, W. and Boonstra, J. (1992) Epidermal growth factor induced signal transduction in A431 cells is influenced by altered gravity conditions. Physiologist 35, S19–22.

49. de Groot, R.P., Rijken, P.J., den Hertog, J., Boonstra, J., Verkleij, A.J., de Laat, S.W. and Kruijer, W. (1990) Microgravity decreases c-fos induction and serum response element activity. J. Cell Sci. 97, 33–38.

50. de Groot, R.P., Rijken, P.J., den Hertog, J., Boonstra, J., Verkleij, A.J., de Laat, S.W. and Kruijer, W. (1991) Nuclear responses to protein kinase C signal transduction are sensi-

tive to gravity changes. Exp. Cell Res. 197, 87–90.

51. de Groot, R.P., Rijken, P.J., Boonstra, J., Verkleij, A.J., de Laat, S.W. and Kruijer, W. (1991) Epidermal growth factor-induced expression of c-fos is influenced by altered gravity conditions. Aviat. Space Environ. Med. 62, 37–40.

52. Rijken, P.J., de Groot, R.P., van Belzen, N., de Laat, S.W., Boonstra, J. and Verkleij, A.J. (1993) Inhibition of EGF-induced signal transduction by microgravity is independent of EGF receptor redistribution in the plasma membrane of human A431 cells. *Exp. Cell Res.* 204, 373–377.

53. Rijken, P.J., Boonstra, J., Verkleij, A.J. and de Laat, S.W. (1994) Effects of gravity on the cellular response to epidermal growth factor. Adv. Space Biol. Med. 4, 159–188.

54. Boonstra, J. (1999) Growth factor-induced signal transduction in adherent mammalian cells is sensitive to gravity. FASEB J. 13, S35–42.

55. Hulleman, E., Bijvelt, J.J., Verkleij, A.J., Verrips, C.T. and Boonstra, J. (1999) Nuclear translocation of mitogen-activated protein kinase p42MAPK during the ongoing cell cycle. J. Cell Physiol. 180, 325–333.

56. Hulleman, E. and Boonstra J. (2001) Regulation of G1 phase progression by growth factors and the extracellular matrix. Cell Mol. Life Sci. 58, 80–93.

57. van Rossum, G.S., Vlug, A.S., van Den Bosch, H., Verkleij, A.J. and Boonstra, J. (2001) Cytosolic phospholipase A(2) activity during the ongoing cell cycle. J. Cell Physiol. 188, 321–328.

58. Sato, A., Hamazaki, T., Oomura, T., Osada, H., Kakeya, M., Watanabe, M., Nakamura, T., Nakamura, Y., Koshikawa, N., Yoshizaki, I., Aizawa, S., Yoda, S., Ogiso, A., Takaoki, M., Kohno, Y. and Tanaka, H. (1999) Effects of microgravity on c-fos gene expression in osteoblast-like MC3T3–E1 cells. Adv. Space Res. 24, 807–813.

59. Hammond, T.G., Benes, E., O'Reilly, K.C., Wolf, D.A., Linnehan, R.M., Taher, A., Kaysen, J.H., Allen, P.L. and Goodwin, T.J. (2000) Mechanical culture conditions effect gene expression: gravity-induced changes on the space shuttle [In Process Citation]. Physiol. Genomics 3, 163–173.

60. Hammond, T.G., Lewis, F.C., Goodwin, T.J., Linnehan, R.M., Wolf, D.A., Hire, K.P., Campbell, W.C., Benes, E., O'Reilly, K.C., Globus, R.K. and Kaysen, J.H. (1999) Gene expression in space [letter]. Nat. Med. 5, 359.

61. Lewis, M.L., Cubano, L.A., Zhao, B., Dinh, H.K., Pabalan, J.G., Piepmeier, E.H. and Bowman, P.D. (2001) cDNA microarray reveals altered cytoskeletal gene expression in space-flown leukemic T lymphocytes (Jurkat). FASEB J 15, 1783–1785.

62. Hughes-Fulford, M. (2001) unpublished observation.

63. Forwood, M.R., Kelly, W.L. and Worth, N.F. (1998) Localisation of prostaglandin endoperoxide H synthase (PGHS)-1 and PGHS-2 in bone following mechanical loading *in vivo*. Anat. Rec. 252, 580–586.

64. Forwood, M.R. and Turner, C.H. (1995) Skeletal adaptations to mechanical usage: results from tibial loading studies in rats. Bone 17, 197S–205S.

65. Kumei, Y., Shimokawa, H., Katano, H., Akiyama, H., Hirano, M., Mukai, C., Nagaoka, S., Whitson, P.A. and Sams, C.F. (1998) Spaceflight modulates insulin-like growth factor binding proteins and glucocorticoid receptor in osteoblasts. J. Appl. Physiol. 85, 139–147.

66. Carmeliet, G., Nys, G. and Bouillon, R. (1997) Microgravity reduces the differentiation of human osteoblastic MG-63 cells. J. Bone Miner. Res. 12, 786–794.

67. Chiquet, M. (1999) Regulation of extracellular matrix gene expression by mechanical

stress. Matrix Biol. 18, 417–426.

68. Duncan, R.L. and Turner, C.H. (1995) Mechanotransduction and the functional response of bone to mechanical strain. Calcif. Tissue Int. 57, 344–358.

69. Duncan, R.L. (1995) Transduction of mechanical strain in bone. ASGSB Bull. 8, 49–62.

70. Zhu, X., Ohtsubo, M., Bohmer, R.M., Roberts, J.M. and Assoian, R.K. (1996) Adhesion-dependent cell cycle progression linked to the expression of cyclin D1, activation of cyclin E-cdk2, and phosphorylation of the retinoblastoma protein. J. Cell Biol. 133, 391–403.

71. Bohmer, R.M., Scharf, E. and Assoian, R.K. (1996) Cytoskeletal integrity is required throughout the mitogen stimulation phase of the cell cycle and mediates the anchorage-dependent expression of cyclin D1. Mol. Biol. Cell 7, 101–111.

72. Guignandon, A., Vico, L., Alexandre, C. and Lafage-Proust, M,H. (1995) Shape changes of osteoblastic cells under gravitational variations during parabolic flight—relationship with PGE2 synthesis. Cell Struct. Funct. 20, 369–375.

73. Rijken, P.J., Hage, W.J., van Bergen en Henegouwen, P.M., Verkleij, A.J. and Boonstra, J. (1991) Epidermal growth factor induces rapid reorganization of the actin microfilament system in human A431 cells. J. Cell Sci. 100, 491–499.

74. Carmeliet, G. and Bouillon, R. (1999) The effect of microgravity on morphology and gene expression of osteoblasts in vitro. FASEB J. 13, S129–134.

75. Hughes-Fulford, M. and Gilbertson, V. (1999) Osteoblast fibronectin mRNA, protein synthesis, and matrix are unchanged after exposure to microgravity. FASEB J. 13, S121–127.

76. Hughes-Fulford, M. (2001) Changes in gene expression and signal transduction in microgravity. J. Gravitat. Physiol. 8 (1), 1–4.

77. Hughes-Fulford, M., Tjandrawinata, R., Fitzgerald, J., Gasuad, K. and Gilbertson, V. (1999) Effect of microgravity on osteoblast growth activation: Analysis of transcription, translation and morphology on STS-Biorack on Spacehab. ESA, Noordwijk, The Netherlands:

78. Stein, T.P. (1997) C-peptide and PGE2 activity during human spaceflight. Adv. Exp. Med. Biol. 407, 443–449.

79. Hughes-Fulford, M., Chen, Y. and Tjandrawinata, R.R. (2001) Fatty acid regulates gene expression and growth of human prostate cancer PC-3 cells. Carcinogenesis 22, 701–707.

80. Reshetnikova, G., Barkan, R., Popov, B., Nikolsky, N. and Chang, L.S. (2000) Disruption of the actin cytoskeleton leads to inhibition of mitogen-induced cyclin E expression, Cdk2 phosphorylation, and nuclear accumulation of the retinoblastoma protein-related p107 protein. Exp. Cell Res. 259, 35–53.

81. Kheradmand, F., Werner, E., Tremble, P., Symons, M. and Werb, Z. (1998) Role of Rac1 and oxygen radicals in collagenase-1 expression induced by cell shape change. Science 280, 898–902.

82. Guignandon, A., Genty, C., Vico, L., Lafage-Proust, M.H., Palle, S. and Alexandre, C. (1997) Demonstration of feasibility of automated osteoblastic line culture in space flight. Bone 20, 109–116.

83. Papaseit, C., Pochon, N. and Tabony, J. (2000) Microtubule self-organization is gravity-dependent. Proc Nat. Acad. Sci. USA 97, 8364–8368.

84. Morey-Holton, E.R. and Globus, R.K. (1998) Hindlimb unloading of growing rats: a model for predicting skeletal changes during space flight. Bone 22, 83S–88S.

85. Van Loon, J.J., Bervoets, D.J., Burger, E.H., Dieudonne, S.C., Hagen, J.W., Semeins, C.M., Doulabi, B.Z and Veldhuijzen, J.P. (1995) Decreased mineralization and increased calcium release in isolated fetal mouse long bones under near weightlessness. J. Bone Miner. Res. 10, 550–557.

86. Harris, S.A., Zhang, M., Kidder, L.S., Evans, G.L., Spelsberg, T.C. and Turner, R.T. (2000) Effects of orbital spaceflight on human osteoblastic cell physiology and gene expression. Bone 26, 325–331.

87. Akiyama, H., Kanai, S., Hirano, M., Shimokawa, H., Katano, H., Mukai, C., Nagaoka, S., Morita, S. and Kumei, Y. (1999) Expression of PDGF-beta receptor, EGF receptor, and receptor adaptor protein Shc in rat osteoblasts during spaceflight. Mol. Cell Biochem. 202, 63–71.

88. Toma, C.D., Ashkar, S., Gray, M.L., Schaffer, J.L. and Gerstenfeld, L.C. (1997) Signal transduction of mechanical stimuli is dependent on microfilament integrity: identification of osteopontin as a mechanically induced gene in osteoblasts. J. Bone Miner. Res. 12, 1626–1636.

89. Pavalko, F.M., Chen, N.X., Turner, C.H., Burr, D.B., Atkinson, S., Hsieh, Y.F., Qiu, J. and Duncan, R.L. (1998) Fluid shear-induced mechanical signaling in MC3T3-E1 osteoblasts requires cytoskeleton-integrin interactions. Am. J. Physiol. 275, C1591–1601.

90. Dike, L.E., Chen, C.S., Mrksich, M., Tien, J., Whitesides, G.M. and Ingber, D.E. (1999) Geometric control of switching between growth, apoptosis, and differentiation during angiogenesis using micropatterned substrates. In Vitro Cell Dev. Biol. Anim. 35, 441–448.

91. Huang, S., Chen, C.S. and Ingber, D.E. (1998) Control of cyclin D1, p27(Kip1), and cell cycle progression in human capillary endothelial cells by cell shape and cytoskeletal tension. Mol. Biol. Cell 9, 3179–3193.

92. Ingber, D.E., Prusty, D., Sun, Z., Betensky, H. and Wang, N. (1995) Cell shape, cytoskeletal mechanics, and cell cycle control in angiogenesis. J. Biomech. 28, 1471–1484.

93. Sawyer, S.J., Norvell, S.M., Ponik, S.M. and Pavalko, F.M. (2001) Regulation of PGE(2) and PGI(2) release from human umbilical vein endothelial cells by actin cytoskeleton. Am. J. Physiol. Cell Physiol. 281, C1038–1045.

94. Hughes-Fulford, M., Tjandrawinata, R., Fitzgerald, J., Gasuad, K. and Gilbertson, V.G. (1999) Effect of microgravity on osteoblast growth activation: Analysis of transcription, translation and morphology. In: Biorack on Spacehab: Biological Experiments on Three Shuttle-to-Mir Missions (M. Perry, ed.), European Space Agency, Noordwijk, The Netherlands, pp. 25–32.

95. Baylink, D.J., Finkelman, R.D. and Mohan, S. (1993) Growth factors to stimulate bone formation. J. Bone Miner. Res. (8) Suppl. 2, S565–572.

96. Lanyon, L.E. (1996) Using functional loading to influence bone mass and architecture: objectives, mechanisms, and relationship with estrogen of the mechanically adaptive process in bone. Bone 18, 37S–43S.

97. Neidlinger-Wilke, C., Wilke, H.J. and Claes, L. (1994) Cyclic stretching of human osteoblasts affects proliferation and metabolism: a new experimental method and its application. J. Orthop. Res. 12, 70–78.

98. Fitzgerald, J., Dietz, T.J. and Hughes-Fulford, M. (2000) Prostaglandin E2-induced up-regulation of c-fos messenger ribonucleic acid is primarily mediated by 3′,5′-cyclic adenosine monophosphate in MC3T3-E1 osteoblasts. Endocrinology 141, 291–298.

99. Wang, J. and Walsh, K. (1996) Resistance to Apoptosis conferred by Cdk inhibitors during myocyte differentiation. Science 273, 359–361.

100. Hatton, J.P., Pooran, M., Li, C.-F., Luzzio, C. and Hughes-Fulford, M. A short pulse of mechanical force induces gene expression and growth in MC3T3-E1 osteoblasts via an ERK $\frac{1}{2}$ pathway. JBMR, in press

Cell Biology and Biotechnology in Space
A. Cogoli (editor)
© 2002 Elsevier Science B.V. All rights reserved

Osteogenesis in Altered Gravity

Ranieri Cancedda[ab] and Anita Muraglia[a]

[a]*Centro di Biotecnologie Avanzate, Istituto Nazionale per la Ricerca sul Cancro, Largo Rosanna Benzi 10, 16132 Genova, Italy; and* [b]*Dipartimento di Oncologia, Biologia e Genetica, Università di Genova, 16132 Genova, Italy*

Introduction

Bone is continuously remodelled

Bone is continuously remodelled in the lifespan of any vertebrate, including man. Bone turnover is the result of the balance between the activities of two different cell populations, the osteoclasts and the osteoblasts, responsible for the old bone resorption and for the new bone deposition, respectively.

Osteoclasts are formed through the differentiation and fusion of monocytes (specific white blood cells). The characteristic feature of osteoclasts is a ruffled edge where active resorption takes place. The osteoclasts secrete bone-reabsorbing enzymes, which digest bone matrix.

Osteoblasts are derived from progenitor cells present in the bone marrow, which migrate to the resulting pits, and secrete and organize a new protein matrix where eventually calcium mineral is deposited. In this way the new bone is formed and the old bone is replaced. Osteoblasts secrete a collagen-rich extracellular matrix, defined as osteoid, essential to later mineralization. Alkaline phosphatase is contained in osteoblasts and secreted during osteoblastic activity. Osteoblasts cause calcium salts and phosphorus to precipitate from blood as hydroxyapatite and other crystals; these minerals bond with the newly formed osteoid to mineralize the bone tissue. Osteoblasts that have been trapped in the osteoids are called osteocytes. Osteocytes maintain bones, and play a role in controlling the extracellular concentration of calcium and phosphate.

Physiologically, bone density varies during the whole life. Bone density increases in childhood and adolescence, and reaches its peak at the age of 25–30. Bone density then remains almost the same for 20–25 years. Around the age of 50, bone density starts to progressively decrease and reaches approximately 85% of the initial value in men and 70% in women.

The mode of differentiation, recruitment and inhibition of bone cells is controlled by numerous hormones and growth factors. Some hormones circulating in blood have effects on the whole skeleton, by controlling the number of active

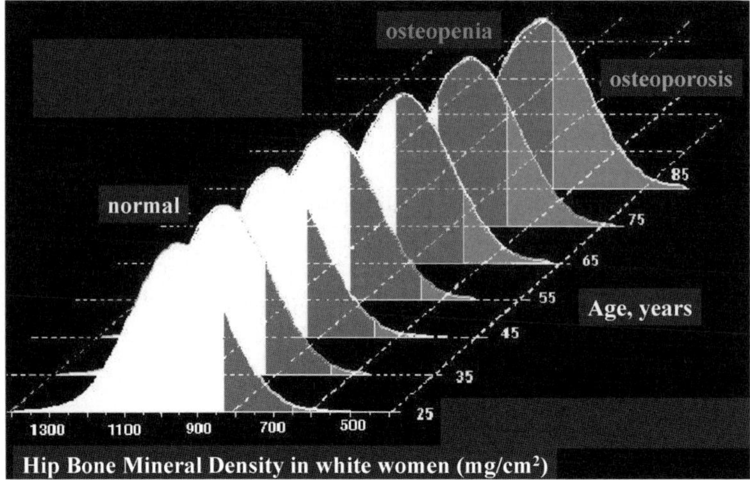

Fig. 1. Permanence in space is like ageing: hip bone mineral density in white women (mg/cm^2).

osteoclasts and osteoblasts. In particular, estrogen plays a major role. Osteoblasts have estrogen receptors; estrogens can increase the number of osteoblasts, thus increasing collagen production. Also osteoclasts have estrogen receptors; indeed estrogens can inhibit their recruitment. A decreased level of estrogen at female menopause can induce a reduced osteoblast bone formation activity and an increased bone resorption, which in turn can lead to osteoporosis (Fig. 1).

Other important hormones are Vitamin D and Parathyroid Hormone (PTH). Vitamin D facilitates osteoclast recruitment, and also plays a role in the mineralization of bone matrix. A lack of Vitamin D results in impaired mineralization (osteomalacia) and too much Vitamin D results in bone loss.

The PTH can increase the recruitment and the activity of osteoblasts and osteoclasts. Therefore, if PTH secretion is too high, an acceleration of the bone turnover occurs. If the increase occurs along a Vitamin D deficiency, the bone cycle is accelerated and entails bone loss. Osteocytes are directly inhibited by PTH, whereas they are stimulated by calcitonin.

Growth factors locally synthesized by skeletal and adjacent cells have effects mainly on the skeletal area where they are released. In some cases circulating hormones act by modulating synthesis, activation, receptor binding, and binding properties of local growth factors.

Bone loss in space

Mechanical loading is essential to the maintenance of a correct bone mass. It is common knowledge that in healthy individuals, a complete immobilisation results in a serious loss of bone mass together with a decreased structural mechanical efficiency.

The first reports of bone loss in humans due to weightlessness during permanence in space came from the Soviet/Russian missions. A significant bone mass decrease was observed after missions lasting up to six months, with the bone loss being roughly proportional to the length of the flight.

When bone loss was examined by measuring bone mineral densities (a measure of bone quality) of different bones such as the forearm (radius) and the heel bone (calcaneus), major differences were observed. Demineralization was always higher in weight-bearing bones. Six months after landing, the recovery was complete in the peripheral region of bones, characterised by low remodelling rate, but it was not yet complete in the central area (cancellous bone), characterised by a high remodelling rate.

Cellular and molecular basis for the observed bone loss

As yet, cellular and molecular mechanisms responsible for the observed bone loss in space are only partially understood. In principle, the changes observed in bone mass could be the consequence either of a decreased osteoblast activity (due to a decreased proliferation and/or differentiation) or of an increased osteoclast activity, or both.

It is worth noting that, in space, several endocrine systems, including the system controlling bone metabolism and turnover, undergo modifications similar to the ones occurring during senescence.

Ground studies (microgravity simulation)

In order to have weightless conditions for prolonged periods of time, it is necessary to go into space. In fact, only in space are "free-fall" conditions present as a result of the existing balance between gravitational and inertial forces. Unfortunately, the number of accesses to orbiting space-vehicles has so far been very limited. Therefore scientists have devised experimental environments where at least some of the conditions found in space can be simulated.

Cells

Different methods, such as hypotonic swelling, stretching or bending of the cell substratum, fluid shear stress, hypergravity and dynamic strain have been used to mechanically stimulate cultured bone cells [1–8]. The cell response is highly dependent both on the extent and frequency of the applied forces and on the nature of bone cells. In several cases, mechanical loading by pulsating fluid flow has been performed. Load-induced compression may result in fluid flow in the marrow cavity, the trabecular network and the osteocytic canalicular system [9–12].

Overall, high strain levels seem to favour proliferation of bone cell progenitors whereas more physiological forces promote osteoblast differentiation and mineralization of the bone matrix [2,13–17]. The mechanisms involved in mechanotransduction in bone remain uncertain. Mechanical loading may induce stretching of bone cells, create extracellular fluid flows and generate cell responses such as intracellular messages playing a role in the releasing of specific growth factors and soluble modulators.

It has been proposed that stretch and compressive forces increase PGE2 synthesis [6,7,18]. PGE2 was found not only to stimulate bone cell proliferation and bone matrix synthesis, but also to increase the recruitment of osteogenic progenitor cells in the bone marrow. PGE2 increases the local expression of insulin-like growth factors (IGFs); thus, PGE2 may play an important role as mediator of the anabolic effects of mechanical strain on osteogenic cells.

Pulsating fluid flow induces a rapid and transient increase in nitric oxide (NO) which is critical to PGE2 release [19]. Apparently, activation of protein kinase C (PKC) mediates the effects of PGE2 release [20]. Activation of phospholipase C, leading to activation of protein kinase C, may be responsible for the increase observed in the intracellular levels of inositol triphosphate (IP3), another important second messenger in cells, and in the release of calcium from intracellular stores [21–23] Strain was found to increase IP3 levels in osteoblastic cells [21] and the inositol phosphate pathway appears to be involved in the mechanical-strain-induced proliferation of bone cells [24]. The finding that the increase in calcium precedes the rise in IP3 in osteoblasts suggests that calcium is the initial cellular signal response to mechanical strain.

Mechanically stimulated bone cells also increase the expression of other factors, such as TGF-β1 [25]. It has been proposed that the strain-induced increased TGF-β affects early expressing genes and subsequently bone cell growth and differentiation (reviewed in Ref. [26]).

Bone cells could respond to cyclic mechanical strain via mechano-sensitive channels. Cation-selective channels, responding to parathyroid hormone and to mechanical strength have been described in osteosarcoma cells [27,28].

Bone cells could respond to mechanical strain also via cell–matrix interactions [14,29–31]. The interconnections between extracellular matrix, integrins, and the cytoskeleton are involved in the transmission of external signals to the nucleus, leading to control cell division and maturation.

In summary, the data available indicate that mechanical strain may have several effects on bone cells. Overall, strain and microgravity appear to affect cell proliferation or differentiation. Multiple associated and coordinated pathways within the cell may be activated in response to strain. Implicated mechanisms include: (i) increased production of prostaglandins and growth factors such as IGFs TGF-βs; (ii) changes in the intracellular levels of second messengers such as cAMP, IP3 and free calcium; and (iii) alterations of the cytoskeleton and of the cell–matrix interactions.

Tissues and organs

Cultured fetal mouse long bones and calvariae exposed to intermittent compressive forces or to hypergravity field show an increased mineralization. The mechanisms involved are not fully understood, but this is most probably the consequence of both a

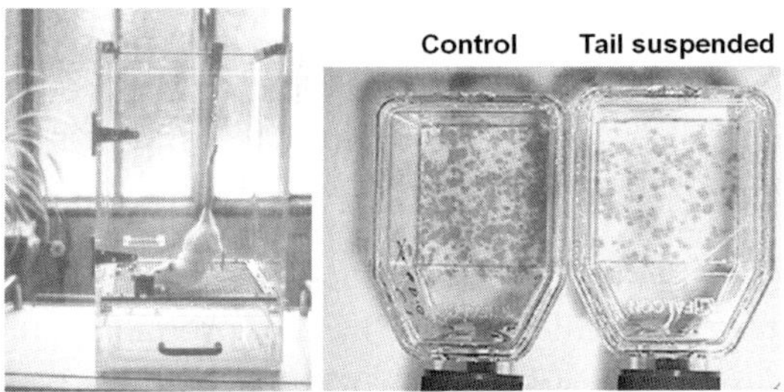

Fig. 2. The number of osteoblast progenitors is decreased in unloaded bones.

decrease in mineral resorption and osteoclast number and an increase in bone formation [32–35]. Mechanical forces promote the release or activation of important factors involved in the control of bone metabolism such as Transforming Growth Factor-β, prostaglandin E and prostacyclin PGI2 [36–39]. It is worth noting that mechanical loading and estrogens in combination have more than additive effects on cell division and collagen synthesis in organ cultures of rat bones, suggesting interactions between strain and estrogens [40,41].

Animals

Bone changes in hind-limb unloaded rats have been reviewed [42]. A sequence of structural changes and molecular events has been described [43–46]. In hind limbs of tail-suspended rats, a decrease in bone formation is observed after one week. This is the consequence of both a decrease in the number of osteoblast progenitors at trabecular and periosteal levels, and an increase in bone resorption (Fig. 2). After six weeks, a significant bone loss and an altered bone structure (reduced trabecular thickness and number) are observed. This results in a decrease in the skeletal structural strength. These alterations are partially reverted by reloading, but the recovery time was always longer than the time of tail suspension and recovery was never complete [47,48].

In tail-suspended animals also changes in the serum levels of hormones such as $1,25(OH)_2D$ and in the expression of growth factors such as of IGF-I and TGF-β2 have been observed. In some cases these changes have been related to the observed changes in bone mineralization and structure; nevertheless none of the data can be considered conclusive.

With the bone mass decrease, impaired biomechanical properties are expected. Hindlimb unloading for more than two weeks results in bone fractures during torsion or bending at lower loads than controls [49,50]. This bone fragility is determined by both the geometry and the material properties of the bone.

Humans

In human bed-rest experiments a progressive loss of mineral from bone, as well as an increase in calcium excretion, were observed during the first four weeks. The calcium

164

loss continued in the following weeks with values corresponding to an average calcium loss of about 0.5% per month with a daily negative balance of 200 to 254 mg/d [51]. At the end of 3–4 month experiments, little change in bone mass was observed, but the bone trabeculae were thicker and less numerous, thus suggesting hyperactive osteoclasts. Reduction in trabecular interconnections may adversely affect bone strength.

The increases in urinary calcium and hydroxyproline excretion were generally interpreted as a sign of increased bone resorption. Indeed, two studies showed increases in bone resorption markers during the first four days of bed rest, with no change in bone formation markers, except for increased osteocalcin levels [52,53].

Spaceflight studies

Cells

Several reports exist on a possible direct effect of microgravity on osteogenic cells. Unfortunately it is very difficult to derive general conclusions since, in most cases, experiments were performed with different cells types, for different periods of time, using different hardware and with completely different experimental design. In addition, not all necessary controls were always included in the experiments. A comprehensive review of the topic is in Ref. [26].

In some cases (MC3T3-E1 mouse calvaria; ROS 17/2.8 rat osteosarcoma; etc), cells showed an abnormal morphology, such as reduced stress fibers and smaller nuclei or cytoplasmic retraction and rounding, as compared to ground controls, see for example Refs. [54] and [55], but in other cases (MG-63 osteosarcoma cells) no differences were found in cell morphology between unit-gravity, microgravity and in-flight unit-gravity cultures [56].

ALP activity and collagen type I synthesis, measured biochemically in response to hormones and growth factors, were generally slowed down, thus suggesting that microgravity reduced the differentiation of osteoblasts [56], but in the case of ROS 17/2.8 rat osteosarcoma cells, ALP activity may have increased [57].

Microgravity was reported to increase PGE2 and IL-6 production and to alter the expression of other growth factors and related proteins, such as insulin-like growth factor binding proteins and glucocorticoid receptor [58]. The precise sequence of events and the involved intracellular signaling remain unclear.

In summary, proliferation of osteogenic cells seems not to be altered in microgravity conditions, whereas their differentiation (i.e. production of bone matrix protein) seems to be affected.

Tissues and organs

Some studies have been performed to examine the effects of microgravity exposure on explanted fetal mouse cartilaginous long bones. Bones were cultured during three Space Shuttle flights, to study cartilage growth, the differentiation to form bone and the process of mineralization [59]. Exposure to gravity resulted in a decreased growth in length. Calcium incorporated into the mineralized areas was significantly lower and bone resorption was increased.

Animals

Most animal observations have been made with rats returning from 4-day to 21-day space flights. Only small groups of animals were studied and, since rats were sacrificed at various time-periods after flight, they were exposed to additional stresses related to landing.

It is difficult to interpret these experiments; in fact in addition to the above, results of these studies were limited by the differences in strain and age of the animals and in the type of cage. An alteration in the bone trabeculae organisation occurred as early as the end of the first week [60–63]. Nevertheless some general conclusions can be derived. After two weeks, a significant loss of bone was observed, mostly due to a sudden and transient increase in bone resorption. After three weeks, also bone formation was decreased [64,65]. In all animals, recovery of bone formation occurred after return to Earth, although the altered bone trabeculae organisation often persisted [64,66].

Few studies present data on gene expression. Bikle et al. [67] showed an increased mRNA level for IGF-I and IGF-IR and alkaline phosphatase, but a decreased mRNA level for osteocalcin after a six-day flight. In an 11-day flight, the levels of TGF-β were reduced relative to the ground controls in the hind-limb periosteum, but were not significantly altered in the tibia growth zone [68].

The combined effect of lack of estrogen and microgravity has been studied in experiments in which ovariectomized rats were subjected to orbital spaceflight [69]. The ovariectomized rats returning from space showed an accentuated bone loss that was not observed in estrogen deficient rats submitted to treadmill running.

In a few Russian experiments monkeys were flown. After a two week space flight a severe bone loss in the primate iliac bone was observed, as well as a decrease in mineralization activity and an altered mineralization pattern.

Humans

Since the first observation of an increase of urinary calcium made in 1962 during the 2–3 hour Vostok III and Vostok IV flights [70], several reports have been published showing a reduced bone density and a decreased bone mass in astronauts and cosmonauts remained in space for either short (few days) or long (several months) duration missions. For a more detailed report on the resulting observations and for a more complete reference list, see Ref. [71].

Often quantitative computerized tomodensitometry (QCT) was used to determine peripheral (cortical) and central (trabecular) bone mass. When it was investigated the effect of microgravity was different at different skeletal sites. Maximal response was observed in bones that are gravity loaded on Earth.

In most cases, the published data have been related to a limited number of cosmonauts and large individual variations have been observed. More recently, bone mineral density measurements in 15 cosmonauts who had stayed in the Russian Mir space station for one month (two cosmonauts), two months (two cosmonauts), and six months (11 cosmonauts), have been reported [72]. Measurements were performed both of the forearm (radius) and of the shin bone (tibia). Bone deterioration, especially shin bone, continued to increase during flight.

After a recovery period comparable to the duration of the space mission, forearm bone density had not significantly changed. On the contrary, the weight-bearing shin bone, regardless of the cosmonauts' physical training during the missions, still showed a central bone loss after one month and a peripheral bone loss, after two months of recovery.

Biotechnology and application perspectives

Osteoporosis research

Over the last century, we have experienced an extraordinary extension of the average human life. Old age is associated with a decreased bone mass and damaged bone structure.

Osteoporosis, defined by reduced bone loss with a high propensity of fracture, is indeed a major public health problem and represents a huge medical and social cost: more than 9 billion Euros per year in European countries excluding Russia. Osteoporosis occurs when the body fails to form enough new bone and/or when too much bone is reabsorbed. This bone loss occurs over several years, a fracture of vertebrae, wrists, or hips being the first sign of a pathology of which the patient was not aware.

There is a genetic and a gender predisposition to osteoporosis. Almost 25% of white and Asian women over the age of 50 have osteoporosis and about 50% have osteopenia, a low bone density situation that may eventually result in osteoporosis. More than 50% of women over 50 will suffer from a fracture of the hip, wrist or vertebra. The fracture risk for an age matched male population is about 12–15%.

The prime cause of osteoporosis is the reduction in estrogen due to loss of ovarian function, in post-menopausal women. Other causes include endocrinological disorders, immobilisation, bone malignancies, some genetic diseases, a low amount of calcium in the diet, eating disorders, heavy alcohol consumption, smoking, and the use of certain drugs, such as steroids.

Alterations similar to the ones associated with osteoporosis are observed after prolonged exposure to microgravity conditions. However, in the latter case, the changes occur at a highly accelerated rate and in healthy individuals. It has been proposed that experimentation in microgravity could be performed to investigate mechanisms responsible for old age bone loss and osteoporosis occurrence, both at the molecular/cellular level and at the whole organism level.

Development of study models

An objective of future space research in osteobiology will be to provide new tools for the diagnostic, prevention and treatment of osteoporosis and related bone diseases.

Specific goals will be the understanding of their underlying mechanisms, including genetics. Animal-based, and possibly cell-based, research models should be developed, taking advantage of the accelerated bone loss process in microgravity, which offers an excellent tool for such studies.

Engineered tissue models

Tissue engineering can be defined as "an interdisciplinary field that applies the principles of engineering and the life sciences toward the development of biological substitutes that restore, maintain or improve tissue function." [73]. Recent advances in cell biology and biomaterial science are giving new impetus to this discipline. Cells and extracellular matrix are the main components for successful tissue engineering. Biomaterials mimic the extracellular matrix, support cell growth and differentiation and play a major role in delivering growth factors.

Several laboratories are trying to reconstruct *in vitro* skeletal tissues to be eventually used in space as a model for investigating microgravity effects. So far, the best results have been obtained with tissue engineering of cartilage.

Tissue engineering of cartilage, i.e., the *in vitro* culture of cartilage cells on synthetic polymer scaffolds, was studied on the Mir Space Station and on Earth. Three-dimensional cell–polymer constructs consisting of articular chondrocytes and polyglycolic acid scaffolds were grown in rotating bioreactors, first for three months on Earth and then for four additional months on either the Mir Space Station or Earth (1 *g* control). Both environments yielded cartilaginous constructs, but, compared with the Earth group, Mir-grown constructs were more spherical, smaller, and mechanically inferior [74].

In vitro reconstruction of bone appears to be a much more difficult task, especially due to the difficulty of reproducing *in vitro* an efficient vascularization of the newly formed bone.

We have proposed an experimental model for human bone formation in unloaded conditions which combines some *in vitro* and some *in vivo* features. Bone formation has been assessed by implanting human Bone Marrow Stromal Cells (BMSC), previously expanded *in vitro* and adsorbed on porous hydroxyapatite (HA) bioceramics, subcutaneously in immunodeficient mice (nude mice) (Fig. 3). In this system human bone formation and remodelling occurs and can be studied in unloaded conditions i.e. with no influence of muscle tension [75].This system allows bone formation and resorption to be monitored in conditions virtually free of mechanical stress and to discriminate between the effects on bone metabolism of the animal being in a microgravity environment and of the absence of muscle tension (present also on ground).

A feasible goal for the development of an exclusively *in-vitro* model should be limited and may be defined as follows: (i) the model shall allow for a 3D co-culture on resorbable biomaterials of the bone-forming osteoblasts and of the bone-resorbing osteoclasts; (ii) in order to mimic some of the vascularization aspects, the model shall allow the controlled flow of fluids both outside and inside the biomaterials.

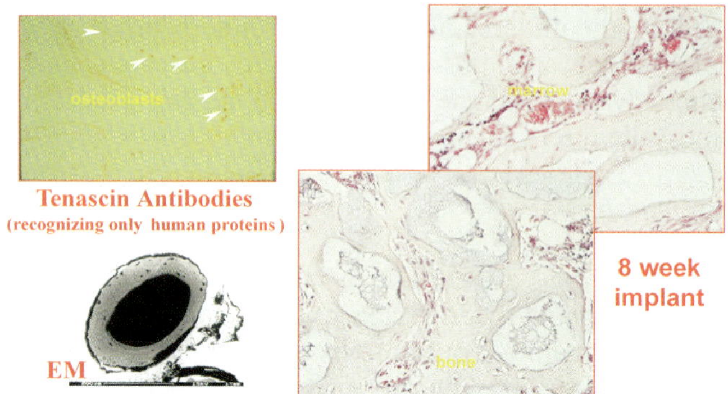

Fig. 3. Human bone formation in the immunodeficient mouse.

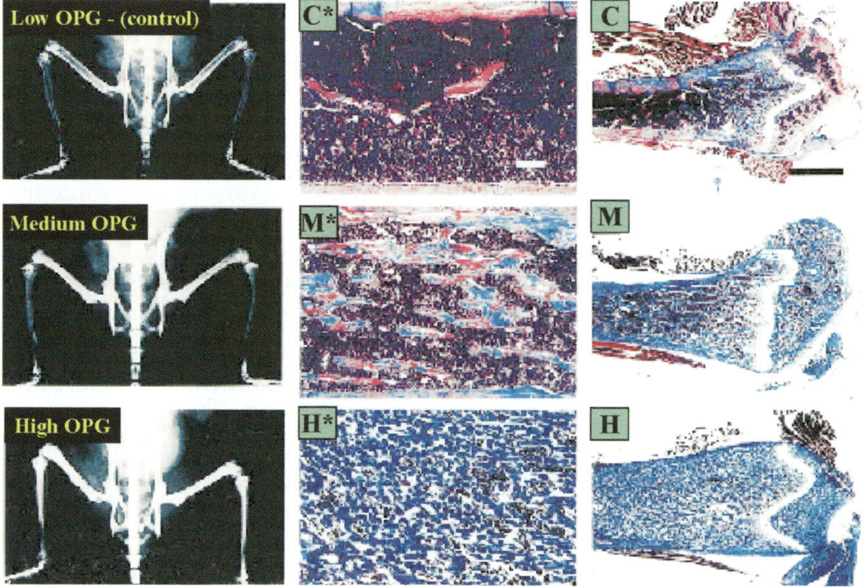

Fig. 4. Increased bone density in mice overexpressing osteoprotegerin (OPG). (From Simonet et al., 1989).

Once developed, and the potential impact of the microgravity proven, the model will be very useful to understand the basic biology of the bone cells not exposed to mechanical load, i.e. the mechanisms of cell growth, cell interaction and cell sensitivity to the mechanical strength.

In addition, a model like this will allow study of the drug efficiency either in enhancing osteoblast activity or in restoring a correct osteoblast/osteoclast activity balance

Animal models

The possibility of manipulating the mouse genoma and of introducing mutations in specific genes is one of the most important advances in modern molecular genetics. These technologies have increased the number of available vertebrate mutants. This approach is becoming the main research tool for determining the potential activities of the new genes discovered by the human genome sequence (genomic-functional analysis).

Thanks to the extraordinary advances in molecular biology and genetics, it is possible today to create animals where either additional copies of certain genes (transgenic genes) are introduced into the genoma, or specific genes are inactivated (knock out genes). Several 'transgenic' and 'knock-out' mice have been obtained that mimic human bone and cartilage pathologies, and where a particular gene responsible for bone development and remodelling has been either modified or inactivated. Some of these mice are characterized by an increased or decreased bone density. Examples of such mice are given in Table 1 and Fig. 4. Studies on the transgenic and knock-out phenotype of these mice, maintained in microgravity, will help revealing mechanisms controlling bone formation and turnover, both in normal and in osteoporotic situations.

Table 1

Altered bone density in "transgenic" and "knock-out" mice

Gene	Type of mutant	Reference
Osteopetrotic phenotype		
Osteoprotegerin (OPG)	transgenic	Simonet et al. (1989) Cell 89: 309–319
TNF receptor-associated factor 6 (TRAF6)	knock-out	Lomaga et al. (1999) Genes Dev. 13: 1015–1024
c-fos	knock-out	Johnson et al. (1992) Cell 71: 577–586
c-sarc	knock-out	Boyce et al. (1992) J. Clin. Invest. 90: 1622–1627
microphthalmia (mi) locus	mutant	Hodgkinson (1993) Cell 74: 395–404
myeloid- and B-cell-specific transcription factor PU.1	knock-out	Tondravi et al. (1997) Nature 386: 81–84
Osteosclerotic phenotype		
Leptin (lep)	knock-out	Ducy et al. (2000) Cell 100: 197–207
cell matrix- associated heparin-binding growth-associated molecule (HB-GAM)	transgenic	Imai et al. (1998) J. Cell Biol. 143: 1113–1128
Cbfa1	transgenic	unpublished
Osteoporotic phenotype		
Cbfa1	transgenic	Ducy et al. (1999) Genes Dev. 13: 1025–1036
Osteoprotegerin (OPG)	knock-out	Bucay et al. (1998) Genes Dev. 12: 1260–1268

Development of countermeasures for astronauts

Another long-term objective of these studies is to develop new related counter-measures for space flights. This appears particularly important since it is expected that, when the International Space Station is fully operational, many astronauts will remain in space for several months.

The mechanisms of bone loss in space are only partially understood. All astronauts devote several hours each day to physical exercise in the attempt to prevent bone loss. Unfortunately, the increase in muscle activity is only partially effective.

So far, also other countermeasures appear to be unsatisfactory. Treatment with hormones and growth factors, known to play a role in bone formation and turnover, is partially effective [47,48,76,77]. Under certain conditions, high diet calcium can increase bone mass and mineral contents in normally unloaded animal bones, but does prevent a decrease in the mass of unloaded bones, otherwise loaded, as compared to control (remained loaded) bones of animals subjected to the same diet [78]. Administration of inhibitors of bone resorption, such as bisphosphonates, currently used for the treatment of human osteoporosis, does reduce the trabecular bone loss in the unloaded bones, but it results in an alteration of the bone structure (Barou et al. unpublished).

Conclusions

Bones are continuously remodelled as a result of the coupled bone resorption and bone formation activities. When these two activities cease to be in dynamic equilibrium, excess bone may be reabsorbed and/or insufficient new bone is formed. The progressive decrease in the skeletal mass can result in osteoporosis, a pathological condition characterised by a significant reduction in bone density, with an associated deterioration in the bone micro-architecture. The consequence is an increased risk of fractures and skeletal deformation. Osteoporosis is a debilitating disease with devastating health and economic consequences.

Permanence in space induces bone loss in both humans and animals. Bone is mainly lost from the bones that are mostly loaded by gravity when on Earth. Interesting similarities exist between bone alterations observed in osteoporosis and in space. The mechanisms involved in these alterations need to be investigated at the cellular and molecular levels and also at the level of the whole organism.

The data available on the effects of mechanical stress and of microgravity on cultured bone cells indicate that some of these cells may directly respond to stress and to unloading. Nevertheless present data should be considered with caution, due to the limited number of studies so far performed, and the constraints of the experimental design, especially in space.

Microgravity exposure represents the only condition where the absence of gravity acts on the entire body. Nevertheless, hind-limb elevation studies with animals and bed-rest studies with human volunteers have been used as models to investigate mechanisms of bone loss under unloading.

Despite the fact that, so far, experiments with animals in space have been very limited in number, the results obtained have shown the value and relevance of animal use. The present availability of 'transgenic' and gene 'knock-out' mouse mutants, in which particular genes responsible for bone development and remodeling have been modified or inactivated, will be particularly useful for understanding the genetic component of both bone loss in space and osteoporosis.

It is hoped that the information acquired will lead not only to measures to prevent defective bone formation and bone loss in future long-duration space flights but also to develop tools for prevention and treatment of human osteoporosis.

Acknowledgments

Partially supported by funds from the Italian Space Agency (ASI) and the European Space Agency (ESA; ERISTO project).

References

1. Brighton, C.T., Strafford, B., Gross, S.B., Leatherwood, D.F., Williams, J.L. and Pollack, S.R. (1991) The proliferative and synthetic response of isolated calvarial bone cells of rats to cyclic biaxial mechanical strain. J. Bone Joint Surg. [Am] 73, 320–331.
2. Buckley, M.J., Banes, A.J., Levin, L.G., Sumpio, B.E., Sato, M., Jordan, R., Gilbert, J., Link, G.W. and Tran Son Tay, R. (1988) Osteoblasts increase their rate of division and align in response to cyclic, mechanical tension *in vitro*. Bone Miner. 4, 225–236.
3. Carvalho, R.S., Scott, J.E., Suga, D.M. and Yen, E.H. (1994) Stimulation of signal transduction pathways in osteoblasts by mechanical strain potentiated by parathyroid hormone. J. Bone Miner. Res. 9, 999–1011.
4. Nishioka, S., Fukuda, K. and Tanaka, S. (1993) Cyclic stretch increases alkaline phosphatase activity of osteoblast-like cells: a role for prostaglandin E2. Bone Miner. 21, 141–150.
5. Nose, K. and Shibanuma, M. (1994) Induction of early response genes by hypergravity in cultured mouse osteoblastic cells (MC3T3-E1). Exp. Cell Res. 211, 168–170.
6. Ozawa, H., Imamura, K., Abe, E., Takahashi, N., Hiraide, T., Shibasaki, Y., Fukuhara, T. and Suda, T. (1990) Effect of a continuously applied compressive pressure on mouse osteoblast-like cells (MC3T3-E1) in vitro. J. Cell Physiol. 142, 177–185.
7. Yeh, C.K. and Rodan, G.A. (1984) Tensile forces enhance prostaglandin E synthesis in osteoblastic cells grown on collagen ribbons. Calcif. Tissue Int. 36, S67–71.
8. Zaman, G., Suswillo, R.F., Cheng, M.Z., Tavares, I.A. and Lanyon, L.E. (1997) Early responses to dynamic strain change and prostaglandins in bone-derived cells in culture. J. Bone Miner. Res. 12, 769–777.
9. Burger, E.H. and Klein-Nulend, J. (1998) Microgravity and bone cell mechanosensitivity. Bone 22, 127S–130S.
10. Cowin, S.C., Weinbaum, S. and Zeng, Y. (1995) A case for bone canaliculi as the anatomical site of strain generated potentials. J. Biomech. 28, 1281–1297.
11. Duncan, R.L. and Turner, C.H. (1995) Mechanotransduction and the functional response of bone to mechanical strain. Calcif. Tissue Int. 57, 344–358.

172

12. Eriksson, C. (1985) Surface energies and the bone induction principle. J. Biomed. Mater. Res. 19, 833–849.

13. Buckley, M.J., Banes, A.J. and Jordan, R.D. (1990) The effects of mechanical strain on osteoblasts *in vitro*. J. Oral Maxillofac. Surg. 48, 276–282; discussion 282–283.

14. Harter, L.V., Hruska, K.A. and Duncan, R.L. (1995) Human osteoblast-like cells respond to mechanical strain with increased bone matrix protein production independent of hormonal regulation. Endocrinology 136, 528–535.

15. Hasegawa, S., Sato, S., Saito, S., Suzuki, Y. and Brunette, D.M. (1985) Mechanical stretching increases the number of cultured bone cells synthesizing DNA and alters their pattern of protein synthesis. Calcif. Tissue Int. 37, 431–436.

16. Mikuni-Takagaki, Y., Suzuki, Y., Kawase, T. and Saito, S. (1996) Distinct responses of different populations of bone cells to mechanical stress. Endocrinology 137, 2028–2035.

17. Veldhuijzen, J.P., Bourret, L.A. and Rodan, G.A. (1979) *In vitro* studies of the effect of intermittent compressive forces on cartilage cell proliferation. J. Cell Physiol. 98, 299–306.

18. Somjen, D., Binderman, I., Berger, E. and Harell, A. (1980) Bone remodelling induced by physical stress is prostaglandin E2 mediated. Biochim. Biophys. Acta 627, 91–100.

19. Klein-Nulend, J., van der Plas, A., Semeins, C.M., Ajubi, N.E., Frangos, J.A., Nijweide, P.J. and Burger, E.H. (1995b) Sensitivity of osteocytes to biomechanical stress *in vitro*. FASEB J. 9, 441–445.

20. Reich, K.M. and Frangos, J.A. (1993) Protein kinase C mediates flow-induced prostaglandin E2 production in osteoblasts. Calcif. Tissue Int. 52, 62–66.

21. Jones, D.B., Nolte, H., Scholubbers, J.G., Turner, E. and Veltel, D. (1991) Biochemical signal transduction of mechanical strain in osteoblast-like cells. Biomaterials 12, 101–110.

22. Reich, K.M. and Frangos, J.A. (1991) Effect of flow on prostaglandin E2 and inositol trisphosphate levels in osteoblasts. Am. J. Physiol. 261, C428–432.

23. Sandy, J.R., Meghji, S., Farndale, R.W. and Meikle, M.C. (1989) Dual elevation of cyclic AMP and inositol phosphates in response to mechanical deformation of murine osteoblasts. Biochim. Biophys. Acta 1010, 265–269.

24. Brighton, C.T., Sennett, B.J., Farmer, J.C., Iannotti, J.P., Hansen, C.A., Williams, J.L. and Williamson, J. (1992) The inositol phosphate pathway as a mediator in the proliferative response of rat calvarial bone cells to cyclical biaxial mechanical strain. J. Orthop. Res. 10, 385–393.

25. Zhuang, H., Wang, W., Tahernia, A.D., Levitz, C.L., Luchetti, W.T. and Brighton, C.T. (1996) Mechanical strain-induced proliferation of osteoblastic cells parallels increased TGF-beta 1 mRNA. Biochem. Biophys. Res. Commun. 229, 449–453.

26. Marie, P.J., Jones, D., Vico, L., Zallone, A., Hinsenkamp, M. and Cancedda, R. (2000) Osteobiology, strain, and microgravity: part I. Studies at the cellular level. Calcif. Tissue Int. 67, 2–9.

27. Duncan, R. and Misler, S. (1989) Voltage-activated and stretch-activated Ba2+ conducting channels in an osteoblast-like cell line (UMR 106). FEBS Lett. 251, 17–21.

28. Duncan, R.L., Hruska, K.A. and Misler, S. (1992) Parathyroid hormone activation of stretch-activated cation channels in osteosarcoma cells (UMR-106.01). FEBS Lett. 307, 219–223.

29. Salter, D.M., Robb, J.E. and Wright, M.O. (1997) Electrophysiological responses of

human bone cells to mechanical stimulation: evidence for specific integrin function in mechanotransduction. J. Bone Miner. Res. 12, 1133–1141.

30. Shyy, J.Y. and Chien, S. (1997) Role of integrins in cellular responses to mechanical stress and adhesion. Curr. Opin. Cell Biol. 9, 707–713.

31. Toma, C.D., Ashkar, S., Gray, M.L., Schaffer, J.L. and Gerstenfeld, L.C. (1997) Signal transduction of mechanical stimuli is dependent on microfilament integrity: identification of osteopontin as a mechanically induced gene in osteoblasts. J. Bone Miner. Res. 12, 1626–1636.

32. Klein-Nulend, J., Semeins, C.M., Veldhuijzen, J.P. and Burger, E.H. (1993) Effect of mechanical stimulation on the production of soluble bone factors in cultured fetal mouse calvariae. Cell Tissue Res. 271, 513–517.

33. Klein-Nulend, J., Veldhuijzen, J.P. and Burger, E.H. (1986) Increased calcification of growth plate cartilage as a result of compressive force in vitro. Arthritis Rheum. 29, 1002–1009.

34. Klein-Nulend, J., Veldhuijzen, J.P., de Jong, M. and Burger, E.H. (1987) Increased bone formation and decreased bone resorption in fetal mouse calvaria as a result of intermittent compressive force *in vitro*. Bone Miner. 2, 441–448.

35. Klein-Nulend, J., Veldhuijzen, J.P., van Strien, M.E., de Jong, M. and Burger, E.H. (1990) Inhibition of osteoclastic bone resorption by mechanical stimulation *in vitro*. Arthritis Rheum. 33, 66–72.

36. Dallas, S.L., Zaman, G., Pead, M.J. and Lanyon, L.E. (1993) Early strain-related changes in cultured embryonic chick tibiotarsi parallel those associated with adaptive modeling in vivo. J. Bone Miner. Res. 8, 251–259.

37. El Haj, A.J., Minter, S.L., Rawlinson, S.C., Suswillo, R. and Lanyon, L.E. (1990) Cellular responses to mechanical loading *in vitro*. J. Bone Miner. Res. 5, 923–932.

38. Klein-Nulend, J., Roelofsen, J., Sterck, J.G., Semeins, C.M. and Burger, E.H. (1995a) Mechanical loading stimulates the release of transforming growth factor-beta activity by cultured mouse calvariae and periosteal cells. J. Cell Physiol. 163, 115–119.

39. Rawlinson, S.C., el-Haj, A.J., Minter, S.L., Tavares, I.A., Bennett, A. and Lanyon, L.E. (1991) Loading-related increases in prostaglandin production in cores of adult canine cancellous bone *in vitro*: a role for prostacyclin in adaptive bone remodeling? J. Bone Miner. Res. 6, 1345–1351.

40. Cheng, M.Z., Zaman, G. and Lanyon, L.E. (1994) Estrogen enhances the stimulation of bone collagen synthesis by loading and exogenous prostacyclin, but not prostaglandin E2, in organ cultures of rat ulnae. J. Bone Miner. Res. 9, 805–816.

41. Cheng, M.Z., Zaman, G., Rawlinson, S.C., Suswillo, R.F. and Lanyon, L.E. (1996) Mechanical loading and sex hormone interactions in organ cultures of rat ulna. J. Bone Miner. Res. 11, 502–511.

42. Morey-Holton, E.R. and Globus, R.K. (1998) Hindlimb unloading of growing rats: a model for predicting skeletal changes during space flight. Bone 22, 83S–88S.

43. Barou, O., Laroche, N., Palle, S., Alexandre, C. and Lafage-Proust, M. H. (1997) Pre-osteoblastic proliferation assessed with BrdU in undecalcified, Epon-embedded adult rat trabecular bone. J. Histochem. Cytochem. 45, 1189–1195.

44. Globus, R.K., Bikle, D.D. and Morey-Holton, E. (1986b) The temporal response of bone to unloading. Endocrinology 118, 733–742.

45. Machwate, M., Zerath, E., Holy, X., Hott, M., Modrowski, D., Malouvier, A. and Marie,

174

P.J. (1993) Skeletal unloading in rat decreases proliferation of rat bone and marrow-derived osteoblastic cells. Am. J. Physiol. 264, E790–799.

46. Vico, L. (1998) Summary of research issues in biomechanics and mechanical sensing. Bone 22, 135S–137S.

47. Matsumoto, T., Nakayama, K., Kodama, Y., Fuse, H., Nakamura, T. and Fukumoto, S. (1998) Effect of mechanical unloading and reloading on periosteal bone formation and gene expression in tail-suspended rapidly growing rats. Bone 22, 89S–93S.

48. Westerlind, K.C., Morey-Holton, E., Evans, G.L., Tanner, S.J. and Turner, R.T. (1996) TGF-β may help couple mechanical strain and bone cell activity *in vivo*. J. Bone Miner. Res. 11 S, 377.

49. Halloran, B.P., Bikle, D.D., Harris, J., Autry, C.P., Currier, P.A., Tanner, S., Patterson-Buckendahl, P. and Morey-Holton, E. (1995) Skeletal unloading induces selective resistance to the anabolic actions of growth hormone on bone. J. Bone Miner. Res. 10, 1168–1176.

50. Vanderby, R., Jr., Vailas, A.C., Graf, B.K., Thielke, R.J., Ulm, M.J., Kohles, S.S. and Kunz, D.N. (1990) Acute modification of biomechanical properties of the bone-ligament insertion to rat limb unweighting. FASEB J. 4, 2499–2505.

51. LeBlanc, A.D., Schneider, V.S., Evans, H.J., Engelbretson, D.A. and Krebs, J.M. (1990) Bone mineral loss and recovery after 17 weeks of bed rest. J. Bone Miner. Res. 5, 843–850.

52. Palle, S., Vico, L., Bourrin, S. and Alexandre, C. (1992) Bone tissue response to four-month antiorthostatic bedrest: a bone histomorphometric study. Calcif. Tissue Int. 51, 189–194.

53. Zerwekh, J.E., Ruml, L.A., Gottschalk, F. and Pak, C.Y. (1998) The effects of twelve weeks of bed rest on bone histology, biochemical markers of bone turnover, and calcium homeostasis in eleven normal subjects. J. Bone Miner. Res. 13, 1594–1601.

54. Hughes-Fulford, M. and Lewis, M.L. (1996) Effects of microgravity on osteoblast growth activation. Exp. Cell Res. 224, 103–109.

55. Guignandon, A., Usson, Y., Laroche, N., Lafage-Proust, M.H., Sabido, O., Alexandre, C. and Vico, L. (1997b) Effects of intermittent or continuous gravitational stresses on cell–matrix adhesion: quantitative analysis of focal contacts in osteoblastic ROS 17/2.8 cells. Exp. Cell Res. 236, 66–75.

56. Carmeliet, G., Nys, G. and Bouillon, R. (1997) Microgravity reduces the differentiation of human osteoblastic MG-63 cells. J. Bone Miner. Res. 12, 786–794.

57. Guignandon, A., Genty, C., Vico, L., Lafage-Proust, M.H., Palle, S. and Alexandre, C. (1997a) Demonstration of feasibility of automated osteoblastic line culture in space flight. Bone 20, 109–116.

58. Kumei, Y., Shimokawa, H., Katano, H., Hara, E., Akiyama, H., Hirano, M., Mukai, C., Nagaoka, S., Whitson, P.A. and Sams, C.F. (1996) Microgravity induces prostaglandin E2 and interleukin-6 production in normal rat osteoblasts: role in bone demineralization. J. Biotechnol. 47, 313–324.

59. Van Loon, J.J., Bervoets, D.J., Burger, E.H., Dieudonne, S.C., Hagen, J.W., Semeins, C.M., Doulabi, B.Z. and Veldhuijzen, J.P. (1995) Decreased mineralization and increased calcium release in isolated fetal mouse long bones under near weightlessness. J. Bone Miner. Res. 10, 550–557.

60. Patterson-Buckendahl, P., Arnaud, S.B., Mechanic, G.L., Martin, R.B., Grindeland,

R.E. and Cann, C.E. (1987) Fragility and composition of growing rat bone after one week in spaceflight. Am. J. Physiol. 252, R240–246.

61. Roberts, W.E. and Morey, E.R. (1985) Proliferation and differentiation sequence of osteoblast histogenesis under physiological conditions in rat periodontal ligament. Am. J. Anat 174, 105–118.

62. Simmons, D.J., Russell, J.E. and Grynpas, M.D. (1986) Bone maturation and quality of bone material in rats flown on the space shuttle 'Spacelab-3 Mission'. Bone Miner. 1, 485–493.

63. Wronski, T.J., Li, M., Shen, Y., Miller, S.C., Bowman, B.M., Kostenuik, P. and Halloran, B.P. (1998) Lack of effect of spaceflight on bone mass and bone formation in group-housed rats. J. Appl. Physiol. 85, 279–285.

64. Morey, E.R. and Baylink, D.J. (1978) Inhibition of bone formation during space flight. Science 201, 1138–1141.

65. Spengler, D.M., Morey, E.R., Carter, D.R., Turner, R.T. and Baylink, D.J. (1983) Effects of spaceflight on structural and material strength of growing bone. Proc. Soc. Exp. Biol. Med. 174, 224–228.

66. Bourrin, S., Palle, S., Genty, C. and Alexandre, C. (1995) Physical exercise during remobilization restores a normal bone trabecular network after tail suspension-induced osteopenia in young rats. J. Bone Miner. Res. 10, 820–828.

67. Bikle, D.D., Harris, J., Halloran, B.P. and Morey-Holton, E. (1994) Altered skeletal pattern of gene expression in response to spaceflight and hindlimb elevation. Am. J. Physiol. 267, E822–827.

68. Westerlind, K.C. and Turner, R.T. (1995) The skeletal effects of spaceflight in growing rats: tissue-specific alterations in mRNA levels for TGF-beta. J. Bone Miner. Res. 10, 843–848.

69. Westerlind, K.C., Wronski, T.J., Ritman, E.L., Luo, Z.P., An, K.N., Bell, N.H. and Turner, R.T. (1997) Estrogen regulates the rate of bone turnover but bone balance in ovariectomized rats is modulated by prevailing mechanical strain. Proc. Nat. Acad. Sci. USA 94, 4199–4204.

70. Hinsenkamp, M., Burny, F., Bourgois, R. and Donkerwolcke, M. (1981) *In vivo* bone strain measurements: clinical results, animal experiments, and a proposal for a study of bone demineralization in weightlessness. Aviat. Space Environ. Med. 52, 95–103.

71. Vico, L., Hinsenkamp, M., Jones, D., Marie, P.J., Zallone, A. and Cancedda, R. (2001) Osteo biology, strain, and microgravity: part II. Studies at the tissue level. Calcif. Tissue Int. 68, 1–10.

72. Vico, L., Collet, P., Guignandon, A., Lafage-Proust, M.H., Thomas, T., Rehaillia, M. and Alexandre, C. (2000) Effects of long-term microgravity exposure on cancellous and cortical weight-bearing bones of cosmonauts [In Process Citation]. Lancet 355, 1607–1611.

73. Langer, R. and Vacanti, J.P. (1993) Tissue engineering. Science 260, 920–926.

74. Freed, L.E., Langer, R., Martin, I., Pellis, N.R. and Vunjak-Novakovic, G. (1997) Tissue engineering of cartilage in space. Proc. Nat. Acad. Sci. USA 94, 13885–13890.

75. Muraglia, A., Martin, I., Cancedda, R. and Quarto, R. (1998) A nude mouse model for human bone formation in unloaded conditions. *Bone* 22, 131S–134S.

76. Jee, W.S., Wronski, T.J., Morey, E.R. and Kimmel, D.B. (1983) Effects of spaceflight on trabecular bone in rats. Am. J. Physiol. 244, R310–314.

77. Lafage-Proust, M.H., Collet, P., Dubost, J.M., Laroche, N., Alexandre, C. and Vico, L. (1998) Space-related bone mineral redistribution and lack of bone mass recovery after reambulation in young rats. Am. J. Physiol. 274, R324–334.
78. Globus, R.K., Bikle, D.D., Halloran, B. and Morey-Holton, E. (1986) Skeletal response to dietary calcium in a rat model simulating weightlessness. J. Bone Miner. Res. 1, 191–197.

Spaceflight Bioreactor Studies of Cells and Tissues

Lisa E. Freed and Gordana Vunjak-Novakovic

Division of Health Sciences and Technology, Massachusetts Institute of Technology, Cambridge, MA, USA

Introduction

Since the early 1970s, human exploration of space has included cell culture studies that have yielded valuable data describing effects of spaceflight on cell morphology, locomotion, growth and function (reviewed in [1]). Early studies have utilized simple systems in which the cells could be inoculated, cultured, and sampled during spaceflight, but the culture environment could not be precisely controlled. During the 1990s, biological research in space has been advanced by the use of bioreactors, i.e. vessels that provide for gas and medium exchange and for mass transfer, and therefore represent a more precisely controlled tissue culture environment. This chapter briefly reviews some of the landmark spaceflight cell culture studies done using simple systems and then describes four spaceflight bioreactors and their application in five representative spaceflight bioreactor studies of cells and tissues.

Cogoli et al. [2] studied lymphocyte activation in microgravity in a three-day experiment aboard the space transportation system (STS, i.e. space shuttle) mission STS-9 in 1983. Lymphocytes were selected as a model cell system because the exposure of humans to spaceflight causes immune dysfunction (reviewed in [3]), and because ground-based studies showed that lymphocyte proliferation was decreased in simulated microgravity (cells cultured in a rapidly rotating clinostat [4]) and increased in hypergravity (cells cultured in a centrifuge at 10 g [5]). In the 1983 spaceflight study, human peripheral blood lymphocytes were cultured either in microgravity or at 1 g in temperature-controlled, medium-filled chambers developed by the European Space Agency (ESA) [2]. Lymphocyte activation by concanavalin A was assessed by radiolabeled tracer incorporation into DNA. The activation of lymphocytes cultured in microgravity was only three percent that of controls cultured at 1 g. This finding supported the hypothesis that microgravity inhibits cell proliferation, as did other spaceflight studies utilizing a human embryonic lung cell line [6] and human kidney cells [7].

Pellis et al. [8] studied lymphocyte locomotion in microgravity in a two-day experiment aboard STS-56 in 1993. The cell preparation consisted of human peripheral blood lymphocytes sandwiched between two layers of gelled collagen type I and cultured either in microgravity or at 1 g in temperature-controlled, medium-filled chambers developed at the National Aeronautics and Space Administration's Johnson Space Center (NASA-JSC). Cell migration from the cell-collagen interface was assessed using a microscope. The locomotion of lymphocytes cultured in microgravity was markedly reduced compared to controls cultured at 1 g. In a separate ground-based experiment, cell locomotion through collagen type I beads was markedly reduced in rotating bioreactors (the Slow Turning Lateral Vessel, STLV, Synthecon, Houston, TX) as compared to static controls. These findings are consistent with the hypothesis that the culture environment in rotating bioreactors operated at 1 g simulates some aspects of the microgravity environment of space [9,10]. Rotating bioreactors have found extensive ground-based application in mammalian cell culture (reviewed by Unsworth and Lelkes [11]).

Klaus et al. [12] studied bacterial cell growth in microgravity in four to five-day experiments aboard seven space shuttle missions (STS-37, 43, 50, 54, 57, 60 and 62) carried out between 1991 and 1994. The *E. coli* strain ATCC 4157 was selected as a model cell because of its resilience, rapid doubling rate and small cell size which minimizes indirect, hydrodynamic effects of gravity. *E. coli* were cultured either in microgravity or at 1 g in a temperature-controlled enclosure containing one of the following cell culture systems developed by BioServe Space Technologies (Boulder, CO) in collaboration with NASA: (i) chambers within two blocks that were moved relative to each other to provide mixing (the Material Dispersion Apparatus); (ii) a syringe in which two different fluids were contained and mixed by depressing a plunger (the Cell Syringe); and (iii) a syringe-like apparatus in which three different compartments containing cells, medium, and fixative were mixed to start or terminate an experiment (the Fluid Processing Apparatus). Cell cultures were initiated in-flight and fixed at timed intervals for post-flight assessment of cell concentration and the three phases of cell growth (i.e. lag, exponential, stationary). The exponential growth phase was lengthened and final cell concentrations were higher in microgravity cultures than in 1 g controls. The authors proposed that in microgravity, the lag phase was shortened due to the lack of convective mixing that resulted in faster conditioning of the medium surrounding the cells, the exponential phase was lengthened due to inhibition of cell growth as metabolic products accumulated, and the final cell concentration was increased due to a lack of cell settling that resulted in more spatially uniform cell distributions.

The above and other studies demonstrated that cell culture during spaceflight is feasible and that microgravity can profoundly affect cell behavior. However, the type and size of biological specimens and the duration of the experiments have been limited by the capabilities of the available flight hardware, and there is a clear and present need to improve the cell culture systems by more precisely controlling medium composition and providing for mass transfer. We believe that these requirements can be met by the development of advanced spaceflight bioreactors.

NASA-developed spaceflight bioreactors

Gas permeable cell culture bags

Gas permeable cell culture bags (Fig. 1a–c) are the heart of the Biological Specimen Temperature Controller (BSTC), a system designed and built at NASA-JSC and used aboard the space shuttle and the Mir Space Station (e.g. [13,14]). The BSTC measures 48×53×25 cm (to replace one Middeck shuttle locker) and weighs 29 kg (Fig. 1d). The BSTC consists of a chamber with a 12-liter usable volume that provides temperature control and gas exchange to the gas-permeable Teflon bags containing cultured cells. Chamber temperature can be controlled over the range of 2°C above ambient to 40°C, with an accuracy of ±0.1°C. A mixture of air and carbon dioxide is used in conjunction with a carbon dioxide sensor to control the chamber concentrations of carbon dioxide in the range of 0–20%. Mixing of gases in the chamber is achieved through forced convection in conjunction with a barometric pressure sensor and a pressure relief valve. Four 25 ml bags (Fig. 1a,b) or eight 10 ml bags are contained within an assembly (Fig. 1c); up to eight assemblies (i.e. 64 bags) can be accommodated. The bags are permeable to carbon dioxide and oxygen (e.g. a 25 ml

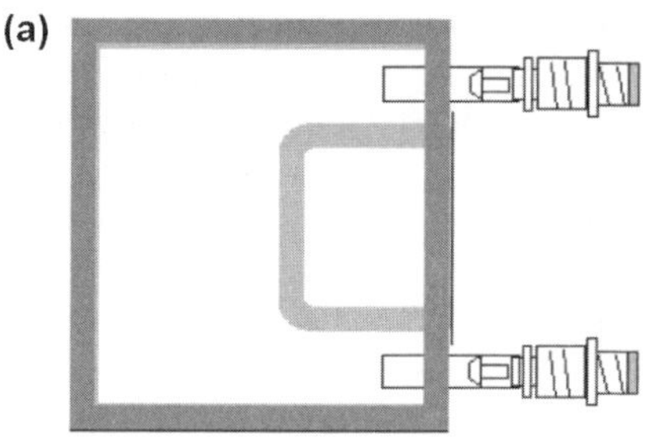

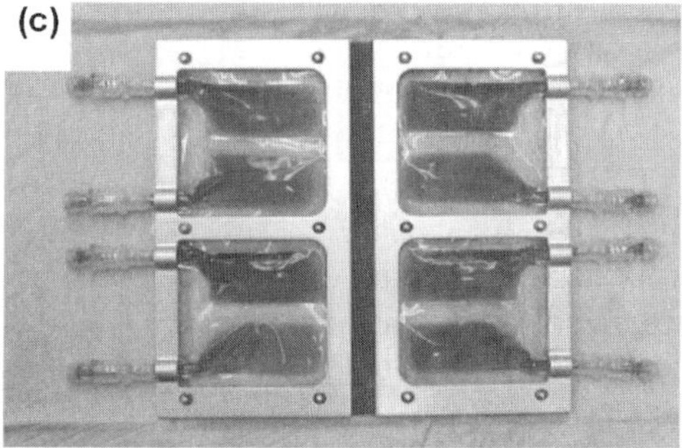

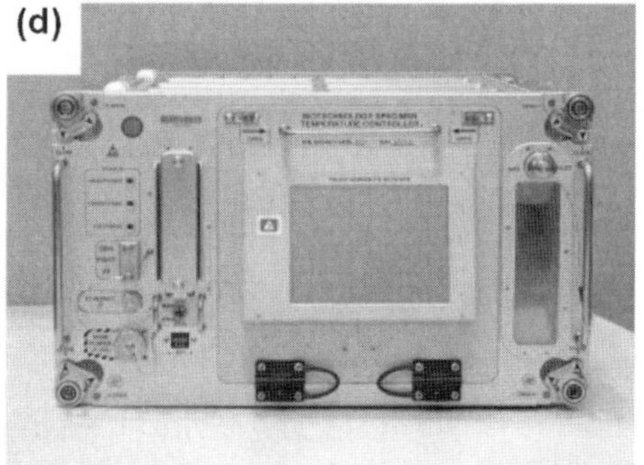

Fig. 1. Gas permeable bags and the Biological Specimen Temperature Controller. (a, b) Schematic and photograph of 25 ml bags; (c) assembly of eight 10 ml bags; (d) photograph of the Biological Specimen Temperature Controller (BSTC). (Courtesy of P.C. Cottingham, Wyle Laboratories, Houston, TX, USA.)

180

bag transfers 0.17 ml carbon dioxide and 0.09 ml oxygen per day). The bags have injection ports for the addition or withdrawal of biological specimens, culture medium, or fixatives.

Perfused cartridges

Perfused cartridges (Fig. 2a,b) are the heart of the Cell Culture Module (CCM), a system formally known as the Space Tissue Loss (STL) module, developed by the Walter Reed Army Institute of Research (WRAIR) through the Department of Defense Space Test Program, and used in numerous flights aboard the space shuttle [15,16]. The system has external dimensions of approximately $48 \times 41 \times 23$ cm (to fit into one middeck shuttle locker) and weighs 26 kg (Fig. 2c). The biological specimens are generally housed in cylindrical cartridges (55 ml in volume, 2.4 cm in diameter and 13 cm long) that are positioned within individual recirculating loops designed to perfuse the cartridges with culture medium containing controlled concentrations of gases and nutrients (Fig. 2d). In one configuration, the cartridge is a hollow fiber

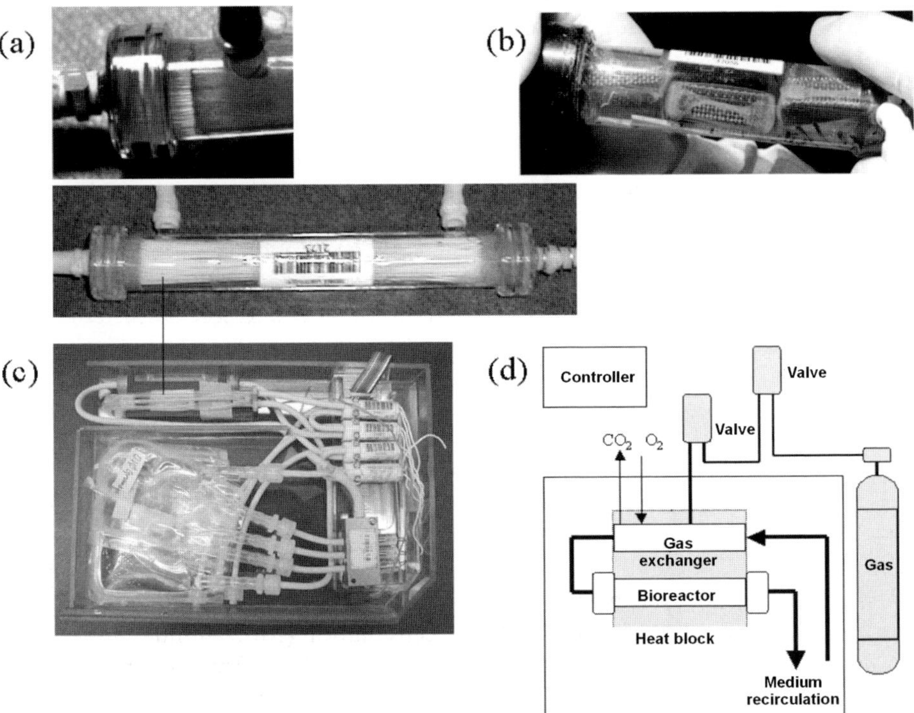

Fig. 2. Perfused cartridges and the Space Tissue Loss Module. (a,b) Photographs of ~55 ml perfused cartridges containing either hollow-fibers or engineered muscle organoids; (c,d) photograph and schematic of the Space Tissue Loss (STL) Module. (Courtesy of C. Waterhouse, Walter Reed Army Institute of Research, Washington, DC, USA.)

bioreactor (Cellco, Spectrum Laboratories, Laguna Hills, CA) in which microcarrier-immobilized cells are cultured in the extracapillary spaces and culture medium is perfused through the fibers (Fig. 2a [15]). In an alternative configuration, the fibers are removed to allow engineered tissues to directly contact culture medium that is perfused through the cartridge (Fig. 2b [16]). Each cartridge is placed within an individual recirculating loop consisting of a pulsatile pump, a gas exchanger (consisting of a coil of silicone tubing), a heat sink (maintained at 37°C), and reservoir bags, 15 to 250 ml in volume, for fresh and waste media. The reservoir bags, as well as medium additives and fixatives, and in some cases biological samples, can be stored in a 6°C cooling chamber (700 ml volume). Systems containing 3, 4 and 8 cartridges have been flown [15,16] (Fig. 2c), and the total number of cartridges can be increased to 24. Specific regimes of recirculation and infusion can be programmed to meet experimental requirements and maintain metabolic parameters within target ranges. For example, in an 11-day study of bone cells, the system was operated with intermittent infusion of fresh medium at a rate of 1.5 ml every six hours [15]. In a ten-day study of engineered muscle, the system employed continuous recirculation of 120 ml medium at a flow rate of 1.5 ml/min [16].

Rotating, perfused bioreactor

A rotating, perfused bioreactor is the heart of the Engineering Development Unit (EDU) (Fig. 3) designed and built at NASA-JSC, and used aboard the space shuttle and the Mir Space Station (e.g. [17,18]). The EDU-M has dimensions of approximately $43 \times 51 \times 64$ cm (to fit into 2 1/2 Middeck shuttle lockers) and weighs 62 kg. The biological specimens (cells on microcarriers or engineered tissues) are housed in a rotating bioreactor (125 ml) that is positioned in a recirculating loop designed to perfuse the vessel with culture medium containing controlled concentrations of gases and nutrients. The bioreactor is comprised of the annular space between two cylinders that can be differentially rotated at up to 40 revolutions per minute. The inner cylinder has a disc attached to one end to enhance secondary fluid flow. Fluid enters the bioreactor at one end and exits through a mesh on the inner cylinder. The recirculating loop consists of a silicone membrane gas exchanger and a peristaltic pump. The recirculation loop is connected via flexible tubing to reservoirs containing fresh and waste media. Specific regimes of recirculation and infusion can be established to meet experimental requirements and maintain metabolic parameters within target ranges. For example, in a four month long study of engineered cartilage, the system was operated with intermittent recirculation (for 20 minutes at a rate of 4 ml/min every six hours) and infusion of fresh medium (50–100 ml, once per day) [17].

A modified rotating perfused bioreactor, the EDU-2, is under development for the International Space Station (ISS) Biotechnology Facility (BTF, see below) [19]. Improvements with respect to the EDU-M described above include reduced weight and power consumption, on-line monitoring and control of culture medium volume, pressure, and pH, and an internal camera and illumination assembly for bioreactor imaging.

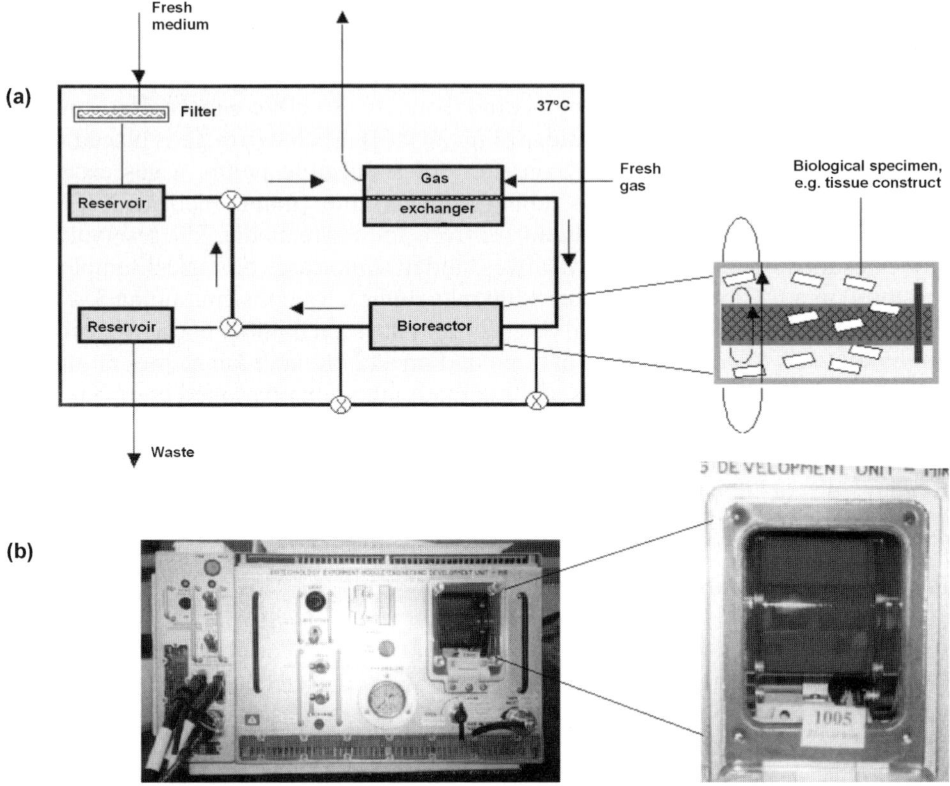

Fig. 3. Rotating, perfused bioreactor and the Engineering Development Unit. (a,b) Schematics and photographs of the Engineering Development Unit used aboard Mir (EDU-M), with insets showing the 125 ml rotating, perfused bioreactor. (Photographs taken at Kennedy Space Center in 1996, prior to the launch of STS-79.)

Perfused chambers

Perfused chambers are the heart of the Cell Culture Unit (CCU) currently under development for the ISS Gravitational Biology Facility (see below) by the NASA Ames Research Center (ARC) in conjunction with Payload Systems Inc. (PSI) and the Massachusetts Institute of Technology (MIT), both of Cambridge, MA [20,21] (Fig. 4). The Middeck configuration of the CCU will have dimensions of approximately 53×46×56 cm (to fit into two Middeck shuttle lockers) and weigh 45 kg; a second (host) configuration will measure approximately 53×48×64 cm and weigh 65 kg. The biological specimens (cell monolayers, cell suspensions, engineered tissues) are housed in perfused chambers (3, 10 or 30 ml) that are positioned within individual recirculating loops designed to perfuse the chambers with culture medium containing controlled concentrations of gases and nutrients (Fig. 4a,b). Culture

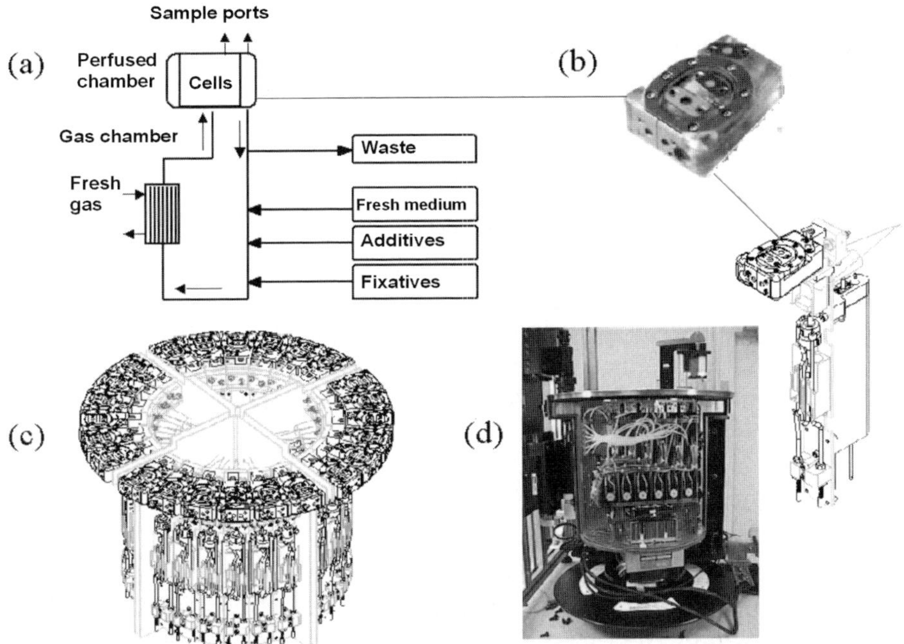

Fig. 4. Perfused chambers and the Cell Culture Unit. (a) Schematic of the Cell Culture Unit (CCU); (b) photograph of a 3 ml perfused chamber; (c,d) schematic and photograph of the prototype Science Evaluation Unit (adapted from [21]).

medium enters into the central portion of the chamber and exits through a porous membrane into the annular cell-free space and back into the perfusion loop. The system can accommodate twenty-four 3-ml chambers, twelve 10-ml chambers, or eight 30-ml chambers. On the chamber bottom is a removable optical window that can be used as a platform for monolayer cell culture. On the chamber top is a condenser window for videomicroscopy and injection ports for the addition or withdrawal of cells, culture medium, or other specific reagents (Fig. 4b).

Individual chambers, each connected to a recirculation loop with a peristaltic pump and a silicone coil gas exchanger, are mounted in a circle such that chambers are positioned above their respective recirculation loops (Fig. 4c). Specific regimes of recirculation can be established to meet experimental requirements. Examples of flow algorithms include: (i) low-rate, intermittent recirculation, to maintain medium gas concentrations while minimizing hydrodynamic shear; (ii) net unidirectional flow with an alternating (forward–reverse) component, to maintain cells suspended in the chamber and to thereby preventing obstruction of its internal membrane; and (iii) net unidirectional flow with an alternating component in conjunction with mechanical stirring, for cultures that require additional means for cell suspension. Each recirculation loop is connected via flexible tubing to a storage compartment, maintained at 4°C, containing fresh and waste media, additives, and fixatives.

184

The CCU is divided into four quadrants to permit the grouping and differential treatment of chambers (e.g., for sequential subcultures or to compare different thermal environments). The chambers are mounted on a turntable that can rotate ±180° in order to allow automated positioning of an individual chamber for on-line videomicroscopy or for sampling. An automated sampling module (ASM) can transfer cells from a chamber into a container containing a fixative. The system can accommodate a total of 60 samples, stored as required at a selected temperature, ranging from +4° to 37°C. The ASM can also be used to transfer cells between chambers or to inject additives. At the time of this writing, three complete prototypes (Science Evaluation Units) have been fabricated (Fig. 4d) and are undergoing formal testing using six model cell types: (i) the C2C12 murine muscle cell line, (ii) primary avian bone cells, (iii) engineered muscle based on primary avian muscle cells in a collagen gel, (iv) yeast cells, (v) tobacco cells, and (vi) *Euglena gracilis*, a motile, unicellular organism.

ISS Biotechnology Facility

The ISS Biotechnology Facility (BTF), a versatile facility designed to support the science requirements of the cellular biotechnology research community, is currently under development by NASA-JSC [22]. The BTF will consist of two racks housing several bioreactors for cell and tissue cultivation (e.g., the BSTC and EDU shown in Figs. 1 and 3, respectively), modularized support hardware, and experiment-specific analytical equipment. The BTF will provide and manage power, research-grade water, gas mixtures, 4°C refrigeration, and computers for payload operation, data acquisition, storage, processing, and downlink. Supplementary elements of the BTF will include an apparatus for quickly freezing experimental samples, and coolers and cryogenic containers for sample storage. Prototype payload racks developed to Expedite the Processing of Experiments to Space Station (EXPRESS racks) have been in use for cell culture research aboard the ISS since August of 2001 [23]; the two BTF racks are scheduled for installation in 2006 and 2008.

ISS Gravitational Biology Facility

The ISS Gravitational Biology Facility (GBF) and the accompanying habitats, host systems and support equipment are currently under development by NASA-ARC in conjunction with International Space Agencies. The GBF will contain habitats for cell and tissue culture (CCUs, described above), other habitats for whole plants and animals, and a centrifuge facility, glovebox, service systems and laboratory support equipment [24]. In particular, a 2.5 meter diameter centrifuge will provide controlled, artificial gravity levels ranging from 0.01 to 2.0 g. The ability to expose biological organisms to controlled g forces aboard the ISS is expected to allow for the first time specific effects of gravity to be distinguished from all other parameters affected by spaceflight.

Five selected NASA flight studies

Landis et al. [15] studied the effects of spaceflight on bone cells in an 11 day experiment carried out in 1994 aboard STS-59. Bone cells were selected as a model system because the exposure of humans to microgravity results in significant bone loss [25]. The cell preparation consisted of embryonic avian bone cells cultured on microcarrier beads in the extracapillary space of hollow-fiber cartridges (Fig. 2a). In particular, cells that were either uncommitted or committed towards osteogenesis were respectively derived from 14 or 17 day old embryos. Cells and collagen-coated microcarrier beads were added to cartridges in which the fibers were pre-coated with fibronectin. Microcarrier-immobilized cells were cultured in cartridges first in re-circulating loops (CellCo, Gaithersburg, MD) and then in STL modules (Fig. 2c) for a total of five days at 1 g. One STL module containing eight cartridges was operated aboard STS-59 for 11 days; a second otherwise identical STL module was operated on the ground. Medium samples were obtained at timed intervals for post-flight analysis of metabolic parameters; cell samples were harvested post-flight for assessments of ultrastructure and expression levels of collagen type I protein and mRNA, and osteocalcin mRNA. Both uncommitted and committed bone cells accumulated less extracellular matrix in microgravity as compared to at 1 g and exhibited reduced expression of type I collagen and osteocalcin [15]. These findings provide evidence that microgravity suppresses osteogenic differentiation and bone development, and imply that the mechanisms underlying bone loss in astronauts may include the down-regulation of type I collagen and osteocalcin expression by osteogenic cells.

Vandenburgh et al. [16] studied the effects of spaceflight on the development and function of engineered muscle in two 9–10 day experiments aboard STS-66 and STS-77 in 1994 and 1996. Muscle was selected as a model tissue because the exposure of humans to microgravity results in significant losses in muscle mass and strength [26]. Engineered muscle based on embryonic avian skeletal muscle cells cultured in type I collagen gel under passive tension was selected based on extensive experience with this tissue culture preparation (e.g. [27,28]). In particular, cells were plated in collagen-coated elastic culture wells with or without an additional layer of type I collagen gel and cultured at 1 g for a total of approximately 14 days to promote the formation of muscle-like organoids consisting of striated, contractile, multinucleated myofibers aligned in the direction of tension. Wells containing the organoids were placed in perfused cartridges (Fig. 2b) within STL modules (Fig. 2c), and modules containing three and four cartridges with six organoids apiece were operated for nine days aboard STS-66 and STS-77, respectively; otherwise identical STL modules were operated on the ground, at NASA's Kennedy Space Center. During the flight, medium was sampled and radiolabeled tracers were injected for post-flight analyses of metabolic parameters and rates of protein synthesis and degradation respectively. Constructs were harvested post-flight for the assessment of histomorphology, amounts of DNA, total protein, myosin heavy chain (MHC) protein, fibronectin and beta 1 collagen. Myofiber diameter and cross-sectional area decreased significantly in microgravity but not in 1 g cultures (Table 1). Construct amounts of DNA and total

Table 1

Properties of engineered muscle organoids cultured in microgravity are compared to those at launch and ground controls (Vandenburgh et al., 1999)

Parameter	At launch (L)	Flight	Ground
Culture time	14 days	L + 9 days	L + 9 days
Myofiber size			
Diameter (mm)	10.5±0.3	9.8±0.3[a,b]	11.1±0.3
Cross-sectional area (mm^2)	145±7	122±5[a,b]	135±6
Organoid composition			
DNA (μg/organoid)		21±3	24±3
Protein/DNA ratio (μg/mg)		22±2	22±2
Myosin heavy chain (μg/organoid)	12±1	12.5±1.5	16.5±1.5[a]
Fibronectin (μg/organoid)	13.5±1.5	8.5±1.5[a]	9.5±1
Noncollagenous protein (μg/organoid)		450±50	550±50
Organoid function			
Total protein synthesis rate (^{3}H dpm/μg protein/h)			
(a) in-flight or at the corresponding time		50±20[b]	190±40
(b) post-flight at corresponding times		350±40[b]	190±20
MHC synthesis rate (post-flight or at corresponding times) (^{3}H dpm/μg protein/h)		85±8[b]	60±5
Fibronectin synthesis rate (post-flight or at corresponding times) (^{3}H dpm/μg protein/h)		45±5[b]	25±2
Collagen synthesis rate (post-flight or at corresponding times) (^{3}H dpm/μg protein/h)		40±2[b]	30±2

Data represent average ± standard deviation of 12 to 15 individual samples.
Statistical significance was assessed using ANOVA (α= 0.05, P < 0.05) in conjunction with Tukey's studentized range test.
[a]Significantly different from organoids at launch.
[b]Significantly different from the ground control at the end of the study.

protein and protein degradation rates were comparable in microgravity and 1 *g* cultures. The amount of MHC did not increase in microgravity but increased in 1 *g* cultures, whereas the amount of fibronectin decreased in microgravity and 1 *g* cultures (Table 1). Total protein synthesis rates measured during microgravity culture were markedly lower than 1 *g* controls; in contrast, total protein synthesis rates measured post-flight were higher than 1 *g* controls (Table 1). Likewise, post-flight synthesis rates of MHC, beta-1 collagen, and fibronectin were all higher than in 1 *g* controls (Table 1). Together, these data imply that in microgravity, muscle fiber atrophy occurs due to a decrease in protein synthesis rates, a change that is apparently reversible upon return to earth. These results imply that skeletal muscle fibers are directly responsive to spaceflight.

Freed et al. [17] studied the effects of spaceflight on engineered cartilage in a four month long experiment carried out aboard STS-79, the Mir Space Station and STS-81, between September 1996 and January 1997. Cartilage was selected as a model tissue because of its resilience and relevance to studies of the effects of spaceflight on skeletal tissues. Engineered cartilage based on primary bovine calf chondrocytes and a synthetic, biodegradable mesh of polyglycolic acid was selected based on extensive experience with this tissue culture preparation in rotating bioreactors (e.g. [29,30]). Engineered cartilage constructs were cultured in rotating bioreactors (the STLV) at 1 g for three months (a time period dictated by the launch schedule), divided into two groups, and transferred into two rotating, perfused bioreactors (the EDU-M) (Fig. 3). One EDU-M containing ten constructs was launched aboard STS-79, transferred to the Priroda module of Mir, and returned to earth via STS-81, and thereby continually operated for a period of four months in microgravity. Fluid mixing was provided by differential rotation of the inner and outer cylinders in the same direction at rates of 10 rpm and 1 rpm, respectively. A second EDU-M containing ten constructs was operated at 1 g at NASA-JSC. Fluid mixing was provided by solid body rotation of the vessel at 40 rpm. Constructs were assessed and compared to native bovine calf articular cartilage with respect to: size, morphology, cell viability, post-flight biosynthesis rates of glycosaminoglycans (GAG) and collagen, amounts of DNA, GAG and total and type II collagen, and mechanical properties (aggregate modulus and hydraulic permeability in radially confined axial compression) at the time of launch (i.e. after three months of culture) and after an additional four months of culture either in microgravity or at 1 g (i.e. after seven months of culture) (Table 2).

Constructs cultured at 1 g tended to maintain their initial discoid shape, whereas constructs cultured in microgravity tended to become more spherical; in both groups constructs appeared histologically cartilaginous throughout their entire cross-sections (5–8 mm thick) (Fig. 5 a,b). The fraction of the total collagen that was the cartilage-specific type II collagen did not decrease significantly between the time of launch and landing, demonstrating good maintenance of the differentiated chondrocytic phenotype in microgravity (Table 2). Over the time interval of 3–7 months, construct cellularity and wet weight increased in both groups, but growth rates were markedly higher at 1 g than in microgravity (Table 2). Likewise, construct mechanical properties improved in both groups over the time interval of 3–7 months, but constructs cultured at 1 g had 2.5-fold higher wet weight GAG fractions and three-fold higher moduli than those of constructs cultured in microgravity (Table 2). Comparable post-flight synthesis rates of GAG and collagen were measured in constructs from microgravity and 1 g cultures (Table 2). Cells isolated from post-flight constructs and 1 g controls were 95–99% viable, and were alive and metabolically active, as demonstrated by attachment to petri dishes, proliferation, and enzymatic conversion of a tracer substrate (Fig. 5c). Engineered cartilage cultured in microgravity was thus viable and mechanically functional, but was smaller and mechanically inferior to that grown at 1 g; the latter findings were attributed to the reduction in physical forces in microgravity and/or to the related differences in construct cultivation conditions (i.e., free-floating aboard Mir vs. gravity settling on Earth).

Jessup et al. [18] studied the effects of spaceflight on cellular apoptosis and differentiation in a preliminary experiment aboard STS-70 in 1995 and a definitive,

Table 2

Properties of engineered cartilage constructs cultured in microgravity are compared to those at launch, ground controls, and native cartilage (Freed et al., 1997)

Parameter	At launch (L)	Flight	Ground	Native cartilage
Culture time	3 months	L + 4 months	L + 4 months	0
Construct size (mg wet weight)	250±38 (20)	330±25 (5)[1,2]	429±14 (5)[1]	
Construct composition				
Cells (millions per construct)	13±1 (3)	14±0.08 (3)	19±0.2 (3)	
Cells (% of wet weight)	0.64±0.03 (3)	0.40±0.01 (3)[1,3]	0.46±0.02 (3)[1,3]	0.66±0.09 (7)
GAG (% of wet weight)	6.03±0.84 (3)	3.59±0.22 (3)[1,2,3]	8.83±0.93 (3)[1,3]	7.05±0.56 (6)
Collagen (% of wet weight)	2.70±0.75 (3)[3]	3.42±0.17 (3)[3]	3.68±0.27 (3)[3]	10.7±0.91 (6)
Water (%)	89.2±0.89 (3)[2,3]	89.2±0.31 (3)[2,3]	86.3±0.53 (3)[3]	82.8±0.93 (7)
Total of above components (% of wet weight)	98.6±0.07 (3)[3]	96.6±0.68 (3)[1,2,3]	99.3±0.60 (3)[3]	101±0.46 (6)
Type II collagen (% of total collagen)	91.6±19.1 (2)	78.0±4.1 (4)	75.3±7.8 (4)	90.3±17.9 (5)
Construct function				
Aggregate modulus (MPa)	0.108±0.047 (2)[3]	0.313±0.045 (4)[1,2,3]	0.932±0.049 (3)[1]	0.949±0.021 (3)
Hydraulic permeability ($\times$ 10^{-15} m⁴/Ns)	8.25±1.94 (2)[3]	6.73±3.02 (4)[1]	3.72±0.167 (3)[1]	2.72±0.641 (3)
Dynamic stiffness (MPa, 1 Hz)	2.23±0.12 (2)[3]	3.80±0.39 (4)[2,3]	7.75±0.30 (3)[3]	16.8±1.14 (3)
Sulfate incorporation rate (ng/μg DNA/day)*	80.8±37.2 (3)[3]	88.9±8.0 (3)[3]	83.4±13.4 (3)[3]	332±36.3 (7)
Proline incorporation rate (ng/μg DNA/day)*	82.0±20.6 (3)[3]	93.0±3.0 (3)[3]	87.9±22.7 (3)[3]	205±46.8 (7)

Data represent average ± standard deviation. Parentheses indicate the number of samples analyzed per group.

Statistical significance was assessed using ANOVA ($\alpha = 0.05$, $P < 0.05$) in conjunction with Tukey's studentized range test.

[1]Significantly different from constructs at launch.

[2]Significantly different from the ground control at the end of the study.

[3]Significantly different from the native cartilage.

*Post-flight radiolabeling and assessment.

12 day experiment aboard STS-85 in 1997. Microgravity cell culture was studied because it allowed a reduction of mechanical forces during suspension culture to otherwise unobtainable ranges [31,32], whereas at 1 g the hydrodynamic forces needed to balance the gravitational settling of cells increase directly with aggregate size and can cause cellular de-differentiation and damage (e.g. [33]). The human cell line MIP-101 was selected since these primitive colon carcinoma cells provide a model for differentiation studies; cells were cultured in the presence of collagen coated microcarrier beads to promote aggregation. Cells and microcarriers were cultured in rotating perfused bioreactors in an EDU like that shown in Fig. 3 for a total of four days at 1 g. One EDU was operated in microgravity; fluid mixing was

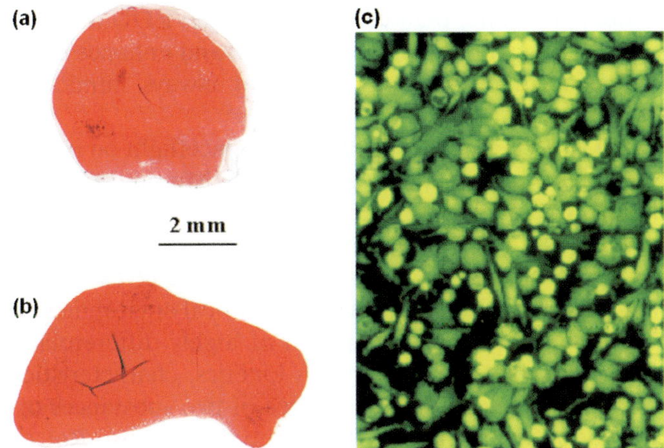

Fig. 5. Microgravity affects engineered cartilage without damaging cell viability. Spaceflight affects the size and structure of engineered cartilage but maintains cell viability. Histological cross sections of engineered cartilage cultured (a) in microgravity or (b) on the ground and stained for glycosaminoglycans with safranin-O. (c) Cells isolated post-flight from tissues grown in microgravity, cultured for two days in a dish, and stained for intracellular esterase activity (adapted from Freed et al. [17]).

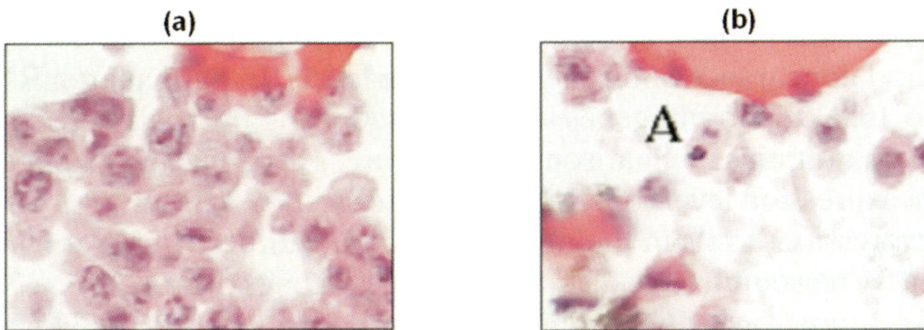

Fig. 6. Microgravity affects the cell cycle. Spaceflight affects the apoptosis rate in human colon carcinoma cells cultured (a) in microgravity, (b) on the ground. An apoptotic cell is indicated by "A" (from Jessup et al. [18]. Copyright © 2000 by the Society for In Vitro Biology (formerly the Tissue Culture Association). Reproduced with permission of the copyright owner.

provided by differential rotation of the inner and outer cylinders in the same direction at rates of 11.5 rpm and 0.5 rpm, respectively. A second EDU was operated at 1 g at NASA-JSC; fluid mixing was provided by solid body rotation of the vessel at 25 to 30 rpm. As an additional control, cells and microcarriers were cultured at 1 g in static Teflon bags. Apoptosis and the expression of carcinoembryonic antigen (CEA) were measured at timed intervals. Apoptosis rates measured for cells grown in the EDU in microgravity (5–18%) and in Teflon bags at 1 g (2–12%) were lower than in the corresponding rates measured for cells grown in the EDU at 1 g (15–60%) (Fig. 6). Likewise, cells grown in the EDU in microgravity expressed higher levels of CEA,

190

a marker for colon carcinoma, than cells grown in the EDU at 1 g. These findings show that (i) spaceflight can reduce apoptosis rate and enhance cell differentiation and (ii) cell behavior in microgravity does not always mimic that in rotating bioreactors operated at 1 g.

Hammond et al. [13,14] studied the effects of spaceflight on gene expression in a preliminary experiment carried out aboard STS-86, Mir, and STS-89 in 1997–1998, and in a definitive, six day experiment aboard STS-90 in 1998. The goal of the study was to quantify effects of different mechanical environments on gene expression in a three dimensional cell culture system. In particular, the investigators hypothesized that microgravity would provide an optimal mechanical environment for the generation of three dimensional cultures composed of highly differentiated cells. Renal cortical cells were selected for their growth kinetics, characteristic differentiated features, and because these cells are the source of two substances of major clinical and commercial importance: vitamin D_3 and erythropoietin. Rat cells were used aboard Mir, and human renal cortical cells were used aboard STS-90; in both cases cells were cultured in the presence of collagen coated microcarrier beads to promote aggregation.

Four different mechanical environments were compared: (i) microgravity cultures in gas permeable bags (25 ml bags in the BSTC) (Fig. 1); (ii) 1 g cultures in rotating bioreactors (STLVs); (iii) 3 g cultures in gas permeable bags (carried out in the NASA-ARC short-arm centrifuge, designed to apply artificial gravity as well as control over culture temperature, dissolved gases and humidity); and (iv) 1 g control cultures in static gas permeable bags. Cells were evaluated post-flight with respect to ultrastructural appearance and gene expression levels. In particular, mRNAs corresponding to 10,000 different genes were assessed using microarray technology that allowed the quantitative comparison of the nature, grouping, and extent of changes. In addition, expression levels of selected proteins were assessed.

Gene expression levels in the microgravity bag culture, 1 g rotating bioreactor culture, and 3 g bag culture are displayed against 1 g bag controls in Fig. 7a, b, and c, respectively. In microgravity bag cultures, expression levels of 1,632 genes (16%) differed by at least three-fold from those in 1 g bag cultures. In rotating bioreactor cultures, expression levels of 914 genes (9%) differed from 1 g bag controls. In 3 g bag cultures, expression levels of 15 genes (1.5%) differed from 1 g bag controls. Genes exhibiting changes in expression levels included differentiation mediators, apoptosis genes, and a variety of signaling molecules. Specific transcription factors (Wilms' tumor zinc finger protein and the vitamin D receptor) underwent large changes in microgravity. Substantial changes in gene expression were induced by reducing the gravity from 1 g to microgravity (Fig. 7a). Significant changes in gene expression were also induced in 1 g rotating bioreactor cultures (Fig. 7b), and the patterns of changes observed were distinctly different from those observed in microgravity. Few changes in gene expression level were induced by increasing the gravity level from 1 g to 3 g (Fig. 7c). These findings demonstrate that genes respond to mechanical regulatory signals and that microgravity induced large changes in gene expression.

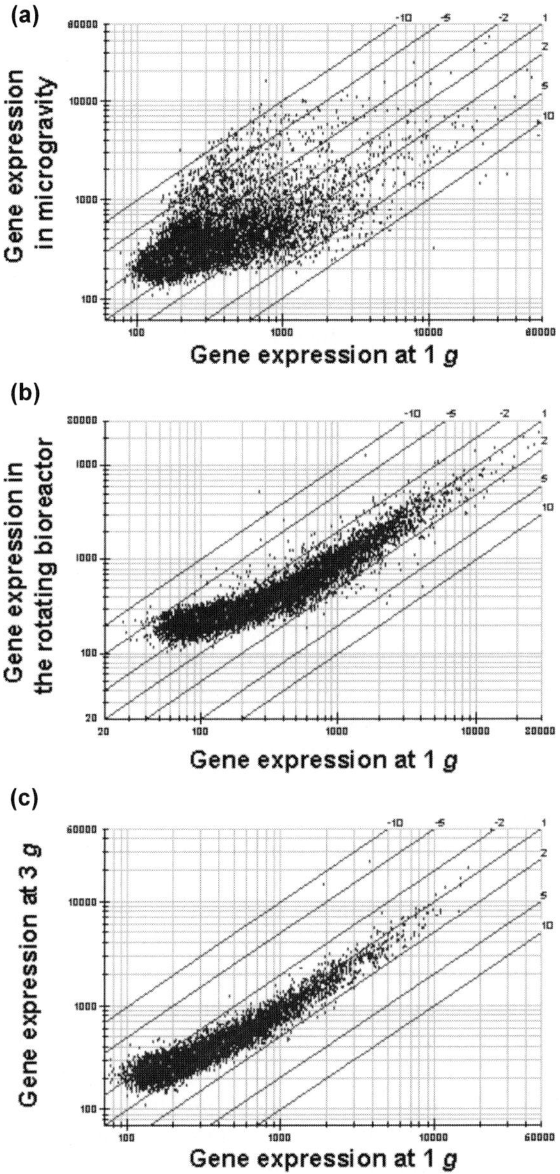

Fig. 7. Microgravity affects gene expression Gene expression levels in (a) microgravity bag cultures, (b) 1 g rotating bioreactor cultures, and (c) 3 g bag cultures, each displayed against 1 g bag cultures. Each dot represents an individual gene; the distance from the dot to the origin represents the level of gene expression; the distance from the dot to the line denoted by "1" represents the difference in gene expression between the culture systems compared in the graph. Changes of three-fold or more were considered to be above background. (From Hammond et al. [14], reproduced with permission from the copyright owner).

Summary and conclusions

Studies of the fundamental role of gravity in the development and function of biological organisms are a central component of the human exploration of space. Microgravity affects numerous physical phenomena relevant to biological research, including the hydrostatic pressure in fluid filled vesicles, sedimentation of organelles, and buoyancy-driven convection of flow and heat. These physical phenomena can in turn directly and indirectly affect cellular morphology, metabolism, locomotion, secretion of extracellular matrix and soluble signals, and assembly into functional tissues. Studies aimed at distinguishing specific effects of gravity on biological systems require the ability to: (i) control and systematically vary gravity, e.g. by utilizing the microgravity environment of space in conjunction with an in-flight centrifuge; and (ii) maintain constant all other factors in the immediate environment, including in particular concentrations and exchange rates of biochemical species and hydrodynamic shear. The latter criteria imply the need for gravity-independent mechanisms to provide for mass transport between the cells and their environment.

Available flight hardware has largely determined the experimental design and scientific objectives of spaceflight cell and tissue culture studies carried out to date. Simple culture vessels have yielded important quantitative data, and helped establish *in vitro* models of cell locomotion, growth and differentiation in various mammalian cell types including embryonic lung cells [6], lymphocytes [2,8], and renal cells [7,31]. Studies done using bacterial cells established the first correlations between gravity-dependent factors such as cell settling velocity and diffusional distance and the respective cell responses [12].

The development of advanced bioreactors for microgravity cell and tissue culture and for tissue engineering has benefited both research areas and provided relevant *in vitro* model systems for studies of astronaut well-being (loss of muscle and skeletal tissues [15–17]) and gene- and cell-level responses to the mechanical environment [13,14,18]. All five of the spaceflight bioreactor studies described above utilized three-dimensional cell culture systems in which the cells were associated with biodegradable polymer scaffolds [17], collagen gel [16], or microcarrier beads [13–15,18] in order to promote the expression of differentiated cell function. In four of the five spaceflight bioreactor studies [15–18], cells were cultured in perfused vessels (cartridges or rotating bioreactors) within recirculating loops designed to maintain medium composition within target ranges by a combination of gas exchange and fresh medium supply. Future spaceflight studies of cells and tissues are likely to involve a three-dimensional culture system, to promote cellular differentiation, and perfusion with or without rotation, to provide a gravity-independent mechanism for fluid mixing and mass transport.

Previous spaceflight studies have guided the ongoing development of NASA flight hardware for the ISS (e.g. the EDU-2 and the CCU). This next generation of hardware will have extended operational capabilities including on-line microscopy, in-line sensors for the monitoring and control of metabolic parameters, modular design for replicate cultures, and, perhaps most importantly of all, compatibility with the ISS centrifuge. The latter will permit in-flight, 1 *g* control cultures, and thereby allow the experimental variable to be gravity itself rather than the more general "spaceflight environment". Technical limitations of spaceflight studies (e.g. allowable size, mass, and power) continue to motivate a creative approach to system design and to result in "spin-off" technologies (e.g. the STLV) for ground-based cell and tissue culture research. The increasing scientific and medical relevance of this work is evidenced by the growing number of publications in which advanced bioreactors are used for *in vitro* studies in physiologically relevant cell and tissue models.

Acknowledgements

We are indebted to T.G. Hammond (Tulane University), J.M. Jessup (George Washington University) and H.H. Vandenburgh (Brown University) for contributing their published data, to P.C. Cottingham (Wyle Laboratories), J. de Luis (PSI), D. Klaus (University of Colorado), J. Shansky (Brown University) and C. Waterhouse (WRAIR) for providing information about hardware and spaceflight experiments, and to N.R. Pellis (NASA-JSC) and N. Searby (NASA-ARC) for their critical reviews of the manuscript. We also thank S. Kangiser for her expert help with manuscript preparation. This work was funded by NASA grants NCC8-174, and NAS2-96001.

References

1. Lewis, M. and Hughes-Fulford, M. (1997) Response of living systems to spaceflight. In: Fundamentals of Space Life Sciences (Churchill, S.E., ed.) Vol. 1, pp. 21–39.
2. Cogoli, A., Tschopp, A. and Fuchs-Bislin, P. (1984) Cell sensitivity to gravity. Science 225, 228–230.
3. Gmunder, F.K., Cogoli, A., Fregly, J.J. and Blatteis, C.M. (1996) Effect of spaceflight on lymphocyte function and immunity. In: Handbook of Physiology (Gmunder, F.K., Cogoli, A., Fregly, J.J. and Blatteis, C.M., eds.). Oxford University Press, New York, pp. 799–813.
4. Cogoli, A., Valluchi-Morf, M., Mueller, M. and Briegleb, W. (1980) Effects of hypogravity on human lymphocyte activation. Aviat. Space Environ. Med. 51, 29–34.
5. Tschopp, A. and Cogoli, A. (1983) Hypergravity promotes cell proliferation. Experientia 39, 1323–1329.
6. Montgomery, P., Cox, J., Reynolds, R., Paul, J., Hayflick, L., Stock, D., Shuiz, W. and Kimzey, S. (1977) Biomedical results from Skylab. Technical paper N77-33780, US Government printing office, Washington DC, pp. 221–234.
7. Todd, P.M., Kunze, E., Williams, K., Morrison, D.R., Lewis, M.L. and Barlow, G.H. (1985) Morphology of human embryonic kidney cells in culture after space flight. Physiologist 28, S183–S184.
8. Pellis, N.R., Goodwin, T.J., Risin, D., McIntyre, B.W., Pizzini, R.P., Cooper, D., Baker, T.L. and Spaulding, G.F. (1997) Changes in gravity inhibit lymphocyte locomotion through type I collagen. In Vitro Cell. Dev. Biol.-An. 33, 398–405.

9. Schwarz, R.P., Goodwin, T.J. and Wolf, D.A. (1992) Cell culture for three-dimensional modeling in rotating-wall vessels: an application of simulated microgravity. J. Tiss. Cult. Meth. 14, 51–58.

10. Goodwin, T.J., Prewett, T.L., Wolf, D.A. and Spaulding, G.F. (1993) Reduced shear stress: a major component in the ability of mammalian tissues to form three-dimensional assemblies in simulated microgravity. J. Cell. Biochem. 51, 301–311.

11. Unsworth, B.R. and Lelkes, P.I. (1998) Growing tissues in microgravity. Nat. Med. 4, 901–907.

12. Klaus, D., Simske, S., Todd, P. and Stodieck, L. (1997) Investigation of space flight effects on *Escherichia coli* and a proposed model of underlying physical mechanisms. Microbiology 143, 449–455.

13. Hammond, T.G., Lewis, F.C., Goodwin, T.J., Linnehan, R.M., Wolf, D.A., Hire, K.P., Campbell, W.C., Benes, E., O'Reilly, K.C., Globus, R.K. and Kaysen, J.H. (1999) Gene expression in space. Nat. Med. 5, 359.

14. Hammond, T.G., Benes, E., O'Reilly, K.C., Wolf, D.A., Linnehan, R.M., Taher, A., Kaysen, J.H., Allen, P.L. and Goodwin, T.J. (2000) Mechanical culture conditions effect gene expression: gravity-induced changes on the space shuttle. Physiol. Genomics 3, 163–173.

15. Landis, W.J., Hodgens, K.J., Block, D., Toma, C.D. and Gerstenfeld, L.C. (2000) Space-flight effects on cultured embryonic chick bone cells. J. Bone Miner. Res. 15, 1099–1112.

16. Vandenburgh, H.H., Chromiak, J., Shansky, J., Del Tatto, M. and Lemaire, J. (1999) Space travel directly induces skeletal muscle atrophy. FASEB J. 13, 1031–1038.

17. Freed, L.E., Langer, R., Martin, I., Pellis, N. and Vunjak-Novakovic, G. (1997) Tissue engineering of cartilage in space. Proc. Natl. Acad. Sci. USA 94, 13885–13890.

18. Jessup, J.M., Frantz, M., Sonmez-Alpan, E., Locker, J., Skena, K., Waller, H., Battle, P., Nachman, A., Weber, M.E., Thomas, D.A., Curbeam, R.L., Baker, T.L. and Goodwin, T.J. (2000) Microgravity culture reduces apoptosis and increases the differentiation of a human colorectal carcinoma cell line. In Vitro Cell. Dev. Biol.-An. 36, 367–373.

19. Casteel, M. (2001) Project specific functional requirements rotating wall perfused vessel engineering development unit #2 (EDU-2) Functional Requirements Document JSC 28738A. Sept. 10, 2001, NASA-Johnson Space Center.

20. Searby, N.D., de Luis, J. and Vunjak-Novakovic, G. (1998) Design and development of a space station cell culture unit. J. Aerospace 107, 445–457.

21. Vunjak-Novakovic, G., Searby, N.D., de Luis, J. and Freed, L.E. (2002) Microgravity studies of cells and tissues. *Ann. N. Y. Acad. Sci.*, in press.

22. Pellis, N.R., Langford, S.D., Love, J.E., Gonzalez, M. and Gonda, S.R. (2001) Cellular biotechnology research utilizing the International Space Station biotechnology facility. In: Proceedings of the American Institute of Aeronautics and Astronautics, October 15–18, Cape Canaveral, Florida.

23. Fortenberry, L., Bartoe, J.F. and Holloway, T.W. (2001) Progress on the International Space Station-we're part way up the mountain. In: Proceedings of the International Astronautical Federation's 52nd International Astronautical Congress, October 1–5, Toulouse, France

24. Kern, V.D., Bhattacharya, S., Bowman, R.N., Donovan, F.M., Elland, C., Fahlen, T.F., Girten, B., Kirven-Brooks, M., Lagel, K., Meeker, G.B. and Santos, O. (2001) Life sciences flight hardware development for the International Space Station. Adv. Space

Res. 27, 1023–1030.

25. Cann, C.E. (1997) Response of the skeletal system to the spaceflight. In: Fundamentals of Space Life Sciences (Churchill, S.E., ed.) Vol. 1, pp. 83–103.

26. Edgerton, V.R. and Roy, R.R. (1997) Response of skeletal muscle to spaceflight. In: Fundamentals of Space Life Sciences (Churchill, S.E., ed.) Vol. 1, pp. 105–120.

27. Shansky, J., Del Tatto, M., Chromiak, J. and Vandenburgh, H. (1997) A simplified method for tissue engineering skeletal muscle organoids *in vitro*. In Vitro Cell. Dev. Biol.-An. 33, 659–661.

28. Chromiak, J.A., Shansky, J., Perrone, C. and Vandenburgh, H.H. (1998) Bioreactor perfusion system for the long-term maintenance of tissue engineered skeletal muscle organoids. In Vitro Cell. Dev. Biol.-An. 34, 694–703.

29. Freed, L.E. and Vunjak-Novakovic, G. (1995) Cultivation of cell-polymer constructs in simulated microgravity. Biotechnol. Bioeng. 46, 306–313.

30. Freed, L.E., Hollander, A.P., Martin, I., Barry, J.R., Langer, R. and Vunjak-Novakovic, G. (1998) Chondrogenesis in a cell–polymer–bioreactor system. Exp. Cell Res. 240, 58–65.

31. Todd, P. (1992) Physical effects at the cellular level under altered gravity. Adv. Space Res. 12, 43–49.

32. Saltzman, W.M. (1997) Weaving cartilage at zero g: the reality of tissue engineering in space. Proc. Natl. Acad. Sci. USA 94, 13380–13382.

33. Cherry, R.S. and Papoutsakis, T. (1988) Physical mechanisms of cell damage in micro-carrier cell culture bioreactors. Biotechnol. Bioeng. 32, 1001–1014.

Cell Biology and Biotechnology in Space
A. Cogoli (editor)
© 2002 Elsevier Science B.V. All rights reserved

Space Bioreactors and Their Applications

Isabelle Walther

Space Biology Group ETH-Technopark, Zurich, Switzerland

Abstract

Space biology is a young and rapidly developing discipline comprising basic research and biotechnology. With the prospect of longer space missions and the construction of the International Space Station several aspects of biotechnology will play a prominent role in space. In fact, biotechnological processes allowing the recycling of vital elements, such as oxygen or water, and the in-flight production of food becomes essential when considering the financial and logistic standpoint. Every kilogram which, having been recycled or produced in space, does not have to be uploaded will drastically reduce the cost of space missions.

In addition, the scientific community is offered a better opportunity to investigate long-term biotechnological processes performing experiments with a duration ranging from weeks to months. Therefore, there is an increasing demand for sophisticated instrumentation to satisfy the requirements of future projects in space biology. The carryover of knowledge from conventional bioreactor technology to miniature space bioreactors for a monitored and controlled cell culturing is one of the key elements for this new dimension in space life science. The first space bioreactors were developed and flown at the end of the last century. It has been demonstrated that cells of different types, from bacteria to mammalian cells, can be successfully grown in this type of culture vessel. This chapter presents different generations of bioreactors developed so far, their performances in space and their potential for the future, as well as the activities of the European Space Agency (ESA) in this domain. A dedicated chapter by Lisa Freed on the rotating wall vessel reactor and the latest NASA bioreactor research is also part of this volume.

Introduction

Centuries ago, mankind learned how to use microorganisms for the production of food and beverages such as cheese, wine or beer. The first "biotechnologists" were probably not even aware that minuscule living systems were assisting them in their task. Even though biotechnology became an important scientific tool in the second part of the twentieth century, it is only in recent decades that it has really flourished,

198

thanks to the use of plants, insects and mammalian cells next to microorganisms. With an expanding role in pharmacology, tissue engineering, environmental protection and agriculture, biotechnology is expected to have a significant impact on our lives in the future. Besides appropriate biological systems, biotechnology requires dedicated instruments for their cultivation. Most biotechnological processes are performed in bioreactors, where physical parameters relevant to cell cultivation such as temperature, mixing speed, aeration and pH are monitored and controlled. The three cultivation modes that can be applied are batch, semi-continuous and continuous. In batch mode a fixed amount of nutritive solution is given to the cells, causing the environment to change over the cultivation period due to the constant nutrient metabolization and waste production operated by the cells. On the other hand, in continuous mode the fresh nutrient solution is continuously supplied to the cells, which allows the system to reach a steady state. Only in these conditions is environment constant over time, so that experiments performed at different points in time can be compared. Such an environment is a prerequisite for the quality and reproducibility of cultures. During the last ten years biotechnology has been increasingly exploited for tissue engineering and mammalian cells growth, for which new bioreactors and cultivation processes had to be developed free of disrupting shear forces. It is to avoid these undesired gravity-driven effects that biotechnologists have begun to show a growing interest in the possibility of operating in a weightlessness environment.

Space biology is a relatively young science that has evolved from the scientists' need to better understand the effects of space on living systems (for a review see Cogoli and Moore [1]). At the beginning of space flight, the main concern was the health of the astronauts; therefore experiments were mainly physiology and medicine-oriented. In the second phase, experiments in space focused more on basic science to achieve a better understanding of the microgravity effects on cell functions and their underlying mechanisms. Instruments allowing cell cultivation, at first in batch mode, were developed and flown. Even though the use of bioreactors in space has been very limited up to now, with the increasing duration of space missions and the assembly of the International Space Station a new need has emerged concerning the development of life support systems allowing for the recycling of expendable materials (i.e., water, air) and the treatment of waste by-products [2–4], and the cultivation of micro-organisms, mammalian cells and tissues for food production and medical purposes.

Space bioreactors: their development

General layout

The main elements necessary for an instrument that one can call a bioreactor are a culture chamber, a gas exchange system, a pump, a fresh medium and a waste reservoir, sensors, and a control unit with feedback. The general layout of a bioreactor is presented in Fig. 1. This is true for terrestrial as well as for space

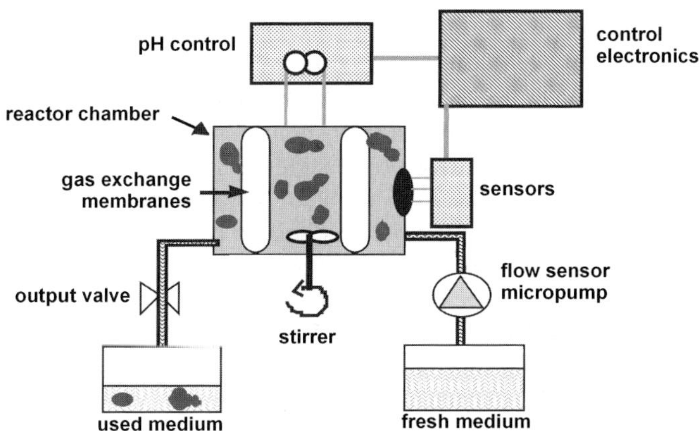

Fig. 1. Layout of a bioreactor system. The main elements are the cultivation chamber, the reservoirs (fresh and waste), the monitoring and the electronic control unit.

bioreactors. To adapt the bioreactor to all types of cells, from bacteria to human cells, all kinds of combinations of these basic elements can be used. One of the key features is monitoring. In fact, for the achievement of biotechnological investigations, the reproducibility of the bioprocess is essential. This can only be made in a bioreactor that allows the control of the physical parameters of the culture such as temperature, mixing, aeration, pH, and nutrient concentration. The usual way is to keep the environmental factors under control by automation and regulation of the cultivation process. The use of a continuous cultivation mode allows a better control of the cells' nutritive environment as in a batch culture in which nutrients are steadily metabolized and waste products accumulated. The growth conditions can be optimized and are reproducible. The cultivation time can be extended to long periods, so that several generations can be grown in the same conditions, allowing the investigation of adaptation processes and comparison between samples withdrawn at different intervals. The time of cultivation is no longer important because the cells grow in a steady-state. While the cultivation conditions are fully controlled, the cells' response to defined changes can be studied.

In space, the bioreactor cultivation allows using optimally the mission time, cells can be produced and harvested at different times without alteration of their quality. As such, bioreactors can operate as cell suppliers for further experiments.

Description and performance of space instruments

Bioreactors, which have been developed for Earth-bound experiments, are not suited to space cultivation. Firstly, most of the materials utilised for the fabrication of a bioreactor are not acceptable in space for safety reasons; no easily flammable plastic and no large pieces of glass are allowed. Secondly, the design of the space apparatus is restricted by limitations of size, weight, and power. Thirdly, the absence of sedimentation in space obliges, for example, to fill completely the cultivation chamber of the reactor (zero headspace) to avoid the presence of a gas bubble in the system. In fact, the bubble will not go to the top of the chamber as at 1 *g* but will float somewhere in the middle of the cultivation chamber creating interference. Finally, as convection movements are also reduced to near zero, nutrient, oxygen and waste products should be efficiently transported by means of medium exchange, perfusion or slow mixing. For these reasons, new types of instruments specifically adapted to space investigations had to be developed. Several types of cultivation systems have been designed (Table 1) or are currently under development. The sophistication grade of the devices presented below varies greatly. Some of the instruments consist of simple cultivation chambers with automatic perfusion system for the exchange of medium. Some are much more elaborate and allow automatic regulation of cultivation parameters (pH, oxygen), sampling or fixation of the cells during flight.

Table 1

Synopsis of the characteristics of different space cultivation systems

Instrument	Flight/Study	Characteristics
Woodlawn Wanderer 9	Skylab (1973)	Perfusion chamber for adherent cells, fixation possible in flight, only thermal regulation
Space Tissue Loss (STL-A) Cell Culture Module (CCM)	Shuttle flights (1992–1996)	Hollow-fiber cartridges for adherent and non-adherent cells perfused with oxygenated medium, spent media sampling, aliquot or whole culture fixation, thermal regulation. CCM is an up-graded STL-A with pH regulation and a 4°C compartment
Space Tissue Loss (STL-B)	Shuttle flights (1994–1996)	as STL-A but with video and microscope
Rotating Wall Vessels (RWV)	Shuttle, several flights Mir Station (1996)	Zero head space bioreactor with rotating vessel wall and gas diffusion membrane, for adherent and non-adherent cells, very low shearing force, sampling port, both fed-batch and perfused systems. In the perfused RWV, dissolved oxygen, pH and temperature are regulated
Carrar	Study	Cultivation chamber with medium exchange (peristaltic pump), no regulation
Dynamic cell culture system (DCCS)	Biokosmos 9 (1989) Shuttle (1992)	Cultivation chamber with medium exchange (osmotic pump), no regulation
Space Bioreactor (SBR I)	Shuttle flights (1994, 1996)	Zero head space bioreactor for yeast cells (3 ml), flexible continuous medium exchange (piezo-electric pump), sampling port, flow rate and pressure sensor, pH regulation, on-line data transfer
MEXSY	STS-84 (1997)	Type I fitting, several containers connected
BIOST-2	ESA Study, no flight	850 ml for mammalian cells
Space bioreactor (SBR II)	STS-107, foreseen 2002	Based on SBR I but larger chamber (7 ml)
MAP bioreactor	Test flight MASER 9 (2002)	Modular bioreactor for mammalian cells
Gradient bioreactor	ESA study (under construction)	Small bioreactor for mammalian cells cultures under specific gradient

The first instrument developed in the USA was the so-called *Woodlawn Wanderer 9* apparatus. It consisted of a fully automated perfusion chamber with devices for light microscopy and a cine-camera. It was installed aboard the US space station Skylab in 1973 [5]. The cells, a strain of diploid human embryonic lung cells, were cultivated 59 days in this instrument. Its size was $40 \times 19 \times 17$ cm^3 and its weight 10 kilograms. No difference was observed between the flight and the ground experiments at the level of cell cycle duration and mitosis. The authors concluded that microgravity did not have any effect on these cells.

Later came the Space Tissue Loss (STL) system, it was developed by the Walter Reed Army Medical Center (Washington DC), based on hollow-fiber technology, providing flexible feeding capabilities, thermal regulation and chemical fixation of the cells. It fitted in a mid-deck locker of the Space Shuttle, had a weight of 26 kg and was used on three flights. NASA at the Johnson Space Center developed another set of bioreactors, all based on the principle of the rotating wall vessel (RWV). Both the fed-batch and the perfusion systems are available, and oxygenation is achieved by diffusion through a silicone membrane. The first flight of a RWV bioreactor took place with mammalian cells in 1991. It was only later that a real regulation of the biological parameters (e.g. oxygen, pH) was implemented in the RWV system. The results obtained with the RWV are specifically described by Dr. Freed in the preceding chapter.

The instruments described above are all of rather large size; in fact the NASA experiment working unit is often the mid-deck locker, the size of which is about $48 \times 43 \times 23$ cm^3. For the scientist working with ESA, the working volume is much more restricted. Figure 2 shows the two standard ESA containers available: the Biorack Type I ($8.1 \times 4.0 \times 2.0$ cm^3) and Type II ($8.7 \times 6.3 \times 6.3$ cm^3). Biorack was one of the most-used ESA facilities dedicated to biological experimentation in space between 1985 and 1997; thus most of the instruments developed in Europe fitted in one of the Biorack standard containers. The Biorack multi-user facility consisted of

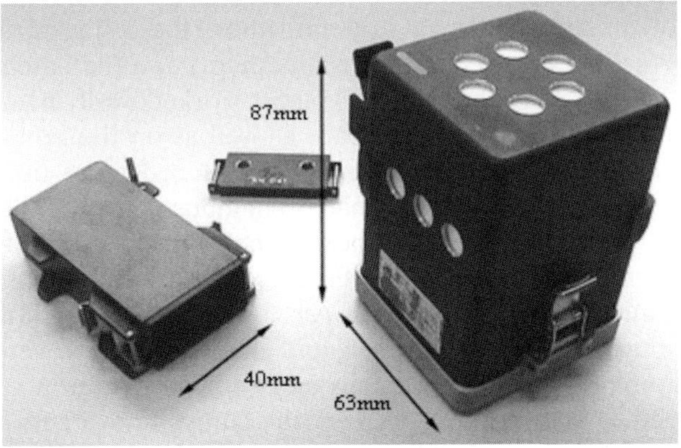

Fig. 2. Standard ESA containers. The small Type I and the large Type II are both available with or without electrical interface connector. Their volume is 65 ml and 385 ml, respectively.

several units: incubators, cooler, freezer, glovebox and stowage. Now the Biorack no longer flies and a new facility, the Biopack, of smaller volume (mid-deck size) will host the next European experiments in the Shuttle. Within the frame of the ESA project, several instruments have been developed for cell cultivation. The hardware for continuous cultivation, which have flown, were developed in Switzerland: the Dynamic Cell Culture System (DCCS) in collaboration with Contraves AG (Zurich, Switzerland) and the Space Bioreactor (SBRI) in collaboration with Mecanex SA (Nyon, Switzerland) and Seyonic SA (Neuchâtel, Switzerland). These two instruments fit in standard ESA Type I and Type II containers, respectively. A third instrument, the MEXSY, developed by Astrium (Germany) flew in 1997.

The Dynamic Cell Culture System (DCCS) is a completely closed system consisting of three main parts: the pump housing, the pump and two culture chambers (200 μl each). The novelty of this system was the self-powered osmotic pump continuously supplying the cells of one culture chamber with fresh medium at a flow rate of 1 μl h^{-1} at 37°C [6]. The osmotic pump principle is relatively simple. The innermost part of the osmotic pump consists of an impermeable reservoir surrounded by an osmotic agent. The reservoir and the osmotic agent are packed in a shell-like semi-permeable membrane. When the filled pump is placed in water, the osmotic agent absorbs water at a rate controlled by the pore size of the semi-permeable membrane. A hydrostatic pressure is created on the flexible reservoir, compressing it slowly and producing a constant flow of culture medium through the outlet. The second culture chamber was not supplied with medium (batch cultivation). This totally automatic cultivation instrument was designed for animal cells and plants cells. Of small size it fitted into an ESA Type I container (Fig. 3). Its biological performance was tested in a 14-day flight on board the Soviet biosatellite Biokosmos 9 in 1989 [7]. In this mission, the growth and development of plant protoplasts were studied. The system worked satisfactorily. It was tested a second time aboard the IML-1 mission in 1992 with hamster kidney cells growing on Cytodex 3 microcarriers [8]. The microcarriers served as a substratum to increase the available area for cell growth. After eight days of cultivation, the cell growth was compared between the perfusion (supplied with fresh medium) and the batch chambers. The results showed clearly that the osmotic pump worked well in microgravity; the volume of fresh medium delivered in space as well as on the ground was between 141 μl and 157 μl. The cells grew better and produced more tissue plasminogen activator in the perfusion than in the batch chamber. This was due to the better growth conditions in the perfused chamber. In the batch chamber, the cells began to die due to the accumulation of waste products and the resulting drop in the pH value. No effect of the microgravity on the cells *per se* was observable. For all parameters measured (pH, glucose, lactate, ammonia, and glutamine) there were no differences between cell culture grown in microgravity and on the 1-g reference centrifuge. The DCCS was proved to be a valuable tool for the cultivation of cells in microgravity. This instrument will be used again aboard a FOTON flight next year. This experiment, very recently selected by ESA, will investigate the growth of chondrocytes in a perfused culture in microgravity.

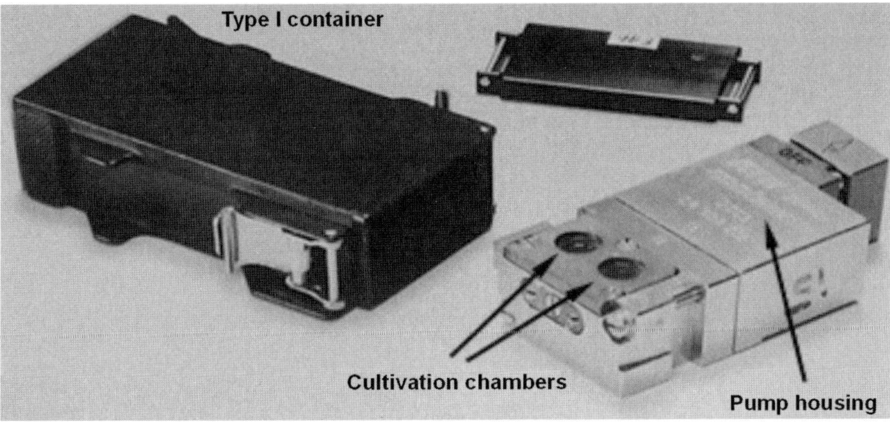

Fig. 3. Dynamic cell cultivation system DCCS. On the left is the container Type I. On the right the DCCS with the two circular windows over the cultivation chambers. The osmotic pump was located inside the metallic bloc.

The promising results observed with the DCCS concerning the perfusion chamber drove us to develop a more sophisticated device: a bioreactor for continuous cultivation in space. In the Space Bioreactor I (SBR I), a 3 ml culture chamber is supplied continuously with fresh medium by means of a piezo-electric silicon micropump of $20\times20\times2$ mm^3 size; the flow rate was variable between 200 and 1200 μl per hour. An additional flow sensor monitored the flow rate (Fig. 4). The micropump is driven with an actuator of 200 V and adjusting the actuating frequency between 0.25 and 1 Hz can set the required flow rate. This instrument was designed for yeast cells which we chose as a model organism [9]. It fits into an ESA standard container Type II as seen in Fig. 6. In both flight experiments (1994 and 1996), the cells were provided with fresh medium at different but fixed dilution rates during eight days. The flow rate was chosen to have dilution rates between 0.08 and 0.35. The data were collected on-line by means of a sensor inserted directly into the cultivation chamber. The chip size was 3.5×3.5 mm^2. This sensor was an integration of: (i) a pH-ISFET (ion-sensitive field-effect transistor) with Al_2O_3 gate insulator (sensitivity 51 mV/pH); (ii) a temperature-sensitive diode in forward bias at 100 μA; (iii) and a thin-film platinum redox electrode. The regulation of the pH was achieved by coulometric generation of hydroxyl ions at a titanium electrode in the bioreactor. The principle is schematically explained in Fig. 5. The counter electrode contained a chlorinated silver wire in a potassium chloride gel. The counter electrode was separated from the chamber by a cation-selective membrane. This type of pH control allows avoiding the use of concentrated NaOH. The pH sensor measured the level between pH 2 and pH 9 with an accuracy of $\pm$ 0.05 pH units. Its only disadvantage was its limited capacity due to the consumption of both the silver anode and the KCl in the counter electrode. The culture was, if wanted, agitated by means of a magnetic stirrer.

204

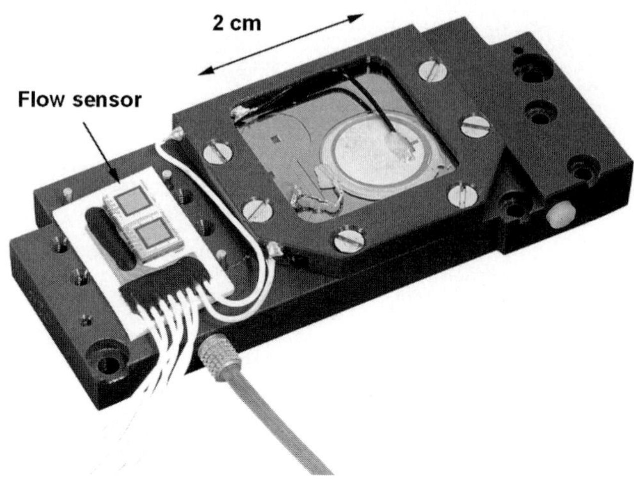

Fig. 4. Micropump. The piezo device is clearly visible as a disk on the right of the pump. The left part of the structure contains the flow sensor with the electrical connector and the tube bringing the medium to the culture chamber.

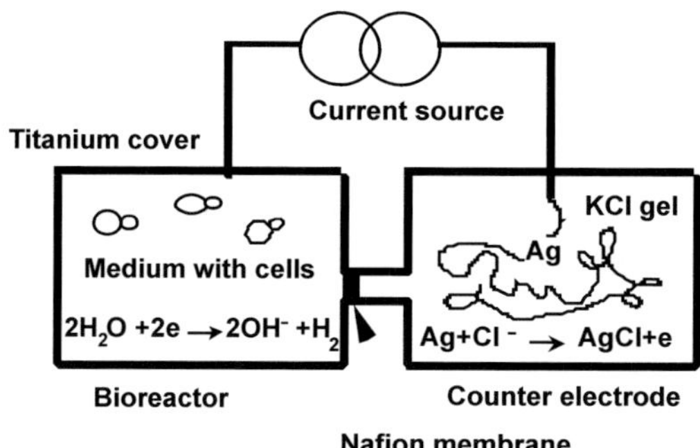

Fig. 5. Principle of the electrochemical pH regulation. K^+ ions are formed in the counter electrode; they pass through the Nafion membrane to the bioreactor chamber and combined with the hydroxyl ions to form KOH.

Technically, the bioreactor functioned according to expectations. The flow rate was stable and the pH was regulated accordingly to the expectations (Fig. 7). Biological data were gathered by means of samples, taken at preset intervals during the cultivation. From the biological standpoint, the yeast cells grew well in microgravity [10,11]. Their metabolism was comparable to that on Earth as they consumed glucose and produced alcohol in similar ways. Morphologically, no differences were observed between flight and ground samples (Fig. 8). A difference between the flight and the ground samples was observed at the level of bud scar positioning. Normally, the bud scars left on the mother cell after separation of the daughter cell are located bipolarly in diploid cells. We observed that this specific positioning was altered under

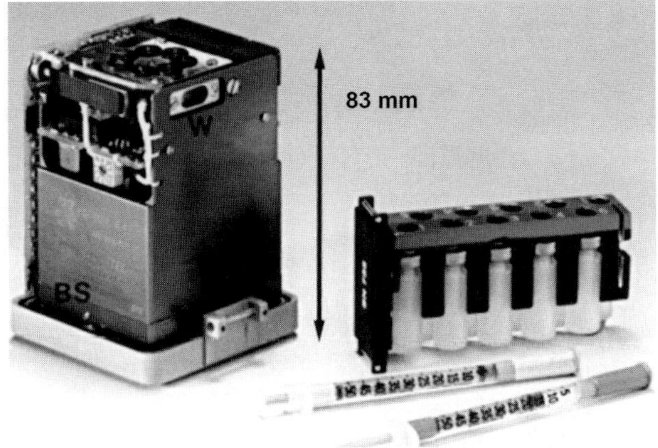

Fig. 6. Miniaturized bioreactor with two syringes for the sampling and the samples bottles. The cultivation chamber is on the top. The inspection window is located on the right upper side (W). The bottom metallic structure (BS) contains the fresh and the used medium. Total height 8.3 cm.

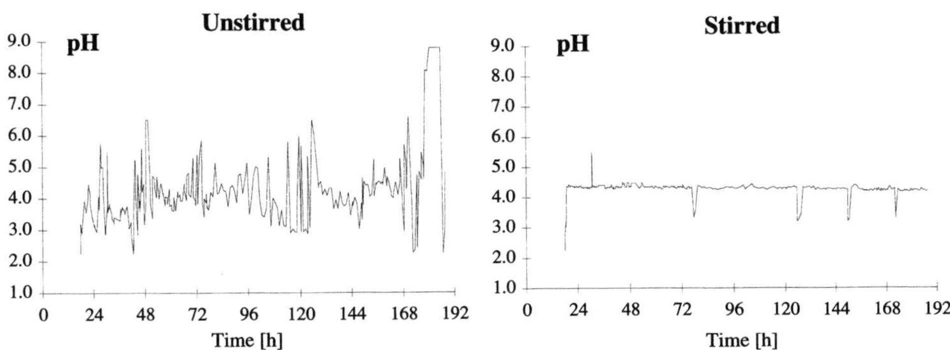

Fig. 7. pH values for both ground bioreactors. In the stirred one, the sampling times are very well visible (drops). About 1 ml of fresh medium at pH 2.5 is delivered to the chamber (pH 4.5) to replace the sampling volume, a pH drop is resulting especially because no compensation of pH is effected during sample withdrawal (no electrical connection of the bioreactor). Similar curve were obtained with the flight units.

microgravity conditions. Interestingly enough, the percentage of cells with bud scars located randomly is much higher in space than on Earth when comparing the same cultivation conditions (unstirred versus stirred) as seen in Fig. 9 [12]).

The MEXSY, modular experiment system developed by Astrium in Germany, offered standardized modules fitting in a Type I ESA container. The complete system consisted of 3–4 containers filled with the different elements of the cultivation system: the growth chamber with sensor, the sampling unit or used medium reservoir, the fresh medium reservoir, the control electronics. The fluidic containers were fitted out with special connectors for the quick, sterile exchange of modules. Based on this

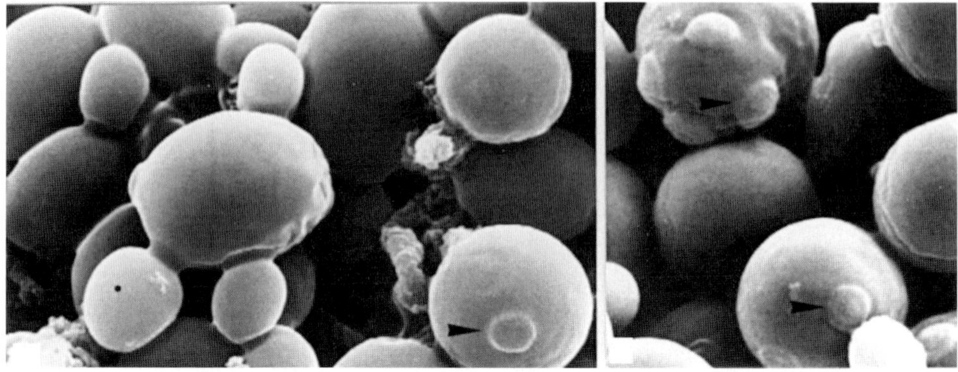

Fig. 8. Scanning electron microscope of the baker's yeast *Saccharomyces cerevisiae*. The arrows show on the budding scars. The left panel shows cells grown in microgravity, the right panel those grown at 1 *g*.

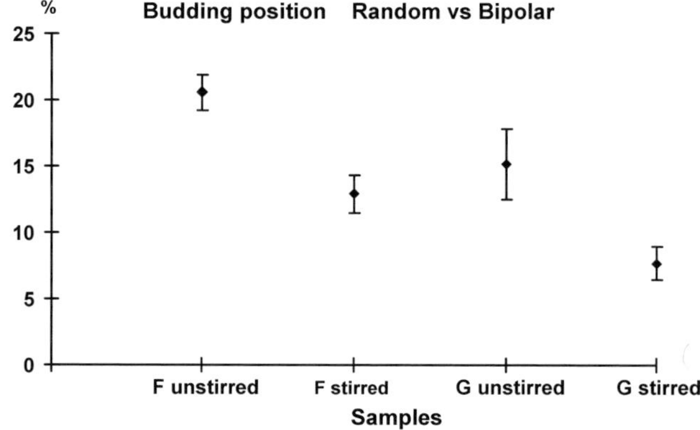

Fig. 9. Percentage of the randomly located bud scars vs. the normally bipolar positioning. The values were obtained counting 500 cells per sample; the standard deviation was calculated on the four samples available per bioreactor. F: Flight, G: Ground.

concept, a special mini-aquarium consisting of one transparent chamber fitting in one Type I container was developed and flew during the Shuttle mission STS-84 (1997). In this experiment, the "vestibulo-ocular reflex of tadpoles and fish" was successfully investigated. The survival rate of the *Xenopus laevi* embryos was over 90% after the incubation time of the experiment. The results obtained showed that long periods of exposure to microgravity affect the roll-induced vestibulo-ocular reflex significantly in a long-term manner, while short periods are almost ineffective [13,14]. No results or flight experiments have been presented with the complete system allowing continuous cultivation and automatic sampling to date.

The experiments performed up to now with bioreactor cultivation in real space are quite scarce, but the possibility of using the international space station as a research platform in the future will allow a large number of experiments to be performed.

Engineering studies and future instruments

Several types of bioreactors have been submitted for study in Europe. These studies were conducted under the direction of the ESA or of national space agencies, such as CNES (France), or ASI (Italy).

The *Carrar cell culture system* has been developed for the IBIS hardware (CNES, France). Its external dimensions are $62\times50\times28$ mm^3. It consists of a 2 ml culture chamber with two medium reservoirs of 2 ml each and a micro-peristaltic pump. The speed of medium exchange (up to 500 μl h^{-1}) is variable. No data are available concerning the use of this device.

The *BIOST-2 BB* has been developed in Switzerland by the same company that developed the *SBR I*. The study was conducted for ESA. This quite large bioreactor had a volume of 850 ml and a mass of 5 kg. It was perfused, sterilisable and the agitation was performed by means of an interesting system: spiral balancing. Both upper and lower plates of the bioreactor chamber, flat with cylindrical rims, rotated partially in opposite directions creating a circular motion, thanks to the special spiral inner geometry. The culture was very gently mixed, thus avoiding excessively high shear forces which could damage sensitive mammalian cells. A portion of at least one plate is porous allowing the oxygen to diffuse into the culture (Patent FR 2724280, 1996). No further result from culture with this bioreactor has been published elsewhere.

Another ESA study (General Support Technology Program), with a larger bioreactor for cell culture, has been conducted with bone cells and human hepatogenic cell line. A bread-board of this instrument was developed and built by Mecanex (Nyon, CH) in collaboration with the Institute of Microtechnology (Neuchâtel, CH) and Logica (B). The cylindrical culture chamber had a total volume of 250 ml; it was equipped with three optical access ports, an agitator with permeable tubing providing the culture with gas and nutrient and removing gaseous metabolic products. Peltier elements were used to provide the desired temperature. Sensors for temperature, pH, dissolved and undissolved oxygen as well as for CO_2 were integrated. The totality was regulated by Labview software and an electronic box for the interface and the data logging. Cultivation tests were performed and results showed that this instrument is suitable for short-term culture (up to four days); long-term cultures were difficult to perform, principally because of contamination problems. Small design changes should alleviate these difficulties. Its total mass is 30 kg including pumps and electronic box (ESA communication).

The next project using a bioreactor, which is currently in the ultimate testing phase, is the YSTRES project. This experiment, planned for the shuttle flight STS-107 (foreseen end-2002), will fly in the newly developed multi-user facility Biopack. Biopack is smaller than the previously used Biorack facility. It consists of one incubator equipped with static and centrifuge rack. The novelty is that on this

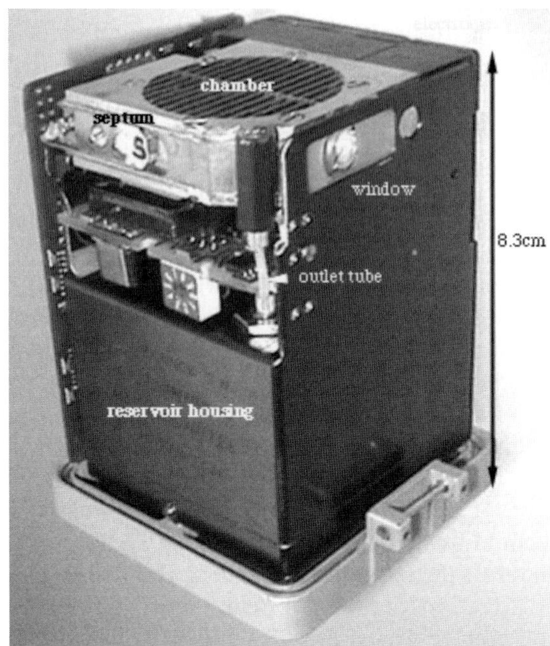

Fig. 10. SBR II: The new version of the SBR I bioreactor has 7 ml working volume in the chamber. Sampling can be achieved by puncturing the septum by a needle (dispenser or syringe). The bioreactor has a height of 8.3 cm.

model the centrifuge can also accommodate Type II containers. In this project a new version of the SBR I bioreactor, called *SBR II*, has been just developed (Fig. 10). This bioreactor is based on the same principle as SBR I but has a working volume of 7 ml instead of the 3 ml of SBR I. This allows the withdrawal of larger samples for further investigation during flight. In this experiment, the sampled cells (in our case the baker's yeast cells, *Saccharomyces cerevisiae*) are then submitted in another hardware unit comprised of six small investigation chambers to several stresses: a heat and an osmotic shock will be applied (Fig. 11). The heat shock is performed by rapidly heating (within a minute) the sample from the cultivation temperature of 27°C to the shock temperature of 37°C. This is achieved by small Peltier elements inserted in the investigation chamber. The current is furnished by the Biopack facility. The osmotic shock is applied on the cells by the adjunction of salt in the culture medium. After an incubation of 30 minutes the reaction is stopped by adding a fixative solution or by freezing the chambers. The analyses of the stress reactions, at the RNA expression and cytoskeleton level, will show if the cells are able to correctly react to stress and if the different stress reactions (osmotic versus heat) are comparable under space conditions.

Further, new types of bioreactors are currently under design and development within the frame of ESA studies, for example, a gradient bioreactor and a modular

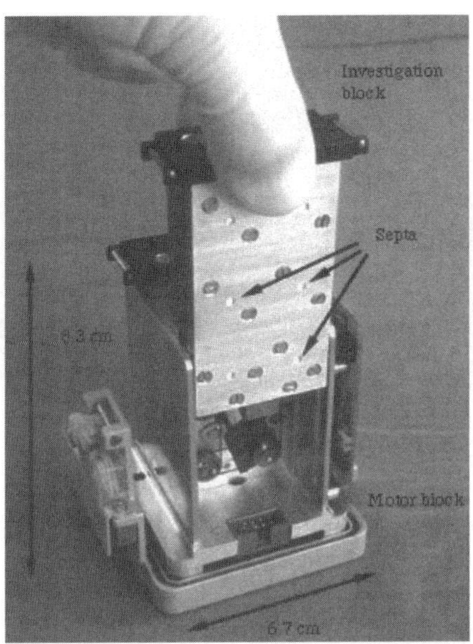

Fig. 11. Insertion of an investigation block into the motor block. Two investigation blocks each with six investigation chambers (under the septa) can be inserted into this motor block. The lowest chamber is electrically connected to the motor block and will be heated. The small motors are driving magnetic agitators in each chamber to avoid sedimentation of the cells on the centrifuge or on the ground.

bioreactor for tissue engineering. The gradient bioreactor principle consists in a cultivation chamber for animal cells in which a gradient can be formed and maintained. The idea was based on the fact that gradients have an enormous influence on cell growth and if one wants to study their effect it is important that they can be monitored in the cultivation chamber. This project is conducted by Mecanex SA (Nyon, Switzerland) in collaboration with Seyonic SA (Neuchâtel, Switzerland) and our group in Zurich. It is in the design phase and the first hardware tests are foreseen in the second half of 2002. It is planned to use different types of mammalian cells to test the stability and the linearity of the gradient within the culture chamber. Visualization will be performed at macro and micro levels.

The second ESA bioreactor project currently under development deals with a modular bioreactor for tissue engineering. This development is part of a MAP project (Microgravity Application Program) financed by ESA and by the participation of a non–space industry, in this particular case Centerpulse (previously Sulzer Medica) (Winterthur, Switzerland). ESA grants funds to universities of an equivalent amount to that invested by associated companies. Three modular cultivation system units have been developed by Fokker (The Netherlands): one for thyroid cells in collaboration with Dr. Ambesi-Impiombato (University of Udine, Italy), one for vascular cells in collaboration with Dr. Bader (Hanover) and one for cartilage cells in collaboration with our group. This latter unit is called CODI (Fig.

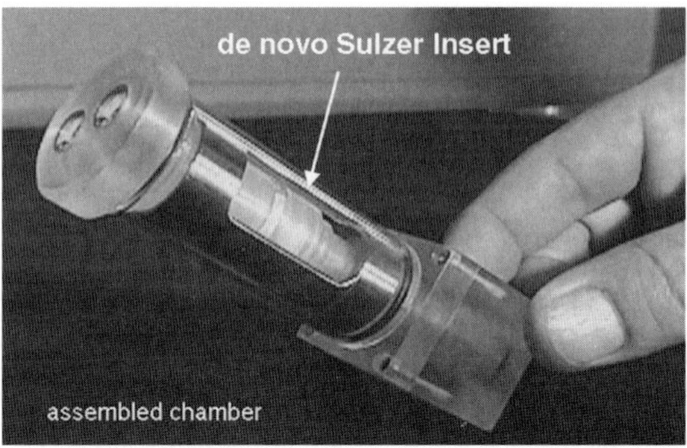

Fig. 12. Chondrocyte module "Codi" for the experiment on MASER 9. The *de novo* insert developed by Sulzer Medica is indicated by an arrow.

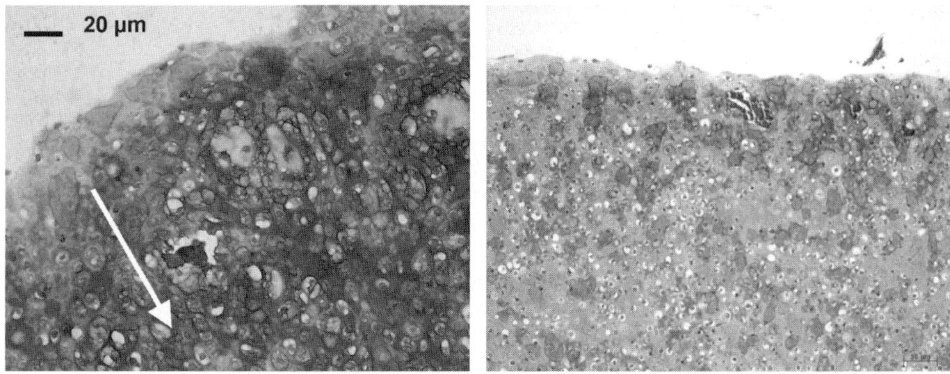

Fig. 13. Cartilages formed from pig chondrocytes cultured for 21 d (50×10^6 cells/ml, *de novo* insert 5 mm deep) under conditions simulating microgravity in the random positioning machine (left panel) and at 1 *g* (right panel).

12). The *de novo* insert, in which the cartilage cells are growing, has been designed and developed by Sulzer Medica. The first test of these growth modules was performed aboard the sounding rocket MASER 9 in March 2002. Preliminary results have shown that solid and compact specimen of cartilage can be obtained after two or three weeks of culture either at 1 *g* or under conditions simulating low-*g* in the random positioning machine (RPM). As shown in Fig. 13, there are clear morphological differences between the static (right panel) and RPM-samples (left panel). Firstly, the cartilage from the RPM is round in shape, almost spherical, whereas that grown at 1 *g* reflects the geometry of the *de novo* insert. Secondly, the histological analysis reveals a more compact cellular structure in the RPM than in the static 1 *g* samples. Thirdly, the arrangement of the cells in the cartilage appears in a more "ordered" geometry in the RPM. In fact, whereas at 1 *g* the cells are distributed at random in the intercellular matrix, in the RPM the cells appear arranged in lines (see arrow in left panel of Fig. 13).

Although these data are preliminary and only qualitative, and need further verification, they indicate that the structure of the cartilage formed under simulated low-g is somehow different from that obtained at 1 g. At this point it cannot be said, however, whether this finding may lead to improved quality of transplants for humans or not. Nevertheless, such experiments will contribute to the understanding of the biological mechanisms of chondrocyte differentiation [15].

Bioreactors: their future

Until now bioreactors have been used in space for basic research and to carry out a preliminary screening of biological system candidates for bioprocessing in microgravity. While interesting observations have been made, no biotechnological breakthrough has been achieved so far. Nevertheless bioreactors will be required in future to supply fresh living cells for basic experiments in space stations and they will be of great importance as elements of the life-support system to deliver and/or recycle oxygen or biomass to be used as food. Now that the use of bioreactors has been proved adequate for the continuous cultivation of cells in space, the era of improvements and new adaptations has begun. Further developments have started to build improved instruments for the cultivation of mammalian cells and tissue engineering in microgravity conditions. The fact that mammalian single cells are undergoing profound alterations in microgravity has nourished hypotheses and speculations about possible commercial and medical applications. Bioprocessing in space has become one of the interesting themes of the exploitation of the ISS. A few pharmaceutical companies manifested their interest in joint application research programs with national and international space agencies. For this reason ESA has started a Microgravity Application Program (MAP) to support application-oriented projects with participation of non-aerospace industries. ESA is also supporting the activity of "Topical Teams" of scientists who are discussing the potential industrial and commercial applications in space.

An example of such activity is the first MAP project which started in May 2000 and is aimed at developing instruments (bioreactor) and technologies for tissue engineering. Team members from academic institutions are A. Bader, Hanover; S. Ambesi-Impiombato, Udine; P. Bruckner, Münster; R. Pörtner, Hamburg; A. Cogoli and I. Walther, Zurich; P. Bittmann is the industrial partner from Centerpulse, Winterthur. The objectives of the project are: to develop procedures of *in vitro* organogenesis of pancreatic islets, thyroid tissue, liver, vessels and cartilage; to study the mechanism of organogenesis in low-g; to define the requirements of a modular space bioreactor for medically relevant organ-like structures; and to set up procedures for the production of implants for medical applications. It is believed that low-g may contribute in two ways to progress in this field. Firstly, as a useful and non-invasive tool to study important and still obscure biological events such as signal transduction, gene expression, and cell proliferation. Secondly, low-g may favor the

212

mass production of single cells by obtaining higher cell densities per unit culture volume as well as a smooth cell–cell aggregation and three-dimensional organogenesis in the absence of sedimentation and shear forces.

Terrestrial spin-offs of space bioreactors are also important. The development of micro-sensors, pumps, and valves has served the medicine of today; and fluid management for titration at the nano level is also a possibility. Newly conceived bioreactors and current research areas include: the effects of reduced levels of mechanical and hydrodynamic shear; the effects of spatial co-location of participating cell populations; the role of mass transport on cellular propagation and tissue assembly; the effects of culture media (e.g., factors) on cellular metabolism and waste accumulation; the development of technologies (biosensors for pH, glucose, and oxygen); new tissue culturing methods and strategies for drugs testing in organo-like tissue; and research into mammalian, plant, and insect culture.

In conclusion it can be said that—independently of their design, origin or mode of function—bioreactors will be increasingly used in space in the future. In cooperation with the medical community, they will be used to prepare better models of different types of tissue, for the investigations of cell growth, and probably for the engineering of larger three-dimensional cell constructs. Moreover, they will be a mandatory component of the system for the production of food and of the recycling of consumable material for longer missions or future space stations.

Acknowledgements

The developments of the DCCS, of the SBR I and SBR II performed in our laboratory have been supported by grants from the Prodex Program of ESA, the ETH Zurich, Contraves AG (Zurich, Switzerland), and Bio-Strath AG (Zurich, Switzerland). The MAP Project is supported partially by ESA and partially by Centerpulse (Winterthur, Switzerland).

References

1. Moore, D. and Cogoli, A. (1996) Biological and Medical Research in Space. Springer Verlag, Heidelberg, pp. 1–106.
2. Skoog, A.I. and Broullet, A.O. (1981) Trends in space life support. Acta Astronautica 8, 1135–1146.
3. Wolf, L. (1996) Bioregeneration in space. Adv. in Space Biol. and Med. 5, 341–356.
4. Muller, C.N. and Porter, R.J. (1996) Evaluation of a bioreactor system used to treat human waste streams on space station. Abstract of the General Meeting of the American Society for Microbiology, 400.
5. Montgomery, P.O.B., Cook, J.E., Reynolds, R.C., Paul, J.S., Hayflick, L., Stock, D., Schulz, W.W., Kimsey, S., Thirolf, R.G., Rogers, T. and Campbell, D. (1978) The response of single human cells to zero gravity. In Vitro 14, 165–173.
6. Gmünder, F.K., Nordau, C.G., Tschopp, A., Huber, B. and Cogoli, A. (1988) Dynamic cell culture system a new cell cultivation instrument for biological experiments in space. J. Biotechnol. 7, 217–227.

7. Rasmussen, O., Gmünder, F.G., Tairbekov, M., Kordyum, E.L., Lozovaya, V.V., Baggerund, C. and Iversen, T.H. (1990) In: Proceedings of the Fourth European Symposium on Life Sciences Research in Space, Trieste, ESA SP-307, 527–530.

8. Lorenzi, G., Gmünder, F.K. and Cogoli, A. (1993) Cultivation of hamster kidney cells in a dynamic cell culture system in space. Microgravity Sci. Technol. 6, 34–38.

9. Walther, I., van der Schoot, B., Jeanneret, S., Arquint, P., de Rooij, N.F., Gass, V., Bechler. B., Lorenzi, G. and Cogoli, A. (1994) Development of a miniature bioreactor for continuous culture in a space laboratory. J. Biotechnol. 38, 21–32.

10. Walther, I., Bechler, B., Müller, O., Hunzinger, E. and Cogoli, A. (1996) Cultivation of *Saccharomyces cerevisiae* in a bioreactor in microgravity. J. Biotechnol. 47, 113–127.

11. Walther, I., van der Schoot, B., Boillat, M. and Cogoli A. (2000) Performance of a miniaturized bioreactor in space flight: Microtechnology at the service of space biology. Enzyme Microb. Technol. 27 (10), 778–783.

12. Walther, I., van der Schoot, B., Boillat, M., Müller, O. and Cogoli, A. (1999) Microtechnology in space bioreactors. Chimia 53, 75–80.

13. Kuebler, U.M. and Kern, P. (2000) MEXSY—A modular tool kit for life science experiments in space. Gravitat. Space Biol. Bull. 14(1), Abstract 143, 65

14. Horn, E.R. and Sebastian, C.E. (1999) A comparison of normal vestibulo-ocular reflex development under gravity and in the absence of gravity. In: Biorack on Spacehab: Biological Experiments on Three Shuttle-to-Mir Missions (M. Perry, ed.). ESA Noordwijk, The Netherlands, ESA SP-1222, pp. 127–138.

15. Conza, N., Mainil-Varlet, P., Rieser, F., Kraemer, J., Bittmann, P., Huijser, R., van den Bergh, L. and Cogoli A. (2001) Tissue engineering in space. J. Gravitational Physiol. 8, 17–20.

Space Cell Physiology and Space Biotechnology in Russia

Anatoly I. Grigoriev[1], Yury T. Kalinin[2], Ludmilla B. Buravkova[1] and
Oleg V. Mitichkin[3]

[1]*Russian State Research Center–Institute for Biomedical Problems, Russian Academy
of Sciences*
[2]*Russian Stock Company "Biopreparat"*
[3]*Russian Space Corporation "Energia"*

Introduction

The progress made in astronautics and the advancements in knowledge of cell and
molecular biology resulted in the establishment of a new scientific discipline: space
biotechnology. Biotechnology is commonly listed among the three global main-
stream industries, along with the Internet and signal communications. Biotechnology
has attained its leading position because we are learning the biological laws of the
evolution of living systems, including the genetic role of DNA, and are developing
the biochemical techniques for research; moreover, biology gives us the social
control to settle some of the most difficult problems of health protection and rational
diet.

Technologies utilizing living organisms have always attended human life. Milk
processing, cheese-making, brewing and vinification, manufacturing of therapeutic
preparations are, in effect, biotechnologies, too. However, it is only in recent years
that space biotechnology processes have became the products of not only empirical
data but also of telic scientific investigations, i.e. of a scrutiny of possible effects and
analysis of action factors in space. Technology is interpreted as a totality of processes
and control procedures securing a pre-determined result. This can be applied in full
measure to space biotechnology, the basis of which has been laid by fundamental
research into the effects of space flight (primarily microgravity) on the biological
processes in living systems on varying levels of development. Made in an attempt to
simultaneously gain insight into the underlying events during space flight and to find
realistic solutions of theoretical predictions, the initial experimental steps in cell
biology and biotechnology are very tentative.

Space biotechnology is a complex of novel technologies aimed at the production
in microgravity of bio-objects and unique biological materials, among them
diagnostic, therapeutic and preventive preparations. This is one of the major areas of
research aboard piloted orbital stations that will allow the use in microgravity of new
methods for producing useful biosamples.

216

Cell cultures

Cultivated cells are widely used in biotechnologies for medicine, pharmacology, agriculture and others. In space biology isolated cells and tissues are a convenient model for studying direct effects of a changed gravity on intracellular processes, including genetic and information flows in living systems. Referred to as adaptive, peculiar shifts in the structure and functioning of cells during space flight can, in some cases, be used as a basis for original biotechnological processes. However, difficulties associated with the design and implementation of space experiments make their results incalculable.

Studies of the biological effects of spaceflight factors, primarily microgravity, on cells *in vitro* have a long history. There have been numerous experiments with various types of cells isolated from tissues of different animals and plants, and with populations of heterotrophic and autotrophic unicellular organisms. These experiments have been performed not only in real space flights on piloted and unmanned space vehicles, but also during simulations of changed gravity on Earth [1–4].

It was shown that neither the number nor the spectrum of chromosomal aberrations in isolated cells were significantly modified by gravity changing in the range up to 10^{-5} g. After exposure in microgravity, cultures did not alter the number of chromosomes, neither did they display any serious deviations in the number of DNA and RNA in cells [5]. It was proposed that microgravity does not affect implementation of the genetic program. For instance, on return from space flight, cultures of somatic polypotent cells were able to form healthy mature plants from the germs that had developed in microgravity [6,7]. The karyotype of cells did not undergo noticeable alterations due to changed force of gravity; in the ensuing years this was confirmed in experimental harvesting of several generations of plants on station Mir [8,9].

However, analysis of experimental samples suggests that changes in the force and vector alter cell morphology and physiology. Abnormal shape and size of cells, rate of protein proliferation and synthesis and shifts in biochemical parameters showed that living systems are not unaffected by gravity [10–15]. At the same time, the main causes for and the molecular mechanisms of the shift observed in microgravity remains unclear. Account should be taken of the implications of space experiment "technology" as some simple laboratory procedures, e.g. stir-up or sedimentation, become impossible. Spaceflight conditions are not a great obstacle to routine procedures with cultivated cells; however, diversion from the experimental protocol may lead to artifacts.

At the same time, control of cell proliferation and differentiation by changing gravity values and vector direction appears quite probable at present. This inspires the hope of future optimization of biotechnological processes for cultivation of isolated cells and unicellular organisms in bioreactors for space life-support systems, and the solution of applied biomedical problems in microgravity.

Analysis of available data in space cell physiology allows the identification of several potential areas of research: firstly, fundamental investigations of molecular mechanisms of the cell gravitational sensitivity and involvement of intracellular structures in adaptation to microgravity; secondly, design of technologies for advanced space life-support systems and optimization of biotechnological processes, specifically development of three-dimensional tissue structures. Thirdly, experiments with cell cultures provide a good opportunity to address important issues of space medicine, such as, for example, immune state of humans, tissue differentiation and oncogenesis, post-traumatic regeneration and wound healing.

The experimental studies on board orbital station Mir have shown that selection of objects and methods for space biology studies should be made with particular care, as attenuation of the gravitational factor or its vector changes did not inflict primary damage on the cell or cell structures, provided all other cultivation conditions (substrate composition and parameters, temperature, etc.) are adhered to. In our opinion, the best choice is the use for space experiments of mechanosensitive cells, e.g. osteocytes, fibroblasts, epithelial and endothelial cells that are highly responsive to gravity. With these models an attempt can be made to analyse the mechanism of gravitational signal reception.

Tissue cultures and malignant cells will allow discern reactions of normal and neoplastic cultures to microgravity. *In vitro* investigations using immunocompetent cells can help bring to light triggers of the immune system suppression in space flight [3,16].

It should be emphasized that most of the above-listed types of cells, as well as plant cell cultures and associations of unicellular plants, have been used in space biology model studies of the microgravity effects and adaptation to the spaceflight factors [17–19]. However, because of technical difficulties and the unpredictability of the results in many cases there have been very few biotechnological experiments in space orbit.

One of the most interesting fields for future investigations is three-dimensional cell cultures. A long period of time has passed in the search for methods that will enable production of implants out of patient's autologous cells to treat or replace damaged or missing tissues. Qualitatively new results were obtained only when consideration was given to the role of gravity in culture growth *in vitro*. It was clearly shown that cell distribution over the surface and cellular interactions are dependent on gravity. This dependence could be used for developing special devices for 3-D cultures [20–22]. Morphology of cell complexes may also be modified by the spatial positioning of cells, mechanic disturbances in their environment, and a matrix composition in microgravity. In addition, three-dimensional cultures appear to be a more adequate model for studies of cellular processes as compared with the traditionally used monolayer.

Bone tissues are one of the objects in this type of investigation. On the one hand, it is unarguable that development of *in vitro* bone implants is a call of the day. On the other, bone losses in long-term space missions constitute a serious threat to the astronaut's health [10]. It is believed that this phenomenon is associated with inhibition of osteogenesis caused by possible disturbances in osteoblast differentiation, matrix maturation and mineralization [23]. During biosatellite missions, bone cells were cultivated in so-called plungers integrated in a Biobox with automatic feed of fresh substrate; gas exchange was not regulated. In preosteoblasts MN7, proliferation was unchanged; however, the differentiation rate estimated by the activity of alkaline phosphatase and expression of the type-I collagen was found to be inhibited [24].

Cultivation of chondrocytes on spongy pads prepared from polyglycolic acid in a rotating bioreactor first for three months in the laboratory and then for four months on Mir resulted in a structure consisting of the polymer material and differentiated cells weighing 0.3–0.4 g. In contrast to the structures obtained on the ground, they had a more regular spherical shape, smaller size, and lower mechanic properties [25].

Laboratory pilot experiments were performed at IBMP in an effort to produce 3D-tissue structures by cultivating fibroblasts and cardiomyocytes on microcarriers [26]. Physical effects in consequence of a change in the position of cell culture relative to the gravity vector included morphology modification of cell organelles, cell functioning and development. Thus, rotation of myocytes in clinostat at 1–50 rot./min increased cell proliferation, the areas of cells and nuclei. Utilization of three-dimensional cultures by cell biology allows us to build a bridge between two disciplines: tissue engineering and cell physiology concerned, specifically, with physiological responses to the environmental factor and pathological processes.

Reactions of cell cultures to space flight are strikingly numerous. This fact is a strong argument for gravitational sensitivity of at least several types of cell cultures. Hence, absence of spaceflight specific alterations in cell structure and physiology in the pioneering experiments [1,27,28] could have accounted for selection of objects and methods.

We suggest the following areas for future investigation:

Cell growth in space flight: intercellular relations, cellular adhesion, and cell proliferation: Previous experiments with cell cultures offered very convincing evidence that they alter function and morphology with changes in gravity. Therefore, examination of the cell cycle (proliferation, differentiation, apoptosis) under conditions of spaceflight and simulated microgravity remains topical.

Changes in the adhesive characteristics of cells and macromolecules in the absence of gravity may assume significance. These changes may be critical for cell migration and intercellular contacts in organisms during, for instance, reparation of vessels and tissues, and tumor development. In addition, cell adhesion to substrate is known to be a signal for cytoskeleton rearrangement which will determine cell shape and spatial parameters of organelles, and influence a large number of regulatory processes in cells. The investigations should be performed with cell cultures and artificial non-cellular model systems.

Development of technologies for production of three-dimensional tissue cultures and studies of organ and tissue regeneration (tissue engineering): Advancement in this area also requires understanding of the differences in cellular interactions at 1 g and microgravity, contribution of different factors (adhesive molecules, intercellular matrix, cytokines, growth factors, oncogenes) to development of three-dimensional cultures, analysis of the cell receptor and signal transduction. Also, it is very important for space medicine to study the mechanisms of cellular interactions during tissue reparation in microgravity.

In the near future, *immunologic investigations in space flight* are going to become one of the main directions of space medicine as soon as techniques enabling immunologic analyses in microgravity are developed. Therefore, a special place is occupied by the perspective of studying microgravity effects on immunological cells responsible for homeostasis control, stability of antibody structures in particular, and constant oversight of the genetic stability of the somatic cells population.

In the past twenty years a series of experiments with isolated immunocompetent cells has studied their growth and development, proliferation, differentiation and other processes in cytoplasm and nucleus in microgravity [16,29]. It was implicitly concluded that the functional activity and structure of the cells had been changed. The investigations were a first attempt to answer the question of whether microgravity modifies cell functions and what are the processes of these changes.

Key stages in the immune reaction are the forms of cell interaction and response such as recognition, activation, proliferation, differentiation and regulation. Therefore, we think it reasonable to focus effort on investigation of these processes in the micro-*g* environment. Appropriate test sets will be needed to get the most objective characteristics of each component of the immune system functioning. One thing is clear: we cannot take the existing immunologic techniques to microgravity for the reason that in the absence of normal gravity isolated cells behave in a different manner. Under this circumstance we need to look for original technologies that will allow us to perform the investigations of isolated immune cells aboard space stations.

The general trend in this area suggests that resources should be primarily directed towards development of the following package of immunologic assays with immunocyte recognition as the first step and immunocyte differentiation as the final step:

1. mixed lymphocytes culture;
2. expression of interleukin-2 receptors on T-lymphocytes activated in culture *in vitro*;
3. analysis of signal transduction systems in mononuclear cell culture;
4. proliferative response of lymphocytes to specific antigens (tuberculin, etc.) *in vitro*;
5. immunization of immunocompetent cells by specific antigens in culture;
6 cytotoxic effect of immune cells on neoplastic target cells *in vitro*;
7. synthesis of specific and nonspecific proteins by immune cells in culture;
8. reactivity of immune cells to immunocytokines *in vitro*;
9. surface antigens of immune cells;
10. metabolic profile of phagocytes.

From our point of view, it might be extremely exciting to investigate also the cytotoxic properties of lymphocyte-killers with respect to neoplastic target cells under conditions of microgravity. Special investigations are necessary to answer the question of how the effects of the lowered gravity of space flight on plasmatic membrane and locomotion influence cytotoxicity of lymphocyte-killers with respect to target cells [30]. At present IBMP investigators are carrying out project "Cell Interactions" aimed at studying the cytotoxic activity of lymphocytes in microgravity [31].

It should be stressed that there are many difficulties in developing immunologic biotechnology in space; however, experiments with isolated immune cells in microgravity not only expand our notion of the functioning of these cells in weightlessness but also lay the basis for designing methods to monitor space-crew immunity immediately during mission.

We could conclude that experiments with isolated immune cells show that a great step forward was made in the past decade, i.e., from relatively simple tests to sophisticated experiments with the use of methods and principles of analysis employed by modern fundamental immunology. This allowed more comprehensive vision of some of the processes taking place in cells and molecules exposed to microgravity. Many questions still remain, but their answers are not too far ahead. As for lymphocytes, we are dealing with a broad spectrum of events associated with cell contacts. These processes may be partially cancelled by microgravity. According to the assumption that cells learn about gravity indirectly when reacting to cytoskeleton deformation (accompanied, possibly, by deformation of the cell recognizing membrane), loss of contact with substrate and other cells as well as disuse of the locomotor functions in microgravity may put lymphocytes isolated in culture in the position of osteocytes of the weight-bearing skeleton of macroorganism during a stay in microgravity.

Biotechnology in life support systems

One of the applications of space biotechnology makes use of biological objects (microorganisms, associations of unicellular organisms) in life-support systems of manned space vehicles designed for long-term orbital and planetary missions. These systems are for regenerating air and water, supplying food, and managing wastes. The demands placed on them are self-sustenance, high reliability and long running time with minimal requirements of power and other resources. Key biological elements of the life support systems will be plants and microorganisms integrated with units realizing physical/chemical technologies [32–35].

Experiments with cultures of free-floating unicellular organisms (autotrophs and heterotrophs) have the purpose of refinement, development and introduction in space biotechnology of cultivation systems and techniques in which the gravity vector will be controlled both in magnitude and direction. The theoretical aspects of the problem have been published by M.G. Tairbekov [36,37]. At present, there is enough design data and methical approaches to research and develop space hardware that will provide adequate support to unicellular organisms cultivated in closed environment [38–41].

One of the very important problems in the field of life-support systems is the influence of space flight on the characteristics of microorganisms in a closed environment. Quantitative and specific monitoring of the microbial environment of the

Russian orbital complexes Salyut-6, Salyut-7, and Mir have been regular procedures over many years [42]. The most in-depth investigations of microbial community changes in space were made during assembly on the orbit and long-term operation of multi-modular space station Mir [43–45]. Dynamic investigations of microflora in air, interior materials, equipment, and water were performed in the period from 1987 to 2000. Microbial samples were collected in the basal and other modules; 234 species of bacteria and microscopic fungi were identified in the Mir environment. The bacterial flora included 108 species, and the fungal flora included 126 species. The greatest diversity (64 species) was demonstrated by the group of technophilic fungi destroying polymers and corroding metals.

The development of a microbial community in habitable compartments of a space vehicle takes place against a specific background of the plural material components which develop in the course of the activities and functioning of crew life and technical means including life-support systems. Along with the main environmental factors of temperature, humidity and microbiology, all biological objects aboard a space vehicle are exposed to the broad spectrum of radiation. Special emphasis should be placed on ionizing radiation of significantly higher doses as compared with terrestrial conditions. It is known that space ionizing radiation has a permanent modifying effect on bio-objects.

It was shown [46] that space station microflora had undergone evolution with peculiarities originating from the conditions of space flight over many years of Mir functioning. For instance, the quantitative dynamics of space vehicle microbial contamination during its long-term exploration in orbit is not linear but fluctuating and displays alternation of the phases of microflora activation and stagnation. This is characteristic of both bacterial and fungal flora. However, it is more expressed in the fungal population; as the time of space flight increases, the swing of the microflora activation amplitude rises and this phase becomes more lengthy, latent by character. On the whole, the trend in quantitative parameters of microflora points to ecological expansion, that is colonization of the Mir interior and equipment by fungi during long-term exploitation of space station.

Another feature of the Mir microbial community is periodic change of genes and species of fungal flora dominating by population density, range and occurrence. This type of event implies establishment of cooperative and competing relations between individual groups of fungi in the process of their expansion on the orbital complex.

Several years' of observations of the specific dynamic of the fungal component on Mir have yielded another valid conclusion. Some representatives of microflora are capable of fixing residence in a space vehicle. Alterability of flight fungal strains and the degree of their similarity/affinity were assessed based on the DNA-polymorphism analysis using polymerase chain reaction with random primer (RAPD: random amplified polymorphs DNA). Certain strains of *Penicillium chrysogenum*, the dominating species in the microflora on Mir, were shown to have resided on the station for a minimum of seven years [46].

The possibility of resident occupation of the space vehicle environment and equipment by molds is very significant since it has been demonstrated that in flight microbial aggressiveness (colonization and biodegrading activity) to decorative and structural materials can essentially increase as compared with reference strains of similar species [47].

Therefore, the following suppositions can be made to characterize the evolution of a microbial community aboard a long-operating space vehicle:

- the environment of a long-operating manned space station may be a peculiar kind of ecological niche for the development and reproduction of bacilli and fungi belonging to particular species,

- bacterio-fungal associations primarily reside on decorative-finish and structural materials of the space station's interior and equipment which gather anthropogenic organic compounds and sufficient air condensate to allow the full vegetative cycle and reproduction of heterotrophic microorganisms, mold fungi *Penicillium*, *Aspergillus*, *Cladosporium* sp.,

- the quantitative and structural dynamic of microflora on the space station Mir is not linear and presents a wave-form cycle of alternating phases of biocenosis activation and stagnation controlled as by internal biological mechanisms of self-regulation and by external cosmophysical factors,

- the phase of microflora activation is connected with medical and technical risks that can significantly impact flight safety and hardware reliability.

The scientists from the Institute of Biomedical Problems have continued the investigation of microorganisms in the closed environment of the International Space Station from the first stage of its construction [42].

Macromolecule biotechnology

The space biotechnological program began in the Soviet Union, firstly from 1982 with the space station "Salyut-7" and since 1986 with the space station Mir. It was the result of a cooperative effort by the Russian Space Corporation "Energia" and Russian Stock Company "Biopreparat", the Institute of Biomedical Problems and some other biomedical institutes. The program united teams of investigators from 25 research institutions and enterprises.

To perform these experiments, 17 space-intended technological units were designed and manufactured. Over an 18-year period (1982–2000) the manned space stations were the venue for more than 150 experiments on different research phases, i.e. demonstration, pilot research and production of biosamples.

The main reason for the space biotechnological experiments was to gain knowledge of the main physical properties of the microgravity environment associated with the absence of various gravity-dependent features of liquids (first of all, concentrating and thermal convection and cell sedimentation) which limit production and quality of bioproducts on Earth. Specifically, the absence of convectional mixing and sedimentation in microgravity significantly raises efficiency of electrophoretic separation of macromolecules in free flow. Throughput, purification and yield of target products (recombinant and natural proteins for medical use, specific cells with pre-determined secretory functions) improve, too.

The process of crystallization of biological molecules like proteins and macromolecular formations, e.g. viral particulates, is very sensitive to environments, particularly to concentrating heterogeneity and microconvective flows in solution. Microgravity provides the optimum for growth of biological crystals as there are no convective flows of solutes at the crystal surface.

On the whole, the aim of the biotechnological research was to obtain new facts concerning the nature of the effects of physical factors of space on various processes of biopreparation production, and biological objects and compounds that might stimulate development of application technologies which on Earth are either difficult or impossible. Scientific data and results of space experiments can also be taken up by different terrestrial trades. That is why expansion of the space biotechnological investigations may, in future, bring in good profits equally to the space industry and advanced technologies on Earth.

Separation and purification of biopreparates by electrophoresis

The main objectives of the experiments associated with electrophoresis of bio-objects in microgravity were as follows:

- to study the baseline principles of protein purification of recombinant and natural preparations with the electrophoresis techniques in space flight;

- to search for novel electrophoretic technologies to separate highly active strains of commercial microorganisms (super-producers);

- experimental testing of the feasibility and efficiency of high-duty technologies for separation of pure proteins;

- the acquisition of test species of new therapeutic drugs.

Pioneering experiments using the techniques of stationary electrophoresis in a column-type unit "Tavria" developed a model of blood protein separation and evaluated the possibility of obtaining super-producers by separation of commercial agricultural antibiotics (tilozin and flavomicin). The principle feasibility of purification of influenza virus haemagglutinin was also tested in these experiments. Analysis of results of the "Tavria" experimental series showed that electrophoresis could be used in space as a core of commercial technologies for separation of bio-objects and processing of very pure biomolecules [48–50].

The electrophoretic investigations continued in the "Genome" unit by separation of a mixture of large polynucleotides (fragments of human DNA) in order to fractionate genic fragments of chromosomes and return separated fractions for laboratory studies. The results of this experiment reaffirmed the possibility of fractionating large macromolecules [51] and were used in part within the framework of the "Human Genome" project.

The next phase (1985–1987) was marked by three experiments in the automated units "Robot" and "Svetlana" built on the basis of the stationary electrophoresis technique. The aim was to gain additional knowledge about the effects of microgravity on the electrophoresis of protein and cell preparations, and to test methods of separating out highly active cell-producers of biologically active substances. It was shown that the purity of proteins could be increased in microgravity using the "Robot" and "Svetlana" space units. The space fraction of hemagglutinine of influenza virus contained more than 20 times less albumin than the control [52]. Data of the experimental fractionation of microorganism-producers attested to the possibility of direct selection of strains by electrophoresis in space flight conditions. Samples of producers of the antibiotics tilozin and flavomicin exhibited the qualities which were 1.5–2 times higher than those of commercial preparations [53].

In 1987–1990, twenty experiments for protein purification were performed in the "Ruchey" ("Stream") unit using high-duty electrophoretic techniques in free flow (Fig. 1). The results justified the theoretical predictions about increasing throughput of electrophoretic purification of protein samples in the microgravity environment with the purity of target product ten times better than its ground-based analog. Another potential application of electrophoresis in space experiments in the "Ruchey" unit is the yield of research samples of recombinant DNA interferon-α_2 free of admixtures and inactive forms undesirable for the human immune system

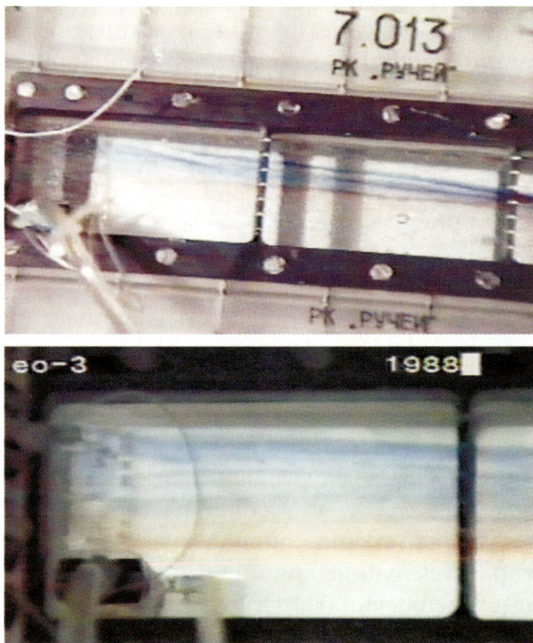

Fig. 1. Results of electrophoretic separation of human albumin and hemoglobin mixture in microgravity by the space device "Stream". Top: Method of isoelectric focusing. Bottom: Method of the voltage step of the electric field.

[54,55]. These data suggest the need to continue experimental research in space using electrophoresis of biopreparations to make them available to pharmacology of the twenty-first century.

Thus the experiments with electrophoresis in microgravity confirmed theoretical predictions of its efficiency in part of the purity of yielded bioproducts and higher throughput. Results of significant practical implications were obtained in tests with commercial bio-objects, i.e. interferon α-2, virus hemagglutinin of influenza, and microorganism-producers of antibiotics tilozin and flavomicin. The results demonstrated the advantage of continuing testing electrophoresis techniques for production of pure pharmaceutical bio-preparations in space.

Crystallization of biomolecules

It is known that microgravity and, consequently, the absence of gravity-driven convection and sedimentation stimulate the development of unique separation techniques and the production of supramolecular structures: crystals and polymers of biomolecules with characteristics (frequency, structure regularity) better than under 1-g gravity. Crystals of biological macromolecules (proteins, nucleic acids) and their complexes are needed for X-structural analysis, the only direct method of 3-dimensional determination of chemicals. In addition to fundamental investigations in biochemistry and molecular biology, data about the d-dimensional structure of biomolecules are used in pharmaceutical, chemical and biotechnological industries, particularly in the development of therapeutic substances that can be recognized by target proteins and thus have the best effects. The presence or absence of gravity is also a determinant for structure of polymers formed in noncellular systems.

Our expectations for the practical use of space biotechnology was justified during the development of protein crystallization in microgravity [54,55]. Monocrystals of functionally important proteins, medical preparations, biosensors and other macromolecules are used to study their spatial structure, the mode of action and development of synthetic analogs of natural macromolecules or their active fragments.

The objectives of experiments with biomolecule crystallization during space flights were as follows:

- to demonstrate the efficiency of growing biocrystals in microgravity;

- to evaluation and thrash out different crystallization methods;

- to test validation of equipment design;

- to produce pharmaceutical and commercial bio-crystals.

Experiments on bio-crystallization were initiated on crystallizer "Ainur" on space station Mir in 1987. Four experiments with fluid diffusion showed indeed the possibility of crystallization of different proteins (lysozyme, concanavalin-A and rhodopsin). Experimental crystals of rhodopsin had the form of a 0.05×1 mm needle and were convenient for X-structural analysis [55,56].

226

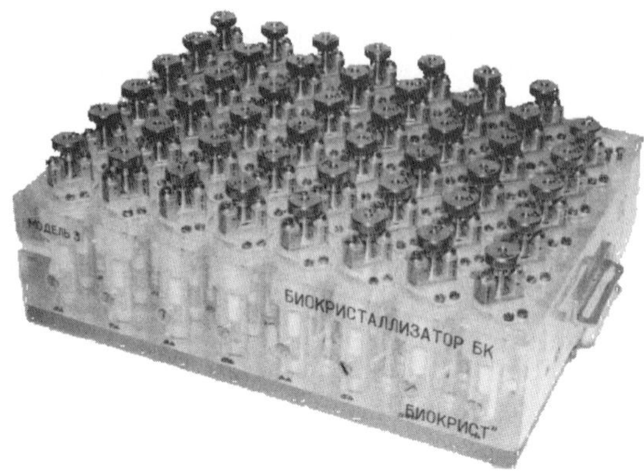

Fig. 2. Space crystallizer "Biocrist".

During 1989–1994, crystallization of bio-objects was performed in original unit "Biocryst" based on a specially developed reinforced-drop method (vapor diffusion) (Fig. 2). The unit was employed in 29 crystallization experiments aimed at testing and enhancing space technology of cultivating crystals for research and commerce, and improving the space unit design. As a result of these experiments in space, crystals of human growth hormone, luciferase, uridine phosphorylase, lysozime, insulin and other important proteins were grown [54,55].

The first experience with crystallization in microgravity gave the possibility to create a new generation of space equipment: the versatile biocrystallizer "Lutch" ("Ray") (Fig. 3). This allowed different crystallization techniques to be used in three experiments aimed at growing monocrystals of valuable pharmaceutical proteins: Fab-1 of a monoclonal antibody to interleukin-2, interleukin-1β and shaperon *Caf*

Fig. 3. Space crystallizer unit "Lutch".

1M Y. Pestis, which were intended for development of immune and anti-viral preparations. Analysis of space samples showed that crystals of all bio-objects grown in microgravity were much better than their control ground-based analogs (Fig. 4). Crystals of fragment Fab-1 of the monoclonal antibody to interleukin-2 synthesized in space flight were up to 0.1×2.0 mm in size, whereas on the ground there were few cases when they reached 0.15–0.2 mm [57,58]. Hence the space experiments have shown that biological crystals grown in space following a special technology surpassed their laboratory analogs by many parameters including large size, more perfect morphology, better resolution of the structure under study.

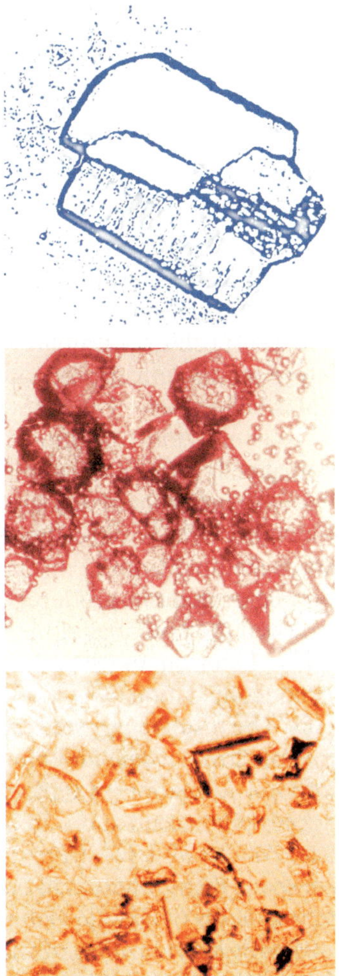

Fig. 4. Results of protein crystal growth in the space crystallizer "Lutch". Top: Space crystals of Fab-1 fragment of the monoclonal antibody to interleukin-2. Middle: Space crystals of interleukin-1β. Bottom: Interleukin-1β crystals after growth in 1-*g* conditions.

228

Production of recombinant strain-producers

One the important goals of space biotechnology is the investigation of cell conjugation and directed transferring of genetic substances in microgravity. The results of these experiments in space could initiate new hybridome technology. Fusion of cell protoplasts resulting in viable hybrids is of considerable use in the production of monoclonal antibodies (affiliating of antibody-producing lymphocytes to myeloma cells) and genetic investigations. The technique can also find many applications in breeding organisms with altered characteristics such as plants with greater adaptability to harsh environments.

Grown in space flight, hybrid cells increase the efficiency of genetic information exchange due to the absence of gravitational factors interfering with the interactions [59]. The arguments for studying cell-producers in microgravity are as follows:

- no gravitation effects on mass-exchange processes during cultivation,

- the influence of the other space flight factors (magnetic fields, space radiation, etc.) on cell morphology and intracellular transmission of information including genetic molecules.

Experiments on hybridization of microbial cell-producers began in the space unit "Recomb" aboard the Mir space station in 1989 with the purpose of studying the effectiveness of different methods of cell fusion and regeneration in space (Fig. 5). Another objective was to determine the effectiveness of intercellular transfer and integration of genetic material in space flight. The following scientific findings were made in four experiments ("Recomb" unit) on intra- and interspecies fusion of protoplasts in different microbial cells [55]. In microgravity, transfer and incorporation of genetic material are steady but without any substantial difference from $1\,g$ conditions. Regeneration of recombinants is more effective in liquid media.

"Recomb" was used on space station Mir in 12 investigations concerning genetic molecule interchange among microorganism-producers based on the conjugation technique. The results have showed the potential of the technology to produce cells with unusual properties in space. Later on, in 1998, the technology of recombinant strain production by bacterial conjugation was evaluated in a modified unit "Recomb-K" in six experiments. The results suggested that there is a high probability of transferring large fragments of chromosome and the possibility of transferring a whole chromosome with temporal formation of diploid hybrid cells. Conjugation (DNA exchange among bacterial cells) occurs more frequently in microgravity than on Earth and depends on time. Almost all DNA from donor cells recombine in the genome of a recipient strain in space flight, whereas in the 1-g environment recombination involves only separate fragments of genetic material (DNA) [60–63].

The results of experiments with hybridization of microbial cells demonstrated high potential of bacterial conjugation as a basis for space biotechnology of synthesizing strain-producers with new properties. High efficiency of conjugation transfer of genetic material in microgravity makes it possible to get sufficient quantities of recombinant cells for directive selection on Earth.

Fig. 5. Space device for microorganism hybridization. Top: "Recomb"; bottom: "Recomb-K".

The experiments with cultured cells producing biological active substances were carried out to fulfil fundamental investigations of the characteristic features of cultivation of various microbial, animal and plant cells, and a differential approach to the role of different factors of space flight in the variation of cultural cell properties. The goal was also a follow-up objective to cultivate improved and novel recombinant strains, and mutant cell of superproducers of prospective biologically active substances.

Special membrane module equipment "Vita" was designed for fundamental and applied purposes in space cell physiology and space biotechnology on the orbital station Mir. Experiments were performed to determine the parameters of the continuous aerobic (with air phase) cultivation of cell cultures, and to study the dynamics of mass-exchange during cultivation of microbial and animal cells in a liquid medium. In addition, the special membrane technology of this unit was tested for yield capacity.

Six experiments provided evidence for the feasibility of continual anaerobic cultivation as a step towards the controlled production of biomass of various cell cultures. As a result, it was discovered that the kinetic parameters of microbial cell growth exceeded those in the ground-based control experiment. Another finding was enhanced stability of the recombinant strain: the loss rate of integrated plasmid with a new gene amounted to 1% as compared to 30% in a ground-based experiment. The

Fig. 6. Space device "Biocont": pre-flight preparation.

results of the experiments with animal cells (myeloma cells of Chinese hamster and hybridoma producing immune-specific antibodies) have shown that cell cultures remained viable in this space equipment [55].

Since 1991, the space station Mir became the venue for more than 20 experiments with cultivation of microorganisms and plant cells in units "Biocont" and "Maksat" (Fig. 6). The purpose of these experiments was to explore the combined space flight effects and to obtain mutant strains of improved and novel producers, as well as strains resistant to extreme environments. Specific attention has been paid to the space flight effects on mutagenesis, cultivation, and morphogenesis of various bio-objects (microorganisms, fungi, plant cells). The experiments involved a variety of biological objects: more than 60 strains of bacteria, yeast, fungi and ten strains of vegetal cells [63–65]. The "Maksat" experiments with cells of farming plants (wheat, maize, potatoes) were carried in cooperation with the Academy of Sciences of the Republic of Kazakhstan [66].

The experimental cultivation of cell-producers on Mir revealed the formation of clones with altered resistance to environmental factors, stable hereditary characters of cell productivity, and better stability of strain with a new gene. Cultures of plant cells distinguished by high productivity of multi-use biologically active substances (shikonin, ginsengsides) were obtained [66]. Moreover, the space flight conditions yielded strain-producers for more effective plant protection and biodegradation of oil-containing pollutants [67].

Analysis of the results of the space experiments revealed changes in the basic morphologic and genetic characteristics of the cells: resistance to external factors, level of mutagenesis, and stability of hereditary characters determining the producing properties of strain. The nature and degree of biological shifts was shown to be dependent on equally inherent properties of cells and the space flight conditions. However, particular space factors responsible for cell mutations have not yet been identified and, therefore, it is still unclear how they actually work, constituting a challenge for investigators in future.

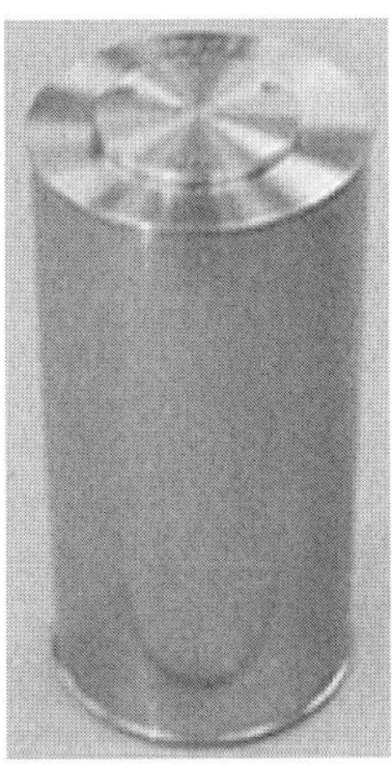

Fig. 7. Space device "Biomagnistat" for screening biological samples from geomagnetic fields.

One of the unknown fields in space cell physiology is the geomagnetic effects on cell functions during orbital flight and the mutagenic effects of space radiation. Space experiments in a "magnetic vacuum" began in 1993 after delivery of the "Biomagnistat" unit to the Mir space station (Fig. 7). The purpose was to determine responses of cell parameters such as growth, segregation stability during selection, and productivity of recombinant strain-producers of biologically active substances to space flight factors without magnetic influences [68].

The unit was designed to screen a container with bio-objects from exposure to the constant electromagnetic field of Earth including the low-frequency component (0–1 kHz), and reduce the intensity 100–1000 times compared with the geomagnetic field. To ensure minimal magnetic intensity within the container, each experiment was preceded by unit "degaussing". Reference cultures were tested in a "Biocont" unit in parallel with the "Biomagnistat" experiments.

Analysis of the results of six "Biomagnistat" experiments led to the conclusion that the geomagnetic field and inversion of its vector (cyclic alterations in the direction of lines of force along the orbit) in space flight significantly influence cell functioning and structure. In some strains, screening from the geomagnetic field may alter the structure of colonies developed in space flight and lead to profound changes in shape and size [69,70].

Thus, the biotechnological experiments on cultivation and exposure of cell-producers using various research equipment demonstrated that space flight factors significantly modify cell properties, giving new opportunities to extend the choice of valuable forms of producers and obtain strains with enhanced yield capacity. Many cultures exposed in the space station contained clones with improved producing qualities and exhibited better viability. However, cultivation and exposure of cells did not allow identification or determination of the individual space factors, including microgravity, that cause changes in cell properties.

It should be mentioned again that the specific factors of space flight, dose and time dependence of their combined effects as well as the triggers of consequent changes are still unclear. This undoubtedly restrains biotechnological research in microgravity and is a subject for future fundamental studies in space biology and biotechnological procedures on the board the orbital station.

Conclusion

Experimental research in the fields of cell physiology and biotechnology carried out by Russian scientists on the board the orbital station Mir has shown that the effects of space flight factors on cells and biotechnology processes are multi-component and significant. They could increase the possibility of developing new strains and substances for medical, pharmaceutical and industrial applications. It could be said that the positive and negative experiences arising from space investigations in this field have brought about not only a new knowledge about the role of gravitational factors in cell processes, but also the creation of a new, young biological discipline—space biotechnology, which will expand on the International Space Station.

References

1. Parfenov, G.P. (1988) Microgravity and elementary biological processes. In: Problems of Space Biology. Leningrad 57, 145–154 (in Russian).
2. Tairbekov, M.G., Parfenov, G.P., Shepelev, E.Ja. and Sushkov, F.V. (1983) Experimental and theoretical analysis of influence of gravity at cellular level. Adv. Space Res. 3 (9), 153–158.
3. Moor, D. and Cogoli, A. (1996) Gravitational and space biology. In: Biological and Medical Research in Space. An Overview of Life Sciences Research in Microgravity (D. Moore, P. Bie and H. Oser, eds.). Springer Verlag, pp. 1–107.
4. Tairbekov, M.G. (2000) Cell as a gravity-dependent biomechanic system. Aviakosm. I Ecolog. Med., 34 (2), 3–17 (in Russian).
5. Gasenko, O.G. and Parfenov G.P. (1967) Results and future perspectives in space genetics. Space Biol. Med., 1 (5) 12–21.
6. Butenko, R.G., Dmitrieva, N.N. and Onko, V.V. (1979) Effects of weightlessness on embryogenesis of somatic cells. In: Biological Investigations on Biosatellites "Kosmos". Nauka, Moscow, pp. 118–124 (in Russian).
7. Stuart, F. and Krikorian A. (1979) Morphogenesis multipotentic cells in weightlessness. In: Biological Investigations on Biosatellites "Kosmos". Nauka, Moscow, pp. 96–117 (in Russian).
8. Levinskikh, M.A., Sychev, V.N., Podolsky, I.G., Padalka, G.I., Avdeev, S.V. et al. (2000) Growth of wheat "from seat to seat" during space flight. Aviakosm. I Ecolog. Med., 34 (4), 43–47 (in Russian).
9. Levinskikh, M.A., Sychev, V.N., Derendyaeva, T.A., Salisbury, F.B., Campbell, W.F., et al. (2000) Analysis of the spaceflight effects on growth and development of Super Dwarf wheat grown on the Space Station Mir. Plant Physiol. 156, 522–529.
10. Grigoriev, A.I. and Gazenko, O.G. (1998) Main areas and results of IBMP scientific investigations from 1963 up to 1998. Aviakosm. I Ecolog. Med., 32 (5), 4–17 (in Russian).

11. Kordyum, E.L. (1994) Effects of altered gravity on plant cell processes: results of recent space and clinostatic experiments. Adv. Space Res. 14 (8), 77–85.

12. Nechitailo, G.S. (2001) Main results of biological experiments at orbital station Mir. Proc. Intern. Symp. "International Scientific Cooperation onboard MIR". Lyon, France, pp. 255–261.

13. Tairbekov, M.G. and Gabova, A.V. (1992) Ecological and morfological peculiarities of growth and movements of unicellular organisms in gravitational field. Aviakosm. I Ecolog. Med., 26 (4), 9–14 (in Russian).

14. Romanov, Yu.A., Kabaeva, N.V. and Buravkova L.B. (2000) Gravisensitivity of human endothelium. Aviakosm. I Ecolog. Med. 34 (4), 23–26 (in Russian).

15. Romanov, Yu.A., Kabaeva, N.V. and Buravkova, L.B. (2000) Simulated hypogravity stimulates cell spreading and woud healing in cultured human vascular endothelial cells. J. Gravit. Physiol. 7 (2), 77–78.

16. Konstantinova, I.V. and Fuchs, B.B. (1991) Experiments on isolated cells in space. In: The Immune System in Space and Other Extreme Conditions (R.V.Petrov, ed.). Harwood Academic, pp. 18–26.

17. Gabova, A.V., Gavrilova, O.V., Rudanova, E.E., Voloshko and L.N. (2000) Peculiarities of growth and morphogenesis of syphoneal alga *Vaucheria sessilis* in microgravity. Aviakosm. I Ecolog. Med., 34 (1), 35–38 (in Russian).

18. Sychev, V.N., Levinskikh, M.A. and Livanskoy, O.G. (1989) Study of growth and development of Chlorella vulgaris on board of Biosatellites "Kosmos—1887". Aviakosm. I Ecolog. Med., 23 (5) 35–39 (in Russian).

19. Mashinsky, A.L., Alehina, T.P., Bogko, A.N., Rumyanzeva, V.B., Schernova, L.S. et al. (1991) First stage development pecularity of *Friticum vulgare* during space flight. Aviakosm. I Ecolog. Med., 25 (1), 39–42 (in Russian).

20. Wolf, D., Sams, C. and Gonda S. (1990) NASA Microgravity Biotechnology Program: microgravity-based cell model and culture system. In Vitro 26 (3), 40A.

21. Ingram, M., Techy, G.B., Saroufeem, R., Yasan, O., et al. (1997) Three-dimentional patterns of various human tumor cell lines in simulated microgravity of NASA bioreactor. In Vitro Cell Dev. Biol. Anim. 33, 459–466.

22. Freed, L.E. and Vunjak-Novakovic G. (1997) Microgravity tissue engineering. In Vitro Cell Dev. Biol. Anim. 33, 381–385.

23. Grigoriev, A.I., Volozin, A.I. and Stupakov, G.P. (1994) Human mineral metabolism under altered gravity (A.B. Rubin, ed.). Nauka, Moscow, 216 pp. (in Russian).

24. Bierkens, J., Maes, J., Ooms, D., Collier, M. et al. (1994) Decreased aquisions of osteoblastic phenotype markers and increased response to interleukin-1 and PTH in osteoblast-like cells under conditions of microgravity. Proc. 5th Europ. Symp. on Life Sciences Research in Space, ESA SP-366, pp. 25–30.

25. Freed, L.E., Langer, R., Martin, I., Pellis, N. and Vunjak-Novakovic, G. (1997) Tissue engineering of cartilage in space. Proc. Natl. Acad. Sci. USA, 94, 13885–13890.

26. Larina, O.N., Fedorenko, L.A. Manuilova, E.S. and Inozemzeva, L.S. (2002) Growth of 3-d cell complex under rotational cultivation. Aviakosm. I Ecolog. Med., 36 (4) in press (in Russian).

27. Sushkov, F.V., Rudneva, S.M., Polikarpova, S.I. and Portugalov, V.V. (1977) Mammaliam cultured cell research on board biosatellite "Cosmos-782" (in Russian). Arch. Anatomy, Histol. Embryol. 73, 28–39.

28. Montgomery, P., Cook, J.E., Reynolds, R.C. et al. (1978) The response of single human cells to zero gravity. In Vitro, 14, 165–173.

29. Cogoli, A. (1996) Gravitational physiology of human immune cells: A review of in *vivo*, *ex vivo* and *in vitro* studies. J Gravit. Physiol. 3, 1, 1–9.

30. Meshkov, D.O. and Rykova, M.P. (1995) The natural cytotoxicity in cosmonauts on board space stations. Acta Astronautica, 36, 719–726.

31. Buravkova, L.B., Rykova, M.P. and Antropova, E.N. (2002) Methodological approaches to study cell interaction natural killers and target cells in vitro on board International Space Station. Proc. XII Conference "Space Biology and Aerospace Medicine", Moscow, pp. 71–72.

32. Gazenko, O.G., Grigoriev, A.I., Meleshko, G.I. and Shepelev, E.Ya. (1990) Biological life support systems. Aviakosm. I Ecolog. Med., 24 (3) 12–17 (in Russian).

33. Meleshko, G.I. and Shepelev, Ye.Ya. (1994) Biological life support systems (closed ecological systems). Moscow, 277 pp. (in Russian).

34. Nazarov, N.M., Bezborodov, A.M., Biber, B.L. et al. (1994) Biocatalisators for water regeneration system in closed environment. Abstr. X Conference on Space Biology and Aerospace Medicine, Moscow. pp. 303–304 (in Russian).

35. Sychev, V.N., Shepelev, Ye.Ya., Meleshko, G.I., Levinskikh, M.A., Popov, V.V. et al. (1999) Biological life support systems: studies on orbital space station "Mir". Aviakosm. I Ecolog. Med., 33 (1) 10–16 (in Russian).

36. Tairbekov, M.G. (1991) Physiological mechanisms of cell adaptation to microgravity. The Physiologist, 34 (1) (suppl.), 62–63.

37. Tairbekov, M.G. (1992) Cell in gravitational field. The Physiologist, 35 (1) (suppl.), 16–18.

38. Abilov, A.A., Alekperov, A.L., Mashinsky, A.L. and Fadeeva, S.N. (1985) Morfological and functional state of photosyntetical plant cells cultivated by different time in space flight. In: Microorganisms in Artificial Ecosystems. Nauka, Novosibirsk, pp. 35–37 (in Russian).

39. Levinskikh, M.A. and Sychev, V.N. (1989) Growth and development of unicelular algae in ecosystem "algobacteria-fish". Aviakosm. I Ecolog. Med., 23 (5), 32–35 (in Russian).

40. Meleshko, G.I., Antipov, V.V., Levinskikh, M.A., Sychev, V.N. et al. (1992) Studies of water ecosystems in microgravity ("Aquarium" investigation). In: Investigation Results on Board of Biosatellites. Nauka, Moscow, pp. 378–383 (in Russian).

41. Sychev, V.N., Levinskikh, M.A., Meleshko, G.I. and Shepelev, Ye.Ya. (2000) The effects of microgravity on active cultures of unicellular algae. Plant Physiol. 47 (5) 1–8 (in Russian).

42. Viktorov, A.N., Novikova, N.D., Deshevaya, E.A. et al. (2001) Main results and future of microbiological investigation of orbital station environment. Proc. of Russian Conf. on Problems of Living in Closed Environment, Moscow, pp. 29–31 (in Russian).

43. Novikova, N.D. and Viktorov, A.N. (1996) Microbial evolution in orbital station environment in condition of multiyear exploration. Abstr. of 31st Scientific Assembly of COSPAR, Birmingham, England, 376.

44. Viktorov, A.N., Novikova, N.D. and Deshevaya, E.A. (1992) Microflora of manned space stations and biodegradation of internal construction materials. Aviakosm. I Ecolog. Med., 26 (3) 41–48 (in Russian).

45. Viktorov, A.N., Novikova, N.D. and Deshevaya, E.A. (1992) Microflora of manned space

stations and biodegradation of internal construction materials. Aviakosm. I Ecolog. Med., 26 (3) 41–48 (in Russian).

46. Viktorov, A.N., Novikova, N.D., Deshevaya, E.A. et al. (1998) Residental expansion of orbital station "Mir" by *Penicillium chrysogenum* and problems of ecological safety during long-term space flights. Aviakosm. I Ecolog. Med. 5, 57–62 (in Russian).

47. Viktorov, A.N., Novikova, N.D., Deshevaya, E.A. et al. (1998) Comparative analysis of biological property of microorganisms from orbital station "Mir" during different stage of exploitation. Aviakosm. I Ecolog. Med. 2, 61–68 (in Russian).

48. Agitsky, G.Yu., Lepsky, A.A., Mitichkin, O.B., Troitsky, G.V. and Fokin, V.E. (1984) Biotechnological experiments on space station "Salyut-7". Proc. of XVIII Symp. In memory of K.E. Tsiolkovsky. Moscow (in Russian).

49. Agitsky, G.Yu., Troitsky, G.V., Mitichkin, O.B., Sharayeva, T.K. and Lepsky, A.A. (1984) Fluid electrophoresis, isoelectric focus and isotachophoresis in microgravity. Proc. Ukrainian Academy of Science, Kiev, 4, 56–60 (in Russian).

50. Mitichkin, O.V., Vavirovsky, L.A., Agitsky, G.Yu., Serebrov, A.A. and Fokin, V.E. (1985) Results of biotechnological experiments "Tavria" on board space station "Salyut-7". Proc. of Conference in Memory of Yu.A. Gagarin, Moscow, pp. 230–233 (in Russian).

51. Mitichkin, O.V. (1993) Results of biotechnological experiments, carried out on board of Mir and Salyut-7 space station. Proc. of Space Station Mir Symp., Washington, pp. 26–42.

52. Krashenyuk, A.I., Krivoshein, Yu.S., Agitsky, G.Yu., Lepsky, A.A., Mitichkin, O.V. et al. (1987) Analysis of the virus heamagglutinine samples, received in microgravity on board space station. Proc. Conference on Act. Probl. Biotechnology, Moscow, pp. 78–79 (in Russian).

53. Krivoshein, Yu.S., Krivorutchenko, Yu.L., Lepsky, A.A., Mitichkin, O.V., Agitsky et al. (1989) Electrophoretic separation and biological objects purification in microgravity conditions. Space Sci. Technol., Kiev 4, 5–10 (in Russian).

54. Vorobyov, D.B., Grigoriev, A.I., Kalinin, Ju.T., Larina, O.N. and Scherbakov G.Ja. (1997) Biological technology in space. Aviakosm. I Ecolog. Med. 5, 13–18 (in Russian).

55. Kalinin, Ju.T. (1998) Achievements and perspectives of space biotechnology. Cosmonautics and Space Ship Devel. 12, 103–107 (in Russian).

56. Kharitonenkov, I.G., Leiver, V.G., Guskov, I.A., Abashev, Ju.P. Borisova, S.N. et al. (1991) Protein crystallization in microgravity on board space station "Mir". Proc. Int. Symposium on Hydromechanic and Heat Exchange in Microgravity, Moscow, pp. 28–30.

57. Zav'yalov, V.P., Mihaylov, A.M., Makintier, Sh., Korpella, T., Abramov, V.M. et al. (1998) Crystallization of molecular peryplasmic shaperon *Caf1M Y. Pestis*. Abstr. XI Conference on Space Biology and Aerospace Medicine, Moscow, p. 275 (in Russian).

58. Kalinin, Yu. T., Pletnev, V.Z., Tsygannik, I.N., Rugeynikov, S.N. (1999) Protein crystallization in microgravity. Proc. 1st Russian Conference on Space Material Technology, Kaluga, p. 56 (in Russian).

59. Kalinin, Yu. T., Krivonogov, S.V., Mitichkin, O.V. and Chorunova, I.V. (1993) Cell hybridization in microgravity conditions. Proc. of XXVIII Symp. K.E. Tsiolkovsky readings, Moscow (in Russian).

60. Zerov, Yu. P., Os'kin, B.V., Mitichkin, O.V. and Scherbakov, G.Ya. (1996) Influence of microgravity and other space flight factors on exchange of genetic substances during bacterial conjugation and protoplast merging. Russia–USA Symposium "Science-

236

       NASA", Korolev, Moscow region (in Russian).

61. Zerov, Yu.P. and Mitichkin, O.V. (1997) Study of the influence of microgravity and other factors of orbital space flight on the transfer and exchange of material in the process of bacterial conjugation and protoplast fusion. Russia–USA Symposium "Science-NASA", Huntsville, Alabama, USA, p. 125.

62. Zerov, Yu.P., Kalinin, Yu.T., Os'kin, B.V., Samarina, M.R., Mitichkin, O.V. and Scherbakov, G.Ya. (1998) Study of microgravity and other space flight factors influence on the genetic material transfer and replacement in bacterial conjugation and protoplast agglomeration. Abstr. XI Conference on Space Biology and Aerospace Medicine, Moscow, p. 276 (in Russian).

63. Zerov, Yu.P., Mitichkin, O.V., Os'kin, B.V., Samarina, M.R. and Scherbakov, G.Ya. (1998) Investigation space flight effects on genetic properties of cells, producing biological active agents. Abstr. XI Conference on Space Biology and Aerospace Medicine", Moscow, p. 278 (in Russian).

64. Voyaekova, T.A. and Tabakov, V.Yu. (1998) Shot-term effects of space flight factors on physiology and genetic property of different microorganisms. Abstr. XI Conference on Space Biology and Aerospace Medicine, Moscow, p. 171 (in Russian).

65. Aithoshina, N.A., Mitichkin, O.V. and Aubakirov, T.O. (1994) Changes of the micellar microorganisms under influence of space factors. J. Appl. Biochem. Microbiol. 4, 420–424.

66. Karabaev, M.K., Lesova, J.T., Kaniev, B.K., Aubakirov T.O., Musabaev T.A. et al. (1995) Growth and development plant cell cultures in space flight. News of Kazahstan Science, Almati, pp. 33–38 (in Russian).

67. Strogov, S.E., Kalinin, Ju.T., Zaizeva G.V., Konstantinova, N.A. et al. (2000) Changes of plant cell productivity in space flight. Aviakosm. I Ecolog. Med. 34 (1), 32–35 (in Russian).

68. Larina, O.N., Kostanyan, I.A., Mitichkin, O.V., Kalinin, Yu, T., Lipkin, V.M. et al. (1996) Geomagnetic field influence on the biological processes in space flight conditions—preliminary results. Russia-USA Symposium "Science-NASA", Korolyov, Moscow region (in Russian).

69. Larina, O.N., Truhanov, K.A., Kostanyan, I.A., Kalinin, Yu, T. et al. (1998) Effects of low magnetic fields on recombinant protein expression in *E. coli*. Proc. IIIrd Int. Conference on Physics and Radioelectronics in Medicine and Biotehnology PRMB98, Vladimir. 310–311 (in Russian).

70. Larina, O.N., Truhanov, K.A., Kostanyan, I.A., Smirnova, E.V., Mitichkin, O.V., Kashicina, G.D. et al. (1998) The problems of low-density magnetic fields on cell systems during space flight. Abstr. XI Conference on Space Biology and Aerospace Medicine, Moscow, p. 4.

Cell Biology and Biotechnology in Space
A. Cogoli (editor)

The Coming Space Science Programme in Animal Cell/Tissue Research

Roger A. Binot

Directorate of Manned Spaceflight and μgravity, ESA-ESTEC, MSM/GA, Noordwijk, The Netherlands

Abstract

Between the genotypical potentials and given phenotypical expressions, cell behaviour is hypothesised largely depending on local distributed control mechanisms resulting from delicate mass flux balances involving biochemical (including cytokine, endocrine and paracrine factors) and mechanical signalling pathways. Better knowledge and exogenous control of these mechanisms is a major objective of our ESA programme in biotechnology through the integrated use of natural sciences and engineering sciences. The control of cell potency and differentiation and of cell regulation mechanisms (including apoptosis and migration triggering events) will have a huge impact on regenerative medicine (tissue engineering), and will also improve our understanding of phenomena important in immune diseases, the ageing process, and cancer. A summary is given of current research projects which have in common the understanding of the cell–environment relationship or the development of analytical and processing methods in the discipline of animal cell technology.

Introduction: from cells to tissues through evolution

Difference in cellular behaviour in a mass of originally identical animal cells undergoing differentiation will depend on local changes in the microenvironment of each individual cell. At the beginning of the process—the cells not being differentiated yet—such changes can only be of a very limited nature and origin.

These changes are, in part, well known and have been studied in many different conditions of cell density and cell metabolism, especially when using living microencapsulated cells. Emphasis was here very early placed on, for example, the establishment of gradients of concentration for oxygen in constructs of encapsulated cells and for the prevention of creation of anoxic conditions or more generally diffusion limitation conditions [1] in the centre of the capsule. The cell–surface properties (expressed in terms of critical surface tension for the material conditioned in physiological medium, i.e. a mechanical chemical interplay) interaction, was also early recognised as being an essential factor in the success of artificial organs grafting in medicine by favouring or reducing the risk for induced necrosis of cells immobilised on the surface [2]. Considerably more complex systems have also been studied for decades by scientists interested in the microbial ecology of natural biofilms or in the formation of well defined and organised micro-colonies and syntrophic microbial associations [3], showing the importance of transport phenomenon on cell–cell interactions and the formation of complex organised tissue-like functional microbial structures. Through evolution, and even long before the appearance of the higher organisms, single cell forming populations or communities had been learning to organise themselves tridimensionally in exopolymeric matrices, synthesizing and exchanging metabolic intermediates, with higher resilience to environmental conditions.

Animal cells must have learned from their ancestors and so mass fluxes, not strictly speaking concentrations, from and towards the cell are hypothesised as being used by the cell as a triggering message not only for growth, but also towards differentiation, apoptosis and migration. Perhaps even more fundamentally than this, cell shape as influenced by the culture conditions, in suspension, immobilised on a petri dish, or interacting with an extracellular matrix, has long since been known to determine their proliferative activity, organisation of the cytoskeleton and differentiation [4]. Mechanical forces, possibly involving some of the above transport or physical shape influences, were also recognised early on as influencing cell activity [5].

Looking for possible space influences on cell behaviour, and even without considering possible direct sensing of gravity at the intra-cellular level, several mechanisms have been identified, known to be gravity dependant and known to influence cell behaviour in normal gravity conditions. Can a better knowledge of these mechanisms, resulting from fundamental investigations in real and simulated space conditions, lead to improved control of such conditions by tissue engineering in classical terrestrial operations?

Tissue engineering; where geneticists, biologists and mechanical/fluid scientists must meet

For tissues and organs, the underlying fundamental mechanisms involved in sensitivity to gravitational forces are still yet to be fully understood. The effect of gravity, among other parameters, must be included in the search for a better understanding of the fundamental principles guiding functional tissue differentiation. Under conditions of weightlessness (microgravity) the formation of three-dimensional aggregates of predefined dimensions and cellular concentrations close to *in-vivo* conditions can *a priori* be better controlled and maintained than on the ground. This is because of the absence of interference due to sedimentation mechanisms proportional to the size and density of the formed aggregates, and the possibility of considerably reducing or maybe even removing polymeric matrix components when so desired. An important aspect here is that matrix structures of any density can be used, providing, for the very low densities, that such matrices are themselves formed and loaded with cells in space. Also, the formation of large-scale structured matrices, playing with chemically defined monomers with functional groups is in principle favoured in a low gravity environment; for exactly the same reasons better quality crystals can be expected from formation in a medium not perturbed by convection patterns induced by medium density evolution and sedimentation.

The precise regulation of the local microenvironment around cellular systems dictates the guidance of cellular proliferation and differentiation towards specific functional tissues. Cell–cell interactions and cellular functions within functional tissues can be influenced or induced by subtle changes in the local biomechanical or biochemical environment. *In-vivo*, mechanical loading is important in tissue formation, remodelling and repair. The regulation of mechanical force on and by connective tissues is accepted as a critical factor in understanding their function. Functional understanding leads to the development of functional, multi-cellular engineered tissues for terrestrial clinical applications. Experimentation in micro-gravity offers the possibility of precise control of subtle modifications of the immediate mechanical and biochemical microenvironment *in-vitro* via the increased stability of fluidic systems and from reduced environmental (i.e. not experimental) mechanical loads.

Some experimental evidence of the influence of gravity-dependent environmental factors such as mass fluxes and mechanical forces in different tissues

Maintaining cell–cell contact during the differentiation stage, in combination with growth factor administration, has been shown to be highly significant in increasing the number of neurons generated from human neurospheres [6]. The most potent factors tested were neurotrophic factors 3 and 4 and platelet-derived growth factor, acting by increasing neuronal survival rather than inducing neuronal phenotype. Also, plating intact, instead of dissociated, neurospheres in the presence of growth factors significantly increased the number of developing neurons. Apart from the prevention of potential mechanical trauma, the authors hypothesised that this improvement could be the result of allowing precursor cells time to establish their fate in a three-dimensional environment while maintaining communication with neighbouring cells.

The importance of the formation of tri-dimensional aggregates of cells has been recognised as being an important step in the cardiogenesis, myogenesis, and neurogenesis [7] from mouse embryonic stem cells, and chondrogenesis [8], and osteogenesis [9] from bone-marrow stromal cells. In the last case, for example, this is observed correlated with an up-regulation of tissue specific proteins (collagen type-I, osteonectin, alkaline phosphatase) and concomitant formation of microcrystalline bone. Whether the change is due to local increased concentration of cytokine factors due to limited diffusion from such "embryoid bodies" or cell spheroids, to cell–cell signalling involving membrane contact, to mechanical factors, or to a cascade mechanism involving several of these conditions is not known at present. Here also, designing biological models to investigate such phenomenon would benefit from low gravity simulation on the ground and validation in real space conditions.

A glimpse of the on-going studies supported by the ESA biotechnology programme in bioengineering and bone remodelling

The Biotechnology programme was initiated under the Microgravity Applications Programme (MAP) of ESA in 1997. The MAP was an original new approach for European Space Research, because it embedded space research in large ground research projects characterised by the involvement of European wide teams with academic and industrial partners, working on experimental programmes rather than on isolated experiments, and with application perspectives. It was conceived from the very beginning that these biotechnology projects would usually use space research to obtain data so as to enable optimisation of earth-based processes. The applications domain is extremely wide, from the improvement and better control of environmental bioprocesses to the genetic improvement of agronomical plants, and understanding the role of the cellular and tissular microenvironment conditions for organotypical artificial bio-organs engineering. The specific instruments necessary for flight experiments and ground preparation are provided by ESA. Financial support is provided by ESA covering up to 100% of additional costs for academic partners. Partner companies are requested to demonstrate interest by contributing in cash or kind to the project. A classical value for the first two-year phase (ground research phase) amounts to one million Euros with one-third covered by ESA and one-third contribution in kind by the academic partners, the remainder being provided by the industrial partners and other financing sources such as national and European authorities. The rights and obligations, including amongst others the confidentiality rules and intellectual property rights, strategic objectives for each Partner, project management structure and composition, are defined and agreed between the Partners signing a "Partnership Agreement". This agreement is part of the Contract.

A major objective of the project "Modular bioreactor for medically relevant organ-like structures" (Fig. 1) is the production of cartilage without using any scaffold structure, and of blood vessels. Because of the extremely high content and organisation of exopolymeric material in cartilage, this may be the only way for *in vitro* production of a functional cartilage analogue. Only microgravity conditions will allow an appropriate cell contact that is stable in position while loose in cohesiveness. The highest effectivity is expected when focusing on those tissues organised primarily by extracellular matrix structures (i.e, cartilage, blood vessel).

The project ERISTO is the second phase of the ERISTO (European Research in Space & Terrestrial Osteoporosis) activity initiated as the first MAP-Biotechnology Project in 1997 (Fig. 2). It is focused on osteoporosis and related bone remodelling issues and takes advantage of the observed accelerated bone evolution in space conditions. The partners are the same as those involved in the first phase. The aims of the first phase were:

AO-99-003 (Initiated May 2000)

Modular Bioreactor for Medically Relevant Organ-Like Structures

International Dimension: CH, D, I

Co-ordinator: *A. Cogoli*, ETH Technopark (CH)

cogoli@spacebiol.ethz.ch

Partners: *F. Ambessi Impiombato*, Univ. Udine (I); A. Bader, Med. School Hannover (D); *P. Bruckner*, Univ. Muenster (D); *W. Mueller*, S.M.-O. (CH); *R. Poertner*, Univ. Hamburg (D); I. Walther, ETH Zurich (CH)

Industry: Contact via Co-ordinator

Assigned flight: MASER 9

Facility: Biotechnol. Mammalian Tissue Cultivation / BMTC (new): Pre-Phase A initiated (Scientific Requirements)

Fig. 1. Project: Modular bioreactor for medically relevant organ-like structures.

AO-99-091 (Phase I initiated in 1997; Phase II initiated 2000)

Osteoporosis - ERISTO

International Dimension: CDN, CH, D, F, FIN, I, NL

Co-ordinator: *A. Berthier* and *L. Braak*, MEDES (F) laurent.braak@medes.cnes.fr

Partners: *R. Cancceda*, IST-CBA Genoa (I); *M. Heer*, DLR Koeln (D); *B. Koller*, Scanco (CH); *S. Pugh*, Millenium Biologix; *P. Ruegsegger*, IBT-ETH Zurich (CH); *P. v.d. Saag*, Hubrecht Lab. (NL); *K. Vaananen* (Part I only), Univ. Turku (FIN); *L. Vico* and *C. Alexandre*, LBTO (F); *A. Zallone*, Univ. Bari (I)

Industry: Millenium Biologix; Scanco Medical (CH), plus additional industrial contacts via co-ordinator

Assigned flight: OSTEO on STS-107

Facility: OSTEO (with CSA)

Biopack (TBC)

Biotechnol. Mammalian Tissue Cultivation / BMTC (new): Pre-Phase A initiated (Scientific Requirements)

AO-LS-99-MED-024

Initiated 2001

ERISTO-Animal Chapter: Transgenic and Normal Mice as Models to Study Osteoporosis Mechanisms and to Test Drugs

Co-ordinator: *idem*

Partners: *R. Cancceda*, IST-CBA Genoa (I); *P. Ruegsegger*, IBT-ETH Zurich (CH); *L. Vico* and C. Alexandre, LBTO (F); *A. Zallone*, Univ. Bari (I)

Industry: Scanco Medical (CH), Sulzer Biologics Inc. (CH) plus additional industrial contacts via co-ordinator

Ground phase

Facility: Mice habitat and *in-vivo* 3D-μcomputer tomograph

Fig. 2. ERISTO project.

- to define a representative three-dimensional *in-vitro* model of bone re-modelling (osteoclasts, osteoblasts, osteocytes) to replace as far as possible and to complement the traditional animal models

- to build an analytical instrument (3D-μCT) for the examination of the bone model and of bone samples

- to propose new/improved animal models for osteoporosis-related research (see ERISTO-Animal Chapter below)

- to study the influence of steroids and cytokines on bone remodelling.

Based on the results and conclusions of this first phase of ERISTO, Phase II extends the objectives of Phase I (experimental 3-D bone artefact, 3D-μCT for *in-vivo* analysis, animal models for space effect simulation, cellular signalling), attracts the interest of the industry to the defined models and instrument, and prepares space-related experiments to be performed in close collaboration with biomedical and pharmaceutical companies.

A prototype version of the *in-vivo* 3D-μCT has now been experimentally validated during a longitudinal study on the rat model.

ERISTO-Animal Chapter (Fig. 2): Within the ERISTO Project, the Animal Chapter deals specifically with animal models for the study of osteoporosis. MED-024 focuses on the influence of microgravity-induced and immobilisation-induced remodelling of bone using animal models (mice). This research is expected to improve the understanding of mechanical constraints in bone remodelling and to provide models to support osteoporosis-targeted drug screening in space and during ground simulations.

Micro-Computer Tomography will be brought to the very limits of this method by hardware refinement and the development of new image reconstruction algorithms by the project "2D & 3D Quantification of Bone Structure and its Changes in Microgravity Condition by Measure of Complexity" (Fig. 3). The method is expected

AO-99-030

New activity initiated in 2000

2D & 3D Quantification of Bone Structure and its Changes in Microgravity Condition by Measure of Complexity

International Dimension: CH, D, DK

Co-ordinator: *W. Gowin*, Univ. Hosp. Benjamin Franklin Berlin (D)
wolf1234@mailer.ukbf.fu-berlin.de

Partners: *A. Boshof, D. Felsenberg, P. Saparin*, Benjamin Franklin (D); *H.-C. Hege*, Konrad-Zuse Zentrum Berlin (D); *J. Kurths*, Univ. Potsdam (D); *J. Thomsen*, Univ. Aarhus (DK)

Industry: Siemens (D) and Scanco (CH)

Facility: TBD

Fig. 3. Project: 2D and 3D quantification of bone structure and its changes in microgravity condition by measure of complexity.

AO-99-122

New activity initiated in 2001

Bone Metabolic Studies in a Combined Perfusion and Loading Chamber

International dimension: B, CH, D, I, UK

Co-ordinator: *D. Jones*, Univ. Marburg–Klin. Der Phillipps (D) jones@post.med.uni-marburg.de

Partners: *B. v.d. Schoot*, Seyonic (CH); *J. v.d. Sloten*, Univ. Leuven (B); *P. v.d. Wal*, Univ. Neuchatel (CH); *U. Zanger*, R. Wolf Endoscope (D); *R.G. Richards,* AO Res. Inst. Davos (CH)

Industry: Seyonic (CH), Visitech (UK), R. Wolf Endoscope (D)

Facility: Biotechnol. Mammalian Tissue Cultivation / BMTC (new): Pre-Phase A initiated (Scientific Requirements)

Fig. 4. Project: Bone metabolic studies in a combined perfusion and loading chamber.

to be of extreme significance in bone diagnosis. This project involves medical doctors collaborating with mathematicians and physicists providing discipline synergy.

The project "Bone Metabolic Studies in a Combined Perfusion and Loading Chamber" (Fig. 4) plans to mimic *in vivo* mechanical and perfusion conditions on bone samples and engineered bone tissue constructs. The test hardware developed will allow synchronous mechanical stimulation, determination of mechanical properties of living bone samples, determination of metabolic parameters, and evaluation of signal transduction phenomena.

Imaging of bone using "Ballistic and Holographic 3-D High-Resolution Imaging" methods (Fig. 5) has great potential in overcoming radiation problems during a series of measurements in space and on the ground. In addition the potential for developing a small, compact and versatile instrument is very high. Since the first submission of the proposal in 1999, the team has managed to attract the interest and participation of additional leading companies in the complementary fields of laser technique, holographic films, and bone microstructure analysis.

AO-99-121

New activity initiated in 2001

Ballistic and Holographic 3-D High-Resolution Imaging of Bone

International dimension: D, IRL

Co-ordinator: *M. Hoffmann*, Univ. Marburg, (D) martin.hofmann@physik.uni-marburg.de

Partners: *D. Jones*, Univ. Marburg (D); *K. Meerholz*, Univ. Munich (D); *J. McInerney*, Univ. Coll. Cork (IRL)

Industry: Sacher Lasertechnik (D), La Vision (D), Soliton (D), Stork/Origin (NL)

Facility: TBD

Fig. 5. Project on Ballistic and holographic 3-D high-resolution imaging of bone.

AO-LS-99-LSS-003

New activity initiated in 2001

Investigation of developmental pathways leading to bone formation and bone homeostasis by genetic dissection and functional analysis of osteoprotegerin in a transgenic fish model on earth and microgravity environment

International dimension: D, N

Co-ordinator: *R. Goerlich* (D) goerlicr@uni-duesseldorf.de

Partners: *P. Ahleström,* Norw. Sch. Vet. Sci. Oslo (N); *P.J. Midtlyng,* Veso (N); *T. Wolffrom,* Veso (N); *E.W. Knapik,* GSF-Neuherberg (D); *M. Schartl,* Univ. Wuerzburg (D); *C. Winkler,* Univ. Wuerzburg (D)

Industry: Contact via VESO (N)

Facility: possibly Aquarack (D)

Fig. 6. Project: Investigation of developmental pathways leading to bone formation and bone homeostasis by genetic dissection and functional analysis of osteoprotegerin in a transgenic fish model on earth and microgravity environment

The project "Investigation of developmental pathways leading to bone formation and bone homeostasis by genetic dissection and functional analysis of osteoprotegerin in a transgenic fish model on earth and microgravity environment" (Fig. 6) proposes a strategy to identify key regulatory elements of conserved eukaryotic signalling pathways controlling bone formation under normal and microgravity conditions. The research team plans to use highly efficient molecular, immunological and cytological techniques in combination with transgenic fish models to identify functions affected by microgravity in bone formation. In combination with pharmacological studies the proposed project may lead to development of novel strategies for better control of osteoporosis.

The project "Vascular endothelial cells in microgravity: gene expression, cellular energy metabolism and differentiation" (Fig. 7) proposed studies of gene expression

---

AO-LS-99-LSS-006

New activity initiated in 2001

Vascular endothelial cells in microgravity: gene expression, cellular energy metabolism and differentiation

International dimension: I, NL

Co-ordinator: *S. Bradamante* (I) brada@icil64.cilea.it

Partners: *J. Maier* (I); *J.W. de Jong* (NL)

Industry: Contact co-ordinator; industrial partnership invited

Facility: TBD

Fig. 7. Project: Vascular endothelial cells in microgravity: gene expression, cellular energy metabolism and differentiation.

and energy metabolism modifications by microgravity conditions are very timely and are supposed to lead to new insights into possibilities of tissue engineering. It also could give leads to the intense and ongoing research effort to tie endothelial tissue development to cancer diseases. The group is for the moment academic, but this does not seriously detract from the potential of the project for health on earth.

Potential applications of space physical sciences in bone biotechnology

In cases of significant gaps between the bone parts to be joined, for example in following severe accidental trauma, infections or tumour, the method of choice is the autologous transplant from e.g. the iliac crest. This is always very painful, presenting real risk of infection, and the amount of autologous bone that can be sampled is also obviously limited. Alternative ways are therefore being proposed, based on the tissue engineering approach, using autologous cells immobilised in a matrix providing for mechanical resistance and also possibly for osteoconductive, osteoinductive and osteogenic properties. Such materials are biocompatible synthetic (ceramics, bio-active glasses) or natural materials like corals. Ceramics produced by powder sintering at high temperature usually possess a crystalline phase and a vitreous matrix. The most frequently used as a substitute for bone is based on hydroxyapatite, the major constituent of bone. Such materials, while osteoconductive, are however extremely stable chemically and cannot be resorbed and replaced by normal bone. Improved hydroxyapatite-based ceramics with higher resorbability can be produced by reducing the purity of the material by the incorporation of more soluble compounds such as tricalcium phosphate and other trace elements. A potentially promising method could involve working on the microcrystalline properties of the hydroxyapatite, close to the bone microcrystallinity, and on the chemical properties of the vitreous mineral or organomineral cement in order to promote its resorbability and osteoinductivity. A contribution from material science to bone bioengineering, investigating for example the crystalline properties of materials formed in organic matrices, would be a good example of interdisciplinary cross-fertilisation.

The existence of osteocompetent cells in the bone marrow, identified as early as the end of the 1960s by researchers from the Gamaleya Institute of Moscow, theoretically allows the inoculation of the implant *in vitro* before implantation. Preosteoblastic cells are, however, rare in the bone stroma and furthermore they decrease with age. Current research focuses on improving the recruitment of pre-osteoblasts and favouring the differentiation of precursor cells present into the bone marrow into osteoblasts (see Fig. 2). A control of the mechanical and biochemical microenvironment for an initial multiplication without differentiation of precursor cells, followed by osteoblastic differentiation is also expected to gain from space research and from the development of specific medium and bioreactors. Control of interstitial fluid velocities in a biocompatible range and mimicking by bioreactor design the cellular/vascular interfaces will need strong involvement of teams dealing with fluid science and bioengineering sciences.

246

In cases of bone fracture, natural cells recruitment from bone tissue boarding the fracture usually provides fast and efficient closure of the gap. However, in the case of large gaps in the bone structure resulting from clinical intervention or from severe trauma, natural bone reformation is usually impossible. Bone substitutes can then be used to close the gap. Two approaches are possible here: a "mechanical stable implant" or a bone substitute biologically functional. Such a bone substitute should resorb to allow the normal bone homeostasis mechanisms to restore biologically and mechanically a functional bone tissue; resorption should proceed at a rate comparable to new bone synthesis and should ideally result largely from the natural process of osteoclastic bone resorption followed by the recruitment of osteoblasts organising exopolymeric matrix deposition followed by mineralisation. Control of this resorption/reconstruction process can certainly gain from attributing biochemical signalling properties to the bone substitute. The use of bone morphogenic proteins in addition to other growth factors is a likely powerful control key, providing that protection against denaturation and controlled release can be achieved.

Controlled release is particularly important both in terms of locally achieved concentrations and time-dependant release. Such molecules usually achieve efficient signalling at concentrations far below the nano-or even femtomolar concentration; at the same time, the effect has to be localised in order to prevent unwanted bone formation at sites different from the site to be repaired. Systemic administration or uncontrolled local release is therefore not acceptable. Dynamic local delivery is thus required and methods involving covalent grafting in/on the bone substitute or encapsulation methods are considered. In the case of encapsulation, the achievement of timely controlled release involves achievement of concentration gradients within the microcapsule at the time of the encapsulation process. In practice, the distribution of active molecules in a forming capsule by solvent evaporation or phase transfer methods is influenced by microcirculations patterns in normal gravity conditions.

The application of fluid sciences methodologies and instruments to the understanding of interfacial transfer, and especially interfacial turbulence and its dependence to gravity, is currently the subject of a topical team supported by ESA in the field of microencapsulation. The objective is to achieve better control of the process of microencapsulation involving emulsion formation followed by mass transfer by solvent extraction at the pellet gas/liquid interface. Better control is necessary to master drug distribution within the microcapsule [10].

Finally, and especially in the case of large gaps in bone repair, one should not only focus on providing local conditions (matrix composition, porosity, biochemical signalling) favourable for bone synthesis, but should also specifically reduce the risk of creating conditions that allow cells from non-bone tissue to settle in the matrix and induce biochemical conditions inhibiting mineralisation. Recent genetic studies indeed suggest that calcification, that is favoured by calcium and phosphate concentrations in extracellular fluids near precipitation, could occur by default, and that inhibition of mineralisation would be an active process involving specific proteins [11]. In cases of large gaps, prevention of the synthesis and filling of the gap by an extracellular matrix actively inhibiting nucleation for mineralisation could be an additional objective. The ways to prevent such inhibiting matrix synthesis and favour nucleation should be discussed, probably in line with nucleation phenomenon in organic matrices.

Conclusion: the art of space science or "the scientist and the (good) cook"

With the great progress made in sequencing the human genome, the past temptation to reduce biology to a sum of yes or no, presence or absence of enzymatic transformation potential, cellular response or not to given signalling molecule, triggering or not of a specific gene expression, or expression or not of a specific gene, is now strongly shifted to a more difficult but necessary integrative approach at system level. This evolution is remarkable and best reflected in Europe by the emphasis placed by the European Molecular Biology Laboratory (EMBL) on functional genomics [13].

The application of mathematics and computation to the understanding and analysis of biological systems was needed because of their extremely complex behaviour in time and space even with a limited set of independent variables. This necessity is now even greater in recent molecular biology when a system-level approach imposes the addition to the previous qualitative functions of a quantitative dimension. Rate of substrate consumption, gradients of intermediate metabolites concentration around the cell, oxygen gradients, rate of cytokines production by the cell, transport phenomenon in the intercellular space, gradients of such cytokines in a tissue due to the dynamic balance of secretion, transport, and inactivation—all these phenomenon will ultimately need to be studied. To quote James E. Bailey in his last commentary paper "Complex biology with no parameters" [12]: "However, such global comprehension of quantitative knowledge, so far beyond our current resources and imagined capabilities in biology, is not usually, if ever, necessary. Qualitative and quantitative understanding and corresponding methodologies for designing desired properties of many complex systems have been successfully achieved in the field of chemistry, physics, and the associated engineering disciplines without knowing all aspects of systems structure and certainly without knowing all parameters value involved. The same must be possible for biology."

This brings me to the title I chose for this conclusion where one will recognise the dualistic aspect of science progression in biology in general, and in medicine and environmental sciences in particular, and the way we may have to plan for experimental research in all these domains when considering the use that life evolution may have made of gravity acceleration. These two approaches are extremely reductionist for truly quantifiable progression in the sound understanding of simplified phenomena, and extremely holistic when there is a need for decisions in matters so complex that summing the bits we know already would be inadequate to help in the mathematical way of demonstrating a unique solution. Not everyone can cook well (good cooking is often referred to as an art); for a medical doctor, a sense for diagnosis is largely a personal gift based on sound professional knowledge; by definition, scientific hypothesis cannot be demonstrated *a priori* but serves as a basis for experimentation that will or will not meet the hypothesis. What then makes the good cook, the medical doctor, and the scientist? It would be presumptuous to propose an answer, but could this have a relationship with the science, the intelligence of global, complex systems—rather like the cook who, not having studied chemistry or olfactory physiology, anticipates the result of the herbs, the tiny piece of garlic and the few drops of olive oil in his newly invented recipe?

References

1. Abbott, B.J. (1977) Immobilized cells. Ann. Rep. Fermentation Proc. 1, 205–233.
2. Baier, R.E. (1980) Substrata influences on adhesion of microorganisms and their resultant new surface properties. In: G. Bitton, and K.C. Marshall (eds.), Adsorption of Microorganisms to Surfaces. Wiley, pp. 59–104.
3. Winter, J.U. and Wolfe, R.S. (1980) Methane formation from fructose by syntrophic associations of *Acetobacter woodii* and different strains of methanogens. Arch. Microbiol. 124, 73–79.
4. Bissell, M.J. and Barcellos-Hoff, M.H. (1987) The influence of extracellular matrix on gene expression: is structure the message? J. Cell Sci. Suppl. 8, 327–343.
5. El Haj, A.J., Minter, S.L., Rawlinson, S.C., Suswillo, R. and Lanyon, L.E. (1990) Cellular responses to mechanical loading *in vitro*. J. Bone Miner. Res. 5, 923–932.
6. Caldwell, M.A., He, X., Wilkie, N., Pollack, S., Marshall, G., Wafford, K.A. and Svendsen, C.N. (2001) Growth factors regulate the survival and fate of cells derived from human neurospheres. Nature Biotech. 19, 475–479.
7. Wobus, A.M., Rohwedel, J, Struebing, C., Shan, J., Adler, K., Maltsev, V. and Hescheler, J. (1997) *In vitro* differentiation of embryonic stem cells. In: S. Klug and R. Thiel (eds.), Methods in Developmental Toxicology and Biology, Blackwell Science.
8. Johnstone, B., Hering, T.M., Caplan, A.I., Goldberg, V.M. and Yoo, J.U. (1998) *In vitro* chondrogenesis of bone marrow-derived mesenchymal progenitor cells. Exp. Cell Res. 238, 265–272.
9. Kale, S., Biermann, S., Edwards, C., Tarnowski, C., Morris, M. and Long, M.W. (2000) Three-dimensional cellular development is essential for *ex vivo* formation of human bone. Nature Biotech. 18, 954–958.
10. Contact: T.L. Whateley, University of Strathclyde (UK) t.l.whateley@strath.ac.uk).
11. Schinke, T., McKee, M.D. and Karsenty, G. (1999) Extracellular matrix calcification: where is the action. Nature Genet. 21, 150–151.
12. Bailey, J.E. (2001) Complex biology with no parameters. Nature Biotech. 19, 503–504.
13. Kafatos, F. (2000) Interesting times—biology, European science, and EMBL. Science 287, 1401–140

List of Addresses

Natalia Battista
Department of Experimental Medicine and Biochemical Sciences
University of Rome Tor Vergata
Via Montpellier 1
I-00133 Rome, Italy

Roger A. Binot
Directorate of Manned Spaceflight and μgravity
ESA-ESTEC/MSM-GA
P.O. Box 299
NL-2200-AG Noordwijk, Netherlands

Richard Bräucker
DLR
Institute of Aerospace Medicine
D-51147 Cologne, Germany

Ludmilla B. Buravkova
Russian State Research Center
Institute for Biomedical Problems
Russian Academy of Sciences
Khoroshevskoye Shosse, Bldg. 76A
Moscow 123007, Russia

Ranieri Cancedda
Centro di Biotecnologie Avanzate
Istituto Nazionale per la Ricerca sul Cancro
Largo Rosanna Benzi 10
I-16132 Genova, Italy;
and
Dipartimento di Oncologia, Biologia e Genetica
Università di Genova
I-16132 Genova, Italy

Augusto Cogoli
Space Biology Group
ETH Zurich
Technoparkstrasse 1
CH-8005 Zurich, Switzerland

250

Jacques Demongeot
Institut d'Informatique et Mathématique Appliquées de Grenoble
Laboratoire de Technique de l'Imagerie de la Modélisation et de la Cognition
Faculté de Médecine
Domaine de la Merci
F-38706 La Tronche Cedex, France

Alessandro Finazzi-Agrò
Department of Experimental Medicine and Biochemical Sciences
University of Rome Tor Vergata
Via Montpellier 1
I-00133 Rome, Italy

Lisa E. Freed
Division of Health Sciences and Technology
Massachusetts Institute of Technology
Building E25 Room 330
45 Carleton Street
Cambridge
MA 02139, USA

Nicolas Glade
Commissariat à l'Energie Atomique
Département de Biologie Moléculaire et Structurale, Laboratoire de Résonance
    Magnétique en Biologie Métabolique
D.S.V, C.E.A. Grenoble
17 Rue des Martyrs
F-38054 Grenoble Cedex 9, France

Ruth Hemmersbach
DLR
Institute of Aerospace Medicine
D-51147 Cologne, Germany

Millie Hughes-Fulford
Laboratory of Cell Growth
Department of Medicine
University of California San Francisco
Department of Veterans' Affairs Medical Center
San Francisco
CA 94121, USA

Anatoly I. Grigoriev
Russian State Research Center

Institute for Biomedical Problems
Russian Academy of Sciences
Khoroshevskoye Shosse, Bldg. 76A
Moscow 123007, Russia

Yury T. Kalinin
Russian Stock Company "Biopreparat"

Marian L. Lewis
250 Hartside Road
Owens Cross Roads
AL 35763, USA

M. Maccarrone
Department of Experimental Medicine and Biochemical Sciences
University of Rome Tor Vergata
Via Montpellier 1
I-00133 Rome, Italy

Oleg V. Mitichkin
Russian Space Corporation "Energia"

Anita Muraglia
Centro di Biotecnologie Avanzate
Istituto Nazionale per la Ricerca sul Cancro
Largo Rosanna Benzi 10
I-16132 Genova, Italy

Cyril Papaseit
Commissariat à l'Energie Atomique
Département de Biologie Moléculaire et Structurale, Laboratoire de Résonance
	Magnétique en Biologie Métabolique
D.S.V, C.E.A. Grenoble
17 Rue des Martyrs
F-38054 Grenoble Cedex 9, France

James Tabony
Commissariat à l'Energie Atomique
Département de Biologie Moléculaire et Structurale, Laboratoire de Résonance
	Magnétique en Biologie Métabolique
D.S.V, C.E.A. Grenoble
17 Rue des Martyrs
F-38054 Grenoble Cedex 9, France